中传学者文库编委会

主　任： 廖祥忠　张树庭

副主任： 蔺海波　李　众　刘守训　李新军　王　晖
　　　　　杨　懿　柴剑平

成　员（按姓氏笔画排序）：
　　　　　王廷信　王栋晗　王晓红　王　雷　文春英
　　　　　龙小农　付　龙　叶　龙　刘东建　刘剑波
　　　　　任孟山　李怀亮　李　舒　张绍华　张　晶
　　　　　张根兴　张毓强　林卫国　郑　月　金　炜
　　　　　金雪涛　周建新　庞　亮　赵新利　徐红梅
　　　　　贾秀清　高晓虹　隋　岩　喻　梅　熊澄宇

中传学者文库

主编／柴剑平
执行主编／龙小农
副主编／张毓强　周建新

电磁超材料及电磁计算

李增瑞团队学术文集

李增瑞 等著

中国传媒大学出版社

·北京·

图书在版编目（CIP）数据

电磁超材料及电磁计算：李增瑞团队学术文集 / 李增瑞等著 . -- 北京：中国传媒大学出版社，2024.8.

(中传学者文库 / 柴剑平主编).

ISBN 978-7-5657-3773-2

Ⅰ. TM271-53；TM15-53

中国国家版本馆 CIP 数据核字第 2024ZG3891 号

电磁超材料及电磁计算：李增瑞团队学术文集
DIANCI CHAOCAILIAO JI DIANCI JISUAN: LI ZENGRUI TUANDUI XUESHU WENJI

著　　者	李增瑞　等
责任编辑	杨小薇
封面设计	锋尚设计
责任印制	李志鹏
出版发行	中国传媒大学出版社
社　　址	北京市朝阳区定福庄东街 1 号　　邮　编　100024
电　　话	86-10-65450528　65450532　　传　真　65779405
网　　址	http://cucp.cuc.edu.cn
经　　销	全国新华书店
印　　刷	北京中科印刷有限公司
开　　本	710mm×1000mm　1/16
印　　张	24.5
字　　数	285 千字
版　　次	2024 年 8 月第 1 版
印　　次	2024 年 8 月第 1 次印刷
书　　号	ISBN 978-7-5657-3773-2/TM・3773　　定　价　122.00 元

本社法律顾问：北京嘉润律师事务所　　郭建平

总 序

媒介是人类社会交流和传播的基本工具。从口语时代到印刷时代，再经电子时代至今天的数智时代，媒介形态加速演变、融合程度深入发展，媒介已然成为现代社会运行的基础设施和操作系统。今天，人类已经迈入媒介社会，万物皆媒、人人皆媒，无媒介不社会、无传播不治理。今天，无论我们怎么用力于信息传播的研究、怎么重视信息传播人才的培养都不为过。

中国传媒大学（其前身为北京广播学院）作为新中国第一所信息传播类院校，自1954年创建伊始，即与媒介形态演变合律同拍、与国家发展同频共振，努力探索中国特色信息传播人才培养模式、构建中国信息传播类学科自主知识体系，执信息传播人才培养之牛耳、发信息传播研究之先声，被誉为"中国广播电视及传媒人才摇篮""信息传播领域知名学府"。

追溯中传肇始发轫之起源、瞩望中传砥砺跨越之未来，可谓创业维艰而其命维新。昔日中传因广播而起，因电视而兴，因网络而盛，今天和未来必乘风破浪、蓄势而上，因人工智能而强。在这期间，每一种媒介兴起，中传均吸引一批志于学、问于道、勤于术的

电磁超材料及电磁计算　李增瑞团队学术文集

学者汇聚于此，切磋学术、传道授业，立时代之潮头，回应社会需求，成为学界翘楚、行业中坚，遂有今日中传学术研究之森然气象，已历七秩而弦歌不断，将传百世亦风华正茂。

自新时代以来，中传坚守为党育人、为国育才初心，励精图治、勠力前行，秉承"系统治理、创新图强、交叉融合、特色发展"的办学理念，牢牢把握高等教育发展大势、传媒业态发展趋势，瞄准"智能传媒"和"国际一流"两大主攻方向，以世界为坐标、以未来为向度，完成了全面布局和系统升级，正在蹄疾步稳、高质量推动学校从传统高等教育向未来高等教育跨越、从传统传媒教育向智能传媒教育跨越、从国内一流向世界一流跨越，全力建设中国特色、世界一流传媒大学。

中国特色、世界一流，在于有大先生扎根中国大地，汇聚古今、融通中外；在于有大先生执教黉门，学高为师、身正为范；在于有大先生躬耕杏坛，敦品积学、启智润心。习近平总书记更强调，高校教师要立志成为大先生，在教书育人和科研创新上不断创造新业绩。中传广大教师素来以做大先生为毕生职志，努力成为新时代"经师"与"人师"的统一者，做真学问、立高品行，践履"立德树人"使命。

2024岁在甲辰，欣逢中传建校70华诞，学校特邀约部分学者钩玄勒要、增删批阅，遴选已公开刊发的论文汇编成集，出版"中传学者文库"，意在呈现学校在学科建设、科学研究、服务行业实践等方面的最新成果，赓续中传文脉，谱写时代新声。

文库汇聚老中青三代学者，资深学者渊渟岳峙、阐幽抉微；中年学者沉潜蓄势、厚积薄发；青年学者踌躇满志、未来可期。文库与五十周年校庆所出版的"北广学者文库"相承接，大致可勾勒中传知识生产薪火相传、三代辉映之概貌，反映中传在构建中国特色新闻传播类、传媒艺术类、传媒技术类学科体系、学术体系和话语体系方面的耕耘与收获，窥见中国特色信息传播类学科知识体系构建的发展脉络与轨迹。

这一构建过程，虽筚路蓝缕，却步履铿锵；虽垦荒拓野，亦四方辐辏。一批肇始于中传，交叉融合、具有中国特色的学科，如播音主持艺术学、广播电视艺术学、传媒艺术学、数字媒体艺术学、政治传播学等，从涓涓细流汇入滔滔江河，从中传走向全国，展现了中传学者构建中国自主知识体系的学术想象力和创新力。文库展示的虽然是历史，实则是呈现今天；看似是总结过去，实则是召唤未来。与其说这套文库的出版，是对既有学术成果的展示，毋宁说是对未来学术创新的邀约。

回首过往，七秩芳华。我们深知，唯有将马克思主义基本原理与中华优秀传统文化相结合，才能推动中华学术创造性转化和创新性发展，推动中国自主知识体系的构建。我们深知，唯有准确把握媒介形态演变的脉动、深刻认知媒介形态变革所产生的影响，才能推动中国信息传播类学科自主知识体系的构建与时俱进。

展望未来，星辰大海。我们深知，以人工智能为代表的产业和科技革命正迅疾而来，媒介生态正在加速重构，教育形态正在全面重塑，大学之使命与价值正在被重新定义；我们深知，唯有"胸怀国之大者"、面向世界科技前沿、面向经济主战场、面向国家重大需求，才能确保中传始终屹立于中国乃至世界传媒教育发展之潮头。

如何应对人工智能带来的深刻变革，对中传而言是一场要么"冲顶"、要么"灭顶"的"兴亡之战"。我们坚信，不管前方是雄关漫道，还是荆棘满途，唯有勇敢直面"教育强国，中传何为？"这一核心命题，奋力书写"智能传媒教育，中传师生有为！"的精彩答卷，才能化危为机，奋力开创人工智能时代中传智能传媒教育新纪元。

功不唐捐，芳华七秩；风帆正举，赓续创新。

是为序。

第十四届全国政协委员，中国传媒大学党委书记、教授、博士生导师

Contents

Quad-Polarization Reconfigurable Reflectarray with Independent Beam-Scanning and
　　Polarization Switching Capabilities ……………………………………………………… 001
Analysis of Scattering Characteristics of Height-Adjustable Phased Array with Ultra-
　　Wideband Dual-Linear Polarized RCS Amplitude Regulation ……………………… 032
Design of an Active Polarizer for Wideband Quad-Polarization Conversion ……………… 052
A Low-RCS Multifunctional Shared Aperture with Wideband Reconfigurable
　　Reflectarray Antenna and Tunable Scattering Characteristic ………………………… 063
Reconfigurable Bidirectional Beam-Steering Aperture with Transmitarray, Reflectarray,
　　and Transmit-Reflect-Array Modes Switching ………………………………………… 085
On the Enforcement of Electric Field Boundary Condition in the Moment Method
　　Solution of Volume-Surface Integral Equation ………………………………………… 114
Using the Interpolative Decomposition to Accelerate the Evaluation of Radome
　　Boresight Error and Transmissivity ……………………………………………………… 125
Extremely Wideband and Omnidirectional RCS Reduction for Wide-Angle Oblique
　　Incidence ………………………………………………………………………………… 136
A Novel Wideband and High-Efficiency Electronically Scanning Transmitarray Using
　　Transmission Metasurface Polarizer …………………………………………………… 153
Design of a Dielectric Dartboard Surface for RCS Reduction ……………………………… 168
Solving the Surface Current Distribution for Open PEC-Dielectric Objects Using the
　　Volume Surface Integral Equation ……………………………………………………… 179
Breaking the High-Frequency Limit and Bandwidth Expansion for Radar Cross Section
　　Reduction ………………………………………………………………………………… 191

Ultrawideband and High-Efficient Polarization Conversion Metasurface Based on Multi-Resonant Element and Interference Theory ············ 212

On the Use of Hybrid CFIE-EFIE for Objects Containing Closed-Open Surface Junctions ············ 228

Ultrawideband Frequency-Selective Absorber Designed with an Adjustable and Highly Selective Notch ············ 240

A Well-Conditioned Integral Equation for Electromagnetic Scattering from Composite Inhomogeneous Bi-Anisotropic Material and Closed PEC Objects ············ 266

Tri-Band Radar Cross-Section Reduction Based on Optimized Multi-Element Phase Cancellation ············ 288

Efficient Triangular Interpolation Methods: Error Analysis and Applications ············ 302

Ultrawideband Monostatic and Bistatic RCS Reduction for Both Copolarization and Cross Polarization Based on Polarization Conversion and Destructive Interference ··· 313

Dual-Polarization Absorptive/Transmissive Frequency Selective Surface Based on Tripole Elements ············ 329

Ultrawideband Radar Cross-Section Reduction by a Metasurface Based on Defect Lattices and Multiwave Destructive Interference ············ 339

A Novel Checkerboard Metasurface Based on Optimized Multielement Phase Cancellation for Superwideband RCS Reduction ············ 362

Quad-Polarization Reconfigurable Reflectarray with Independent Beam-Scanning and Polarization Switching Capabilities[*]

1. Introduction

Reconfigurable reflectarray (RA) antennas with low-cost high-gain beam-scanning characteristics have attracted much attention of antenna researchers in recent years due to their combining feasible features of the reflector and phased array. This type of antenna adopts the spatial feeding technique, which reduces insertion loss caused by the complicated feeding networks leading to the characteristic of low loss, simple structure, low cost, and high gain. Reconfigurable RA aperture can reflect electromagnetic (EM) waves generated from the feed antenna and electrically steer a pencil beam by adjusting the compensation phase of each element. The tunable devices, such as positive-intrinsic-negative (PIN) diodes, varactor diodes, or tunable material are integrated into RA elements to achieve dynamically phase control without using the high-cost radio frequency (RF) phase shifters and transmit/receive (T/R) modules.

Generally, there is a tradeoff between two major factors, i.e., reconfigurability and structure complexity, for a reconfigurable design. For example, in terms of phase quantization, a design with a high phase resolution can minimize the degradation in antenna gain; however, it increases the complexity of the structure and the losses of the tunable devices. Single-bit phase resolution (0°/180°) for beam-scanning is considered a fair balance between performance stability and system complexity, which has been extensively reported in spatially-fed arrays, including transmitarray (TA) and RA.

Nowadays, reconfigurable RAs have not only electronic beam-steering capabilities but also more advanced capabilities in terms of operating frequency and polarization. Among them, simultaneous beam steering and dynamic polarization control of the radiated pattern ensure polarization matching between the transmitter and receiver antennas, which is a key prospect in cognitive radio applications and wireless systems. This type of reconfigurable antenna also has great potential in many scenarios due to its unique features

---

[*] The paper was originally published in *IEEE Transactions on Antennas and Propagation*, 2023, 71 (9), and has since been revised with new information. It was co–authored by Hang Yu, Ziyu Zhang, Jianxun Su, Meijun Qu, Zengrui Li, Shenheng Xu, and Fan Yang.

in enhancing the system performance through frequency reuse, eliminating multipath fading, and avoiding polarization mismatch, as well as the improved capacity for multiple-input multiple-out (MIMO) systems. On the other hand, by using a single polarization-agile antenna instead of multiple antennas with different polarizations, the size, weight, and cost of the RF system are greatly reduced.

However, the studies on reconfigurable RA or TA to date are mostly limited to single linear polarization (LP) or circular polarization (CP). Only a few of the reconfigurable architectures proposed in the literature operate in multi-polarization reconfiguration modes. For example, *Wang et al.* proposed a multi-polarization beam-scanning mechanically reconfigurable RA antenna by manually rotating all-metal rotational elements. However, the mechanical reconfigurable method has the disadvantages of complex hardware control and slow response compared to the electrical reconfigurable strategy in [29]. A dual-linear-polarized reconfigurable TA with 2D beam steering capability is proposed, while the linear polarization conversion is achieved by a single pole double throw PIN diode switch. In [31], *Zhang et al.* present a concept of the reconfigurable RA featuring dual-polarized beam control, where 4 PIN diodes are integrated into each element as the active components for providing independent 1-bit phase control in two orthogonal polarization channels. A circularly polarized reconfigurable TA in the Ka-band has been presented in [32], which enables right-hand CP (RHCP) or left-hand CP (LHCP) switching and beam-steering. It can be seen that these works can only realize electronic beam scanning with one or two polarization modes instead of four polarizations [horizontal polarization (HP), vertical polarization (VP), LHCP, and RHCP]. Actually, the theoretical research on quad-polarization and simultaneous beam steering in the RA design has been studied as early as in pioneer works. However, it is very challenging to design a quad-polarization reconfigurable RA with simultaneous beam scanning capability in an electronically controlled manner.

One of the major challenges in the design of quad-polarized reconfigurable RAs is the dramatic increase in complexity of the element structure and bias feed network, due to the additional polarization or external phase control. To maintain dynamic control for beam-scanning while simultaneously achieving quad-polarization flexibility operation, the element should allow to independently control two-linear polarization with at least a 2-bit phase resolution (0°/90°/180°/270°) for each component in traditional view. Since a 90° or 270° phase shift is needed for conversion from LP to CP based on the principle from the element level; while a 180° phase is needed for conversion from LP to the orthogonal LP.

The increase in phase resolution results in a drastic increase in the number of tunable devices and the complexity of the bias network. For example, a reconfigurable dual linearly-polarized RA element with 2-bit phase resolution is presented , which consists of two orthogonal cross-type printed triangular dipoles loaded by 2×4 PIN diodes. Due to the complexity of the bias circuit and the control circuit, where each PIN diode is connected to a bias line, this work is only limited to the element design, and the array level design is not carried out, thus hindering their further practical applications. The same issues also exist in [37], where one element integrates 4 PIN diodes and 4 varactor diodes. Another important challenge is that similar loss of the element should be achieved in the different phase states, due to the high variation in the losses for different phases will strongly impact on the quality of the polarization regulation.

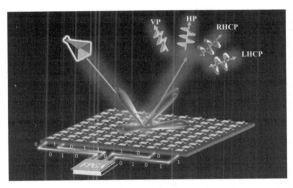

Fig. 1 Schematic diagram of quad-polarization reconfigurable RA antenna radiation modes.

The aim of this work is to propose a quad-polarization reconfigurable RA, which can work in HP, VP, LHCP, and RHCP modes with simultaneous large-angle beam scanning capability, in an efficient and low-cost manner, as shown in Fig. 1. Unlike the traditional view at the element level, where at least 2-bit phases resolution are needed due to this corresponds to a 90° phase shift step required for LP to CP conversion. In this work, we found that the extra polarization conversion phase shifts (90°/-90°) can be achieved by changing the 0/1 coding arrangement in the spatially-fed architecture only using 1-bit phase resolution, which greatly reduces the complexity while increasing the stability of the system. As verification, a 256-element aperture fed by a pyramidal horn antenna is designed, optimized, fabricated, and measured. We have introduced the phase encoding arrangement for quad-polarization beam scanning and the global reference phase optimization for circular polarization using the aperture field approach, which lays a general guide for a 1-bit quad-polarization reconfigurable RA design. The measured results demonstrate the proposed RA antenna can scan pencil beams with dual-linear and dual-circular polarization modes switching in a real-time manner by controlling the PIN diode states using a field programmable gate array. The proposed method of quad-polarization reconfigurable RA may open a new avenue for high-efficient control of the radiation pattern and polarization.

This paper is organized as follows. Section II introduces the analysis of multi-polarization realization and design motivation. The proposed RA element design, encoding arrangement, and array optimization flow of RA performance are described in Section III. Section IV gives the measured multiple polarization beam scanning performance of the proposed RA prototype, and Section V concludes this work.

2. Analysis of Multi-Polarization Realization and Design Motivation

The aim of this work is to propose a quad-polarization (HP, VP, RHCP, and LHCP) electronically reconfigurable RA with beam scanning and polarization switching capabilities. In this section, we first discuss the principle of quad-polarization conversion from the unit cell level, and conclude that at least 2-bit phase resolution is required to realize beam scanning and circular polarization conversion. Then, the influence of the global reference phase on the far-field phase in the spatially-fed array is introduced from the point of array level. The results show that the extra phase shift can be achieved by controlling the array coding distribution only using the 1-bit phase control method. It proves the feasibility of realizing quad-polarization reconfiguration with simultaneous beam scanning capability through a 1-bit reconfigurable unit cell.

2.1 Quad-Polarization Conversion from Unit Cell Level

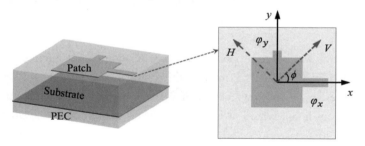

Fig. 2　Configuration of a typical RA unit cell with different phase delays in x and y polarizations.

To realize the proposed quad-polarization reconfigurable RA antenna, the proposed element should simultaneously provide phase control and polarization flexibility from the unit cell level. The configuration of a typical RA element with different phase delays in x and y polarizations is shown in Fig. 2. The unit cell can be characterized by the reflection matrix R, which includes the co-polar and cross-polar reflection coefficients for the two linear polarizations paralleled to the x- and y-axes. The matrix R relating the cartesian complex components of the reflected (E^r) and incident tangential electric field (E^i) on the RA unit cell in a periodic environment. Due to the square patch element with different phase delays in x and y polarizations, matrix R can be written as

$$\mathbf{R} = \begin{bmatrix} R_{xx} & 0 \\ 0 & R_{yy} \end{bmatrix} \tag{1}$$

By counterclockwise rotating the original coordinate axis to 45°, the *V-H* coordinate system can be obtained. The relation between two coordinates is $\begin{bmatrix} \hat{v} \\ \hat{h} \end{bmatrix} = \mathbf{O} \cdot \begin{bmatrix} \hat{x} \\ \hat{y} \end{bmatrix} = \begin{bmatrix} \cos\varphi & \sin\varphi \\ -\sin\varphi & \cos\varphi \end{bmatrix} \begin{bmatrix} \hat{x} \\ \hat{y} \end{bmatrix}$, where **O** is coordinate rotation matrix, $\varphi = 45°$ in this example. Usually, the polarization of the incident wave is fixed; let's assume the polarization of an incident wave is along the *V*-direction. We can get the reflected electric field in the *V-H* coordinate:

$$\begin{bmatrix} E_v^r \\ E_h^r \end{bmatrix} = \mathbf{O} \cdot \mathbf{R} \cdot \mathbf{O}^T \begin{bmatrix} E_v^i \\ 0 \end{bmatrix} = \frac{1}{2} \begin{bmatrix} (R_{xx} + R_{yy})E_v^i \\ (-R_{xx} + R_{yy})E_v^i \end{bmatrix} \qquad (2)$$

From (2), it can be seen that in order to realize the linear polarization conversion between *V* and *H* direction, it should satisfy

$$R_{xx} = -R_{yy} \qquad (3)$$

When $R_{xx} = R_{yy}$, the polarizations of the reflected incident and incident waves are the same. The reflected electric field in the *x-y* component can be expressed as

$$\begin{bmatrix} E_x^r \\ E_y^r \end{bmatrix} = \mathbf{R} \cdot \mathbf{O}^T \begin{bmatrix} E_v^i \\ 0 \end{bmatrix} = \frac{\sqrt{2}}{2} \mathbf{R} \cdot \begin{bmatrix} E_v^i \\ E_v^i \end{bmatrix} = \frac{\sqrt{2}}{2} E_v^i \begin{bmatrix} R_{xx} \\ R_{yy} \end{bmatrix} \qquad (4)$$

Eq. (4) indicates the phases of *Rxx* and *Ryy* should have ±90° phase in order to achieve circular polarization conversion. A phase shift of 90° or -90° between two orthogonal linear polarizations is required for the LP-CP conversion depending on the desired handedness.

Table I
Polarization modes and phases of the reflected polarized waves when the unit cell is with 2–bit phase resolution both in *x*– and *y*–polarization

x-pol. \ *y*-pol.	0°	90°	180°	270°
0°	VP (0°)	LHCP (0°)	HP (0°)	RHCP (0°)
90°	RHCP (90°)	VP (90°)	LHCP (90°)	HP (90°)
180°	HP (180°)	RHCP (180°)	VP (180°)	LHCP (180°)
270°	LHCP (270°)	HP (270°)	RHCP (270°)	VP (270°)

In addition, the unit cell needs at least a 1-bit tunable phase shift, i.e., 0° and 180°, to realize beam scanning in spatially-fed architecture. It can be seen that only a 1-bit phase is needed to realize *V-H* direction polarization conversion and beam scanning. Since ±90° phase shifts are needed for conversion from LP to CP, there is at least a 2-bit phase resolution for each component based on unit cell level. Table I gives the polarization modes and phases of the reflected polarized waves when the unit cell is with 2-bit phase resolution both in *x*- and *y*-polarization. It can be drawn a conclusion that, to realize quad-polarization reconfigurable, the unit cell should allow the independent control of two linear polarizations with at least available phases of 0°, 90°, 180°, and 270°, hence operating as a reconfigurable polarizer with simultaneous phase control capability.

2.2 Far-Field Phase Shift Based on 1-Bit Phase Control in Array Level

There is a trade-off between simplicity in the design of reconfigurable RA elements when choosing the number of bits for phase quantization. A 1-bit resolution design is considered a fair tradeoff between performance and system cost due to its simplicity and low cost. We will discuss how to achieve a 90° phase shift from the perspective of the array level based on the superposition of the aperture fields, which indicates the LP-CP conversion can only be achieved using 1-bit phase unit cells.

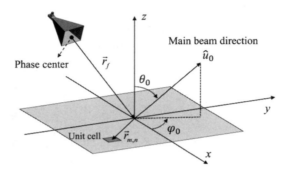

Fig. 3   The coordinate system of RA antenna.

Let us consider a linearly polarized feed horn incident on the RA aperture with M×N elements, the far-field radiation pattern of RA using array theory formulation can be expressed as

$$\mathbf{E}(\theta,\varphi) = \sum_{m=1}^{M}\sum_{n=1}^{N} I(m,n) \cdot \cos^{qe}\theta \cdot e^{jk(\vec{r}_{m,n} \cdot \hat{u})} \quad (5\text{-}1)$$

$$I(m,n) = \frac{\cos^{qf}\theta_f(m,n)}{|\vec{r}_{m,n} - \vec{r}_f|} e^{-jk(|\vec{r}_{m,n} - \vec{r}_f|)} \cdot |R_{m,n}| e^{j\varphi_{m,n}} \quad (5\text{-}2)$$

$$\hat{u} = \hat{x}\sin\theta\cos\varphi + \hat{y}\sin\theta\sin\varphi + \hat{z}\cos\theta \quad (5\text{-}3)$$

where *qe* and *qf* are the unit cell and the feed pattern power factor, respectively. As shown in Fig. 3, θ_f (m, n) is the spherical angle in the feed's coordinate system, and $\vec{r}_f$ and $\vec{r}_{mn}$ are the position vector of feed antenna and the *mn*th unit cell, respectively. *k* is the free space wavenumber. $|R_{mn}|$ is the reflection magnitude of the *mn*th element which is obtained from a unit-cell analysis, and $\varphi_{m,n}$ is the required phase shift of the *mn*th element to set the required beam. To form a pencil-shaped beam in the direction $\hat{u}_0$, the reflection phase of each unit cell should be set as

$$\varphi_{m,n} = k(|\vec{r}_{m,n} - \vec{r}_f| - \vec{r}_{m,n} \cdot \hat{u}_0) + \varphi_0 \quad (6)$$

where φ_0 is a global reference phase indicating that a relative reflection phase rather than the absolute reflection phase is required for RA design. For a 1-bit resolution, the quantified phase can be written as

$$\psi_{m,n} = round(\frac{\varphi_{m,n}}{\pi}) \times \pi \quad (7)$$

where *round* function is used to round off the numeric values. In order to study the influence of the global reference phase on the far-field electric field, according to the array theory in (5), the relationship curves of the far-field amplitude and phase with the reference phase are calculated.

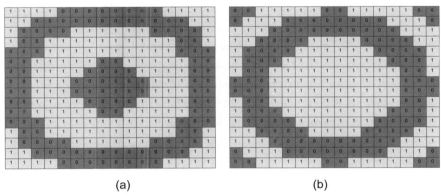

Fig. 4 1-bit compensation phase arrangements for broadside beam at different global reference phases. (a) $\varphi_0 = 50°$. (b) $\varphi_0 = 140°$.

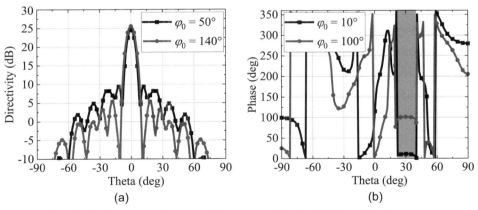

Fig. 5 In the case of the broadside beam, the far-field amplitude and phase with the different global reference phases. (a) Amplitude. (b) Phase.

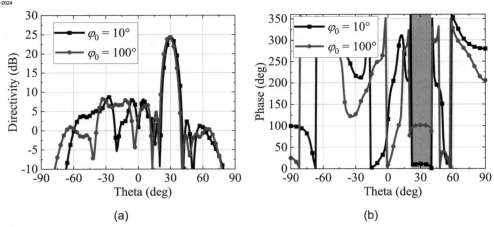

Fig. 6 In the case of scan angle = 30°, the far-field amplitude and phase with the different global reference phases. (a) Amplitude. (b) Phase.

The RA aperture is lying on the *xoy*-plane and the aperture center is located at the center of the coordinate system. The feed antenna is pointing toward the geometrical center of aperture. In first case, we set the array size is N=M=16, the element spacing is λ/2, and the focal-diameter ratio is F/D=1, where F is focal length and D is array dimension. The feed with a $qf = 9$ is incident perpendicular to the center of the array. Using (6) and (7), the 1-bit compensation phase arrangement for the broadside beam with different phase offsets (φ_0 = 50° and 140°) are shown in Fig. 4. The calculated 2-D far-field patterns are shown in Fig. 5 (a). It can be seen that the directivities with the reference phases of 140° and 50° are 25.5 dB and 24.5 dB, respectively, and the side lobe levels (SLLs) are -20.2 dB and -15.1 dB, respectively. The optimization of the reference phase is usually an important parameter of the 1-bit RA or TA to achieve optimal radiation performance. The far-field phase information with two different compensation phase arrangements is also provided in Fig. 5(b), which shows that the phases in 0° direction with the global reference phases of 140° and 50° are 138.7° and 49.9°, respectively. Fig. 6 also shows the far field information of $\varphi_0 = 10°$ and 100° when the scanning angle is 30°. It can be seen that the amplitudes of the far-field in the main lobes are basically the same, and an approximate 90° phase shift in the far-field is achieved through the 1-bit phase arrangement in the array level. This phenomenon indicates that different reference phases can be set in the *x*-direction and *y*-direction to achieve linear to circular polarization conversion. It is worth noting that the value of global reference phase φ_0 directly affects the CP radiation performance since φ_0 is related to quantization errors in the 1-bit case, which can be ignored in the continuous phase. The optimization of the global reference phases will be introduced in Section III-D using the aperture field approach.

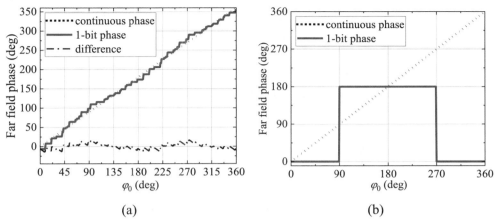

Fig. 7 Influence on the far-field phase in the direction of 0° for the broadside beam with the global reference phase. (a) RA antenna. (b) Conventional phased array antennas.

Fig. 7(a) also gives the influence of the far-field phase in the direction of 0° for the broadside beam with the different global reference phases. It can be seen that in the case of the continuous phase, the two phases show a linear relationship, and the two-phase differences have fluctuated in the range of 17° in the case of the 1-bit phase. This means that choosing a suitable global reference phase can make the CP performance optimal using a 1-bit phase. For comparison, the influence of the far-field phase in the direction of 0° with the global reference phase for the conventional phased array antennas is also investigated in Fig. 7(b). Note that the $|I(m, n)| = 1$ in (5-2), and all elements are uniformly excited for the conventional phased array antennas. Fig. 7(b) indicates that, in the case of the 1-bit phase, the far-field phase is either -180° or 180° with different global reference phases. Benefiting from the spatially-fed frame, where exciting amplitude and phase of each unit are unique, yielding only 1-bit phase controls can realize the continuous far-field phase regulation and pencil-shaped beam scanning.

From the above analysis, it can be seen that it is feasible to use the 1-bit phase controls to realize the linear to circular polarization with simultaneous beam scanning. The utilization of global reference phase differences between two orthogonal polarization waves is the key for the quad-polarization in the main beam and also the reduction from 2-bit to 1-bit phase resolution simplification. The advantages of using the global reference phase control technology to achieve quad polarization reconfigurability include: 1) It reduces the phase resolution of the four-polarization conversion in the element level. By only using a 1-bit phase resolution to realize four kinds of polarization control and beam scanning, which significantly reduces the system complexity. 2) It is easy to realize the broadband design.

At the element level, 1-bit elements are easier to implement broadband design compared with 2-bit elements, because the phases of 1-bit elements need to satisfy two curves with a phase difference of 180°, while the 2-bit elements need four phases with a 90° phase difference. Fewer constraints mean easier implementation of wideband designs. At the array level, since the far-field phase is frequency-independent through code regulation, the broadband circular polarization performance can be achieved by designing a corresponding broadband element.

It should be noted that since the 1-bit phase resolution has a higher phase quantization error than the 2-bit, which inevitably leads to a decrease in antenna aperture efficiency.

3. RA Antenna Design and Optimization

In this section, we will introduce the design and optimization process of quad-polarization reconfigurable RA and beam scanning based on 1-bit phase control.

3.1 Reconfigurable RA Unit Cell Design

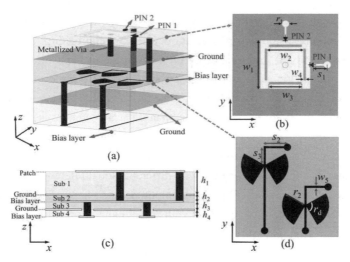

Fig. 8 Geometry of the proposed multi-polarization 1-bit RA element. (a) Schematic. (b) Square patch with four slots. (c) Stack up. (d) Dc biasing circuits.

Table II

Key parameter values

Parameters	Values	Parameters	Values	Parameters	Values
w_1	4.8 mm	s_1	1.3 mm	r_d	90°
w_2	3.5 mm	s_2	2.2 mm	h_1	1.58 mm

Table II Continued

Parameters	Values	Parameters	Values	Parameters	Values
w_3	3.1 mm	s_3	1.8 mm	h_2	0.5 mm
w_4	0.2 mm	r_1	0.7 mm	h_3	0.5 mm
w_5	0.2 mm	r_2	2.3 mm	h_4	0.5 mm

To realize the proposed quad-polarization reconfigurable RA antenna, a new reconfigurable RA element that is capable of independently controlling the reflection phase of x- and y-polarization in a 1-bit manner is introduced. The geometry of the proposed multi-polarization 1-bit RA element is plotted in Fig. 8. This structure geometry is evolved from the classic single-polarization reconfigurable RA. The RA element size is 10 mm× 10 mm, and the structural and geometrical parameters are listed in Table II. It consists of a square patch with four slots, where the center of the patch is connected to the ground and connected to the dc bias layer by two PIN diodes and two metallized vias both in the x- and y-direction. The substrate (Sub 1) is Taconic RF-30 with permittivity 3, loss tangent tanδ = 0.0014, and a thickness of h1=1.58 mm. The PIN diode named MACOM MADP-000907-14020 is used as the RF switch, which is modeled as a series of lumped RLC elements: R = 7.8 Ω and L = 30 pH for the ON state, and C = 25 fF and L = 30 pH for the OFF state. In order to individually turned on or off two PIN diodes both in the x- and y-direction for independent phase controlling, two PIN diodes are individually connected to their own biasing circuits through vias, and then two pairs of open-ended radial stubs to choke the RF signals. Considering that the bias lines are doubled, an additional layer is added at the bottom and is used for the layout of bias lines, which leaves enough room even for a large-sized array. The FR-4 (ε_r=4.4, tanδ=0.02) is used as the substrates (Sub 2, 3, and 4) for biasing circuits.

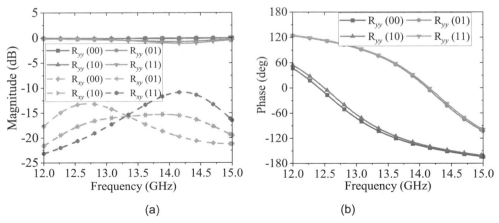

Fig. 9 Simulated element reflection coefficient under y-polarized normal incidence for different PIN diodes states: (a) Magnitude. (b) Co-polarized reflective phase.

In order to represent more intuitively, the two bits binary codes are used to indicate two PIN diode states ("1" denotes ON and "0" denotes OFF). The proposed unit cell has four states: "00" (OFF/OFF), "01" (OFF/ON), "10" (ON/OFF), and "11" (ON/ON). The first code represents the state of the PIN diode along the x-axis, while the second code represents the state PIN diode along the y-axis. The EM performance of the unit cell for four states is simulated using the finite element method (FEM) with periodic boundary conditions. Since the designed element is symmetric for y- or x-polarized incident wave, the performance for y-polarized incidence is only displayed here for brevity. The simulated element performance is shown in Fig. 9. As shown in Fig. 9(a), it can be observed that the co-polarization reflection coefficient loss is < 1.1 dB regardless of any state, the cross-polarization reflection coefficients are < -10 dB in the band of 12-15 GHz, indicating that x-polarized incident wave is totally nearly reflected with low polarization conversion all the PIN diode states. It is shown in Fig. 9(b) that the reflection phase under the y-polarized incident wave only depends on the y-oriented PIN diode state and is almost independent of the orthogonal one. For the cases of "11" and "01", while the y-oriented diode (PIN 2) always is in the ON state, the reflected y-polarized wave has a reflection phase of approximately 65° at 13.5 GHz. For the cases of "00" and "10", while the y-oriented diode always is in the OFF state, the reflected y-polarized wave has a reflection phase of approximately -113°. As a result, an approximate 180° phase difference is obtained for the y-polarized reflection wave when one controls the switch state of the y-oriented diode. When the x-oriented diode PIN state switch, the obtained phase difference is the same as a y-polarized one. Therefore, the reflection phases with low loss performance can be independently controlled in a 1-bit manner both for x- and y-polarization operations.

3.2 Configuration of RA Antenna

The configuration of the RA antenna is shown in Fig. 10, which includes a 16× 16-element RA aperture and an offset-fed horn. The dimension of the RA aperture is 160×160 mm², due to the period of the element being 10 mm. The feed is further tilted by 15° to reduce feed blockage. After a comprehensive parametric study, the feed's phase center is placed at (-30, -30, 158) mm in RA coordinate system, as shown in Fig. 10. The F/D ratio is 1, which is optimized to balance the spillover efficiency and illumination efficiency. In order to simultaneously generate x- and y-polarized components on the RA aperture, the polarization of the feed antenna is set along the H-direction. In other words, the RA aperture is rotated by 45° relative to the feed source coordinate system. A subarray consisting by 2 × 2 elements with the element angular orientation arranged in a 0°, 90°, 180°, and 270° fashion is employed to improve CP quality and suppress cross-polarization level for wide angle scanning beams.

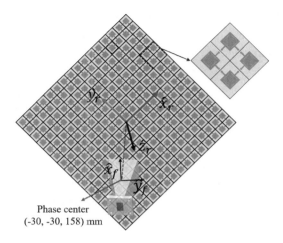

Fig. 10　The configuration of proposed reconfigurable RA antenna.

4. Phase Encoding Arrangement for Four-Polarization Beam Scanning

The phase encoding arrangement of the RA array includes the encoding of the x- and y-polarization. Once the global reference phases of the two, i.e., φ_0^x and φ_0^y, are determined, the phase encoding arrangement ($\psi^{x/y}$) can be obtained accordingly by

$$\varphi_{m,n}^{x/y} = \mathrm{k}(|\vec{r}_{m,n} - \vec{r}_f| - \vec{r}_{m,n} \cdot \hat{u}_0) + \varphi_0^{x/y} \tag{8}$$

$$\psi_{m,n}^{x/y} = round(\frac{\varphi_{m,n}^{x/y}}{\pi}) \times \pi \tag{9}$$

Usually, φ_0^x and φ_0^y are an optimization variable which determine the radiation performance of the array, such as gain, sidelobe, crossover level, etc. If we assume φ_0^y is a parameter related to φ_0^x and can be written as

$$\varphi_0^y = \varphi_0^x + \delta \tag{10}$$

where δ is an undetermined value, depending on which polarizations the RA antenna works in. The polarization of the feed is generally fixed, such as VP, HP, RHCP, or LHCP. For example, if we set the polarization of the feed along the horizontal direction, it can be decomposed into two equal amplitude and opposite phase electric field components, one along the x-direction and the other along the $-y$-direction. Moreover, the reflective phases of each element in the RA aperture can be individually controlled along the x- and y- axis. Thus, the polarization of the reflected wave depends on the phase difference between two field components in the x- and y-directions. The multiple-polarization capability of the RA antenna is described below:

1) *H-polarization beam scanning*: In this case, the operating polarization of the RA antenna is the same as the feed, which means $\delta = 0$ and the phase encoding arrangement in x-polarization and y-polarization are the same indicating the coding on the RA array includes the states "00" and "11".

2) *V-polarization beam scanning*: Since the operating linear polarization of the RA antenna is orthogonal with the feed, the relation of $\delta = \pi$ should be satisfied. The coding on the RA array includes the states "01" and "10" while the element is working at the H-V polarization conversion state.

3) *RHCP beam scanning*: As previously shown, the desired far-field phase shift can be achieved by the 1-bit phase arrangement of the array via the global reference phase. When $\delta = \pi/2$ is met, which results in a 90° phase advance in the x-polarization reflect wave compared to the y-polarization, that enabling RHCP radiation. In the case of circular polarization, the encoding of the RA should include four states, namely "00", "01", "10", and "11". It is worthwhile to emphasize that the global reference phase (φ_0^x) in each beam scan angle needs to be carefully chosen to minimize quantization loss for radiation performance, especially the axis ratio (AR).

4) *LHCP beam scanning*: This case is similar to RHCP beam scanning, while $\delta = -\pi/2$ should be followed.

Table II summarizes the value of δ and the types of RA coding for the four polarizations. Since the ±90° phase shifts x- polarization and y-polarization can be independently achieved by 1-bit phase arrangement via the global reference phase, this means that the RA aperture can be excited by a LP (HP or VP) or a CP (RHCP or LHCP) feed, while reflecting the field in the desired direction with any polarization. The values of δ and types of RA coding for four polarizations under different feed polarizations are also listed in Table III.

Table III

Values of δ and types of RA coding for the four polarizations under different feed polarizations

Feed Polarization	RA Polarization	δ	PIN Diode Working States
HP	HP	0	00, 11
	VP	π	01, 10
	RHCP	$\pi/2$	00, 01, 10, 11
	LHCP	$-\pi/2$	00, 01, 10, 11
VP	HP	π	01, 10
	VP	0	00, 11
	RHCP	$-\pi/2$	00, 01, 10, 11
	LHCP	$\pi/2$	00, 01, 10, 11

Table III Continued

Feed Polarization	RA Polarization	δ	PIN Diode Working States
RHCP	HP	$-\pi/2$	00, 01, 10, 11
	VP	$\pi/2$	00, 01, 10, 11
	RHCP	π	01, 10
	LHCP	0	00, 11
LHCP	HP	$\pi/2$	00, 01, 10, 11
	VP	$-\pi/2$	00, 01, 10, 11
	RHCP	0	00, 11
	LHCP	π	01, 10

5. Global Reference Phase Optimization Using the Aperture Field Approach

In the previous part, the encoding arrangements for four-polarization beam scanning are given. Due to the global reference phases are related to RA radiation performance, the global reference phases in each beam scan angle need to be carefully chosen. The optimization of the reference phase for LP beam scanning is well studied in [5] using 1-bit reconfigurable designs. Since the polarization information of the feed horn and elements are not accounted in the array theory formulation, the CP performance cannot be evaluated. It is worth noting that since the actual physical element still has a cross-polarization level of about -13 dB [see Fig. 9(a)], this item will greatly affect the AR performance. In order to address the above problems, a general optimization technique for the global reference phase using the aperture field approach is shown below.

The field incident at each RA element is determined by the position and radiation pattern of the feed. A directional antenna, i.e., a horn, is usually used as feed, and its radiation pattern is modeled as a *cosine* function in the feed coordinate system. For an *x*-polarized ideal feed, the radiated field is given by

$$E^F(\theta,\varphi) = \frac{jke^{-jkr}}{2\pi r}[\hat{\theta}\cos^{q_E}(\theta)\cos\varphi - \hat{\varphi}\cos^{q_H}(\theta)\sin\varphi] \quad (11)$$

where qE and qH are the E- and H-plane pattern power factors of feed, respectively. For an *y*-polarized feed, the radiated field is given by

$$E^F(\theta,\varphi) = \frac{jke^{-jkr}}{2\pi r}[\hat{\theta}\cos^{q_E}(\theta)\sin\varphi + \hat{\varphi}\cos^{q_H}(\theta)\cos\varphi] \quad (12)$$

A proper superposition of (11) and (12) gives a circularly polarized far field, namely,

$$E^F(\theta,\varphi) = \frac{jke^{-jkr}}{2\pi r}e^{j\gamma}[\hat{\theta}\cos^{q_E}(\theta) + \hat{\varphi}\cos^{q_H}(\theta)] \quad (13)$$

where $\tau=1$ for LHCP, and $\tau=-1$ for RHCP. Using the transformation of the field components from feed to RA coordinate system, the incident field on each RA element can be obtained. The incident electric field on the surface of the RA can be written for x and y linear polarizations in Cartesian coordinates as

$$\vec{E}_{\text{inc}}(x, y) = \hat{x} E_{\text{inc},x}(x, y) + \hat{y} E_{\text{inc},y}(x, y) \tag{14}$$

where $E_{\text{inc},x}$ and $E_{\text{inc},x}$ are the incident electric field for x and y linear polarization, respectively. The relation between the incident and reflected tangential electric field at each element of the RA is given by a matrix of reflection coefficients that expressed as

$$\vec{E}_{\text{ref}}(x_m, y_n) = R^{mn} \cdot \vec{E}_{\text{inc}}(x_m, y_n) \tag{15}$$

where Rmn is the reflection coefficient of the (m, n)-th element and can be written as

$$\mathbf{R}^{mn} = \begin{bmatrix} R_{xx}^{mn} & R_{xy}^{mn} \\ R_{yx}^{mn} & R_{yy}^{mn} \end{bmatrix} \tag{16}$$

The reflection coefficient is obtained in full-wave simulations of unit cell, and the state of each element is determined according to phase encoding arrangement principle.

Once the tangential fields on the RA surface are obtained, the far field radiated by the RA antenna can be computed using the Principle of Equivalence. The radiated far field of the RA aperture in spherical coordinates is given by

$$E_\theta = \frac{jke^{-jkr}}{4\pi r}[P_x \cos\varphi + P_y \sin\varphi - \eta\cos\theta \cdot (Q_x \sin\varphi - Q_y \cos\varphi)] \tag{17-1}$$

$$E_\varphi = -\frac{jke^{-jkr}}{4\pi r}[\eta(Q_x \cos\varphi + Q_y \sin\varphi) + \cos\theta \cdot (P_x \sin\varphi - P_y \cos\varphi)] \tag{17-2}$$

where $\eta=120\pi$ is the impedance in vacuum, P_x, P_y, Q_x, and Q_y are the spectrum functions, which can be obtained using the Fourier transforms of the tangential electric and magnetic fields in the RA aperture and expressed as

$$P_{x/y}(u, v) = K \cdot \sum_{m}^{M} \sum_{n}^{N} E_{\text{ref}, x/y}(x_m, y_n) e^{jk(ux_m + vy_n)} \tag{18-1}$$

$$Q_{x/y}(u, v) = K \cdot \sum_{m}^{M} \sum_{n}^{N} H_{\text{ref}, x/y}(x_m, y_n) e^{jk(ux_m + vy_n)} \tag{18-2}$$

where $H_{\text{ref}}, x/y$ is tangential reflected magnetic field in x or y polarization, which can be computed under local plane wave assumption. $u = \sin\theta \cos\varphi$, $v = \sin\theta \sin\varphi$. K is a constant, and can be written as

$$K = p^2 \text{sinc}(\frac{kp \sin\theta \cos\varphi}{2}) \text{sinc}(\frac{kp \sin\theta \sin\varphi}{2}) \tag{19}$$

where p is the period of element. The RHCP and LHCP components and AR can be obtained as

$$E_{\text{RHCP}} = \frac{1}{\sqrt{2}}(E_\theta - jE_\varphi) \qquad (20\text{--}1)$$

$$E_{\text{LHCP}} = \frac{1}{\sqrt{2}}(E_\theta + jE_\varphi) \qquad (20\text{--}2)$$

$$\text{AR} = \frac{|E_{\text{RHCP}}| + |E_{\text{LHCP}}|}{|E_{\text{RHCP}}| - |E_{\text{LHCP}}|} \qquad (20\text{--}3)$$

The CP performance can be quickly calculated by the superposition of the aperture fields approach without time-consuming complete full-wave simulation. The optimization flow of CP performance is presented as follows.

1) Set the scanning angle and polarization mode.

2) Set the global reference phase in *x*-polarization ($\varphi_0^x = 0°$).

3) Determine the global reference phase in *y*-polarization ($\varphi_0^y = \varphi_0^x + \delta$) according to the polarization mode.

4) Calculate the phase encoding arrangement using (6) and (7).

5) According to the phase encoding arrangement in *Step 4* and the full-wave simulation data of the unit cell, the reflection coefficient in (16) of each element can be obtained.

6) Calculate the reflected tangential electric field in the RA aperture using (15).

7) Compute the far field radiated by the RA antenna according to (17)-(20).

8) Record gain, sidelobe level and the axial ratio in the desired direction. Then turn to *Step 2* with $\varphi_0^x = \varphi_0^x + d\varphi$, where $d\varphi$ is the angle step. When $\varphi_0^x = 360°$ is satisfied, the whole loop is over.

9) The optimal phase encoding arrangement can be obtained by the recorded data.

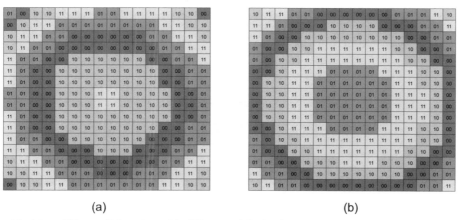

Fig. 11 Two different RA codes with different global reference phases corresponding to the worst and the optimal performance. (a) $\varphi_0^x = 170°$. (b) $\varphi_0^x = 37°$.

In order to show the feasibility of the above process, the optimization with the radiated beam pointing to 15° in RHCP mode is carried out. The RA aperture is excited by a LP horn feed with radiation pattern $qE= qH=6$ as depicted in Fig. 10. According to the data recorded in the optimization process, the values of $\varphi_0^x =170°$ and $\varphi_0^x = 37°$, corresponding to the worst and the optimal performance, can be obtained. Fig. 11 gives two codes in RA aperture with the global reference phases of $\varphi_0^x =170°$ and $\varphi_0^x = 37°$ in x-polarization while the global reference phase in y-polarization $\varphi_0^y = \varphi_0^x + 90°$. Figs. 12 and 13 illustrate the radiation performance computed by the aperture field approach in the cases of two codes, where the cross-polarization levels in the 15° direction are -11.2 dB and -40 dB, and the ARs are 5 dB and 0 dB for two codes, respectively. It can be clearly seen that the selection of the global reference phase is directly related to the circular polarization performance. The full-wave simulations are also given for comparison, which provide the most accurate solution to the RA radiation problem. It's worth to noted that full-wave simulation results have more advantageous than measured ones due to measured results are susceptible to fabrication and measurement errors. The total processing time using the full-wave technique using the transient solver of the CST Microwave Studio was approximately 1 hour and 50 minutes, which runs on a PC with Intel I9 12900 CPU and 64 GB RAM; while the total processing time using the aperture field approach is only approximately 2.2 seconds.

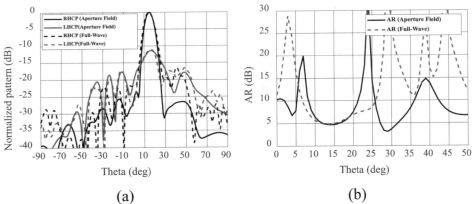

Fig. 12 Radiation performance in RHCP mode when the global reference phase of $\varphi_0^x =170°$. (a) Far-field pattern. (b) AR.

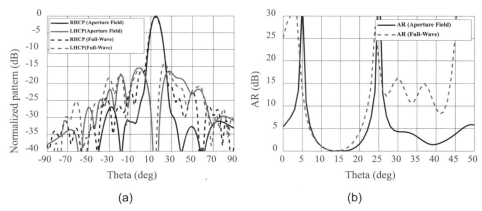

Fig. 13 Radiation performance in RHCP mode when the global reference phase of $\varphi_0^x = 37°$. (a) Far-field pattern. (b) AR.

The comparison shows that the aperture field analysis approach can accurately calculate the radiation patterns and cross-polarization levels in the main beam area. Some discrepancies are observed outside the main beam area since all approximations in RA element (e.g., oblique incidence approximation), mutual coupling, and edge-diffraction effects are not taken into account in the aperture field approach. However, the aperture field approach can quickly and accurately calculate the interested performance parameters, which provides a feasible calculation method for the optimization of the global reference phase or the codes in the RA surface.

6. Antenna Fabrication and Measurement

To experimentally validate the performance of the proposed reconfigurable strategy, a 16 ×16-unit-cell RA is fabricated using printed circuit board (PCB) technology. A pyramidal horn antenna (Ainfoinc LB-62-15-A) is used as the feed antenna. For the purpose of realizing beam steering and polarization reconfigurable, as shown in Fig.14 (a), a dc bias line layer is needed to design for controlling the states of each PIN diode independently. As shown in Figs.14 (b) and (c), the dimension of RA is 220 mm × 220 mm with an effective size of 160 mm × 160 mm, and the extra room is designated for PIN diode dc lines and digital commands. The element is linked to the bias line and FPGA-based digital control circuit to control individually 512 PIN diodes. To verify the capability of the polarization reconfiguration for the designed RA antenna, we investigate the dual-linear and dual-circular polarized beam scanning performance. For convenient measurement, the scanning beam of the RA is measured at H- and V-planes, and the polarization of the feed along the H-direction, where the V- H coordinate system can be obtained by rotating the original X-Y coordinate axis to

45° as shown in Fig. 14(a). The measurement was carried out using the antenna near-field test system in an anechoic chamber, as shown in Fig. 15. The near-field measurements are done in an anechoic chamber where the absorbers are placed on the walls to avoid reflections and reduce the effects of stray signals. A vector network analyzer (VNA) is used to generate the source signal and acts as a receiver, which is connected to the probe and feed horn antenna, respectively. The electric field is measured using a standard Ku-band near-field waveguide probe, which is in front of the aperture. The near E-field value at each pixel can be obtained by scanning the probe along the x- and y-axis. The far-field radiation pattern can be readily obtained from the measured near-field data using the fast-Fourier transformation (FFT).

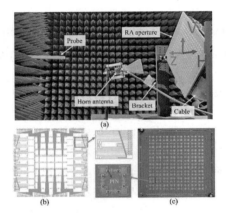

Fig. 14 Photograph of the fabricated RA prototype. (a) Antenna under test in an anechoic chamber. (b) Details on dc bias line layer. (c) RA aperture.

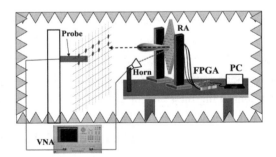

Fig. 15 Measurement in the near-field system.

7. H- and V-Polarization Mode Beam Scanning

In dual-linear polarized beam scanning mode, the codes for the RA include the states "00" and "11" in H-polarization mode while including the states "01" and "10" in

V-polarization mode. The details phase encoding arrangement can be calculated according to the rules introduced in Section II-B.

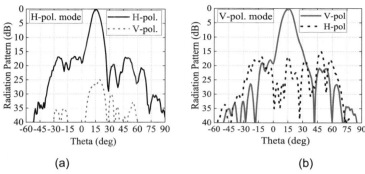

Fig. 16 Radiation patterns of the scanning beam at 15° in V-plane at 13.5 GHz. (a) Measured pattern for H-polarization mode. (b) Measured pattern for V- polarization mode.

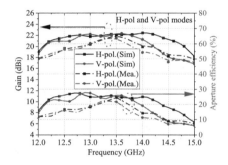

Fig. 17 Measured and simulated dominant polarization gains and aperture efficiencies for H- and V-polarization modes.

The measured radiation patterns of the beam pointing at 15° in V-plane at 13.5 GHz in cases of H- and V-polarization modes are plotted in Fig. 16. It can be seen that the SLL and the cross-polarization level are -16.9 dB and -25.1 dB, respectively, in cases of H-polarization mode. The SLL and the cross-polarization level are -17.9 dB and -15.1 dB, respectively, in the case of the V-polarization mode. The measured half power beam widths (HPBWs) of dominant polarization components in the two modes are 9.1° and 8.7°, respectively. The measured and simulated dominant polarization gains and aperture efficiencies for H-polarization and V-polarization modes are shown in Fig. 17. The measured gains of the proposed RA antenna at 13.5 GHz are 22 dBi and 21.5 dBi (with aperture efficiencies of 24.5 % and 21.7%) corresponding to H- and V-polarization modes.

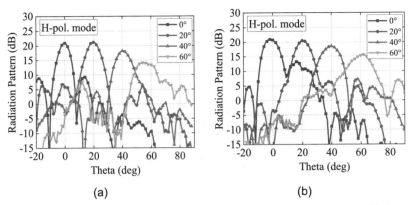

Fig. 18   Measured radiation performance in H-polarization mode with the scanning beams in (a) H-plane and (b) V-plane at 13.5 GHz.

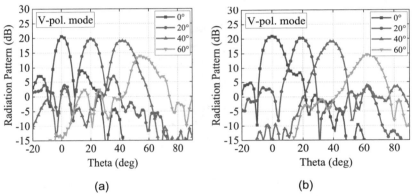

Fig. 19   Measured radiation performance in V-polarization mode with the scanning beams in (a) H-plane and (b) V-plane at 13.5 GHz.

To show the beam scanning ability, the dual-linear-polarization large-angle beam scanning radiation patterns in both H- and V-planes have been shown in Figs. 18 and 19. The measured angles are from 0° to 60° with a 20° increment. Due to the symmetry of feed blockage, the scanning beams in the H-plane at scanning angles of 0° to -60° are similar to 0° to 60°. However, the beam scanning performance is not symmetric in V-plane because of the different feed blockage effects.

8. CP Mode Beam Scanning

In the CP mode beam scanning, the codes for the RA array include the states "00", "01", "10", and "11". The detailed phase encoding arrangement can be calculated and optimized according to the rules introduced in Section II-B and C.

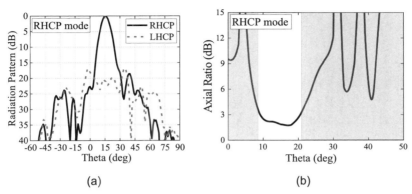

Fig. 20 Measured radiation performance in RHCP mode. (a) Radiation pattern. (b) AR.

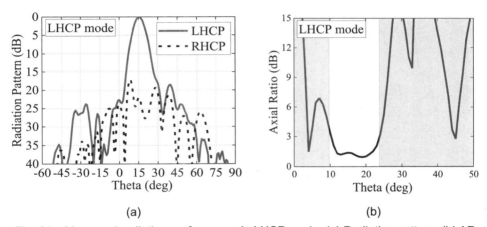

Fig. 21 Measured radiation performance in LHCP mode. (a) Radiation pattern. (b) AR

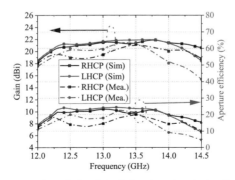

Fig. 22 Measured and simulated dominant polarization gains and aperture efficiencies for RHCP and LHCP modes.

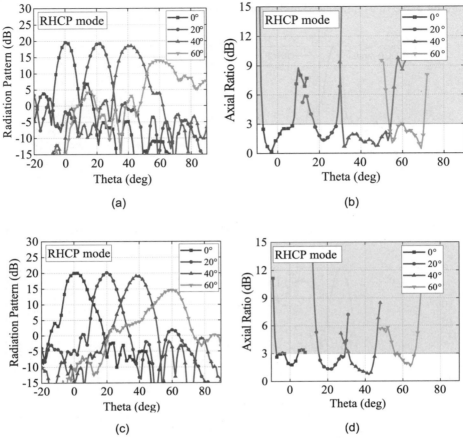

Fig. 23 Measured radiation performance in RHCP mode with the scanning beams in H- and V-plane 13.5 GHz. (a) Radiation pattern and (b) AR in H-plane. (c) Radiation pattern and (d) AR in V-plane.

The measured radiation patterns and AR of the beam pointing at 15° in V-plane at 13.5 GHz in the case of CP mode are plotted in Figs. 20 and 21. It can be seen from Fig. 20(a) that the SLL and the cross-polarization level are -18.5 dB and -16.1 dB, respectively, in the case of RHCP mode. The measured axis ratios are < 3 dB in the range of 9.1-21.5°, as shown in Fig. 20(b), which shows a good circular polarization characteristic.

Similar radiation performance can also be observed in LHCP mode, as shown in Fig. 21. As shown in Fig. 22, the measured gains of the proposed RA antenna at 13.5 GHz are 21.4 dBi and 21.2 dBi (with the aperture efficiencies of 21.3 % and 20.1%) corresponding to RHCP and LHCP modes. Figs. 23 and 24 give the measured radiation performance in RHCP and LHCP modes with the scanning beams in H-and V-plane at 13.5 GHz. It is observed that the RA prototype can achieve excellent dual-CP beam scanning with an angle up to 60° in both H- and V-planes at 13.5 GHz. Finally, the measured gains of four polarization modes at 13.5 GHz with the beam scanning in both H- and V-planes are tabulated in Tables IV and V.

Table IV

Measured gains of scanning beams in H-plane at 13.5 GHz

Scan angle	Gain (dBi)			
	HP	VP	RHCP	LHCP
0°	21.00	20.59	19.65	19.86
20°	21.25	19.76	19.30	20.43
40°	18.22	19.09	18.55	19.88
60°	14.11	13.90	14.08	14.42

Table V

Measured gains of scanning beams in V-plane at 13.5 GHz

Scan angle	Gain (dBi)			
	HP	VP	RHCP	LHCP
0°	21.00	20.93	20.06	19.84
20°	20.51	20.41	20.13	20.13
40°	18.64	19.22	19.17	18.4
60°	15.58	14.57	14.58	15.13

The performance comparison of the proposed antenna with previous works of reconfigurable spatially-fed array is listed in Table VI. It can be seen that the studies in [14], [19-22], [24-26], [28] are mostly limited to single polarization. The reconfigurable architectures proposed in the literature [29-31], [33] operate in multi-polarization reconfiguration modes. Only [29] proposed a quad-polarization beam-scanning mechanically reconfigurable RA antenna. However, the mechanical reconfigurable method has the disadvantages of complex hardware control and slow response compared to the electrical reconfigurable strategy. The advantage and original contribution of this work proposes novel efficient electrical reconfigurable RA with more polarization modes, i.e., the dual-LP and dual-CP modes with simultaneous large-angle beam scanning capability, that cannot be achieved by other works. Moreover, the proposed RA design only needs two PIN diodes in each element to realize a 1-bit phase resolution, which can significantly reduce the cost and the complexity of the bias circuit.

Table VI
Performance comparison of the proposed RA antenna with previous works of reconfigurable spatially-fed array

Ref. and year	Antenna Type	Freq. (GHz)	Number of elements	Number of diodes in one element	Element Loss (dB)	Polarization modes	Max. scanning angle	Max gain (dBi) / Ap. eff.	3-dB gain bandwidth	X-pol. @0° (dB)
[14], 2021	TA	13.5	16×16	2	~1	LP	60°	22.3/25.6%	29%	>21
[19], 2022	RA	12.5	16×16	1	0.5-1.5	LP	60°	20.1/24%	38.4%	>26
[20], 2022	RA	10	16×16	2	<1.1	LP	60°	18.5/13.8%	30%	>18.5
[21], 2022	RA	15	16×16	1	<0.8	LP	60°	21.6/25%	22.5% for 1-dB	>20
[22], 2023	RA	25	20×20	N.A.	N.A.	LP	60°	20.0/11.5%	N.A.	N.A.
[24], 2022	RA	12.5	64	1	N.A.	LP	50°	17.4/17.5%	N.A.	N.A.
[25], 2023	RA	10.1	12×12	2	<1	LP	50°	18.8/20.34%	15.68%	>20
[26], 2022	RA	20.2	16×16	4	<1	CP	60°	20.4/24%	32%	>18

Quad-Polarization Reconfigurable Reflectarray with Independent Beam-Scanning and Polarization Switching Capabilities

Table VI Continued

Ref. and year	Antenna Type	Freq. (GHz)	Number of elements	Number of diodes in one element	Element Loss (dB)	Polarization modes	Max. scanning angle	Max gain (dBi) / Ap. eff.	3-dB gain bandwidth	X-pol. @0° (dB)
[28], 2021	RA	9.5	16×16	2	<1	CP	60°	21.8/20%	9.4%	>18
[29], 2023	RA	10	13×13	N.A.	<2	Dual-LP, Dual-CP	60°	27.2/24.7% for x-pol. 26.7/22.0% for y-pol. 27.1/24.2% for CP	24% 14% 18%	>17
[30], 2022	TA	S-band	8×8	6	<5	Dual-LP	45°	7.9/4.8% for VP 7.4/4.3% for HP	N.A.	N.A.
[31], 2022	RA	7.45	20×20	4	<0.5	Dual-LP	40°	21.13/16.35% for x-pol. 20.89/15.47 for y-pol.	11.0% 11.6%	>25
[33], 2017	TA	29	20×20	2	<1.0	RHCP LHCP	60°	20.8/9.5 for RHCP 20.8/9.5 for LHCP	14.6% 14.6%	>25
This work	RA	13.5	16×16	2	<1.1	Dual-LP, Dual-CP	60°	22/24.5% for HP 21.5/21.7% for VP 21.4/21.3% for RHCP 21.2/20.1% for LHCP	14.4% 16.5% 18.8% 15.4%	>25 >17 >20 >22

| 027

9. Conclusion

In this paper, a novel quad-polarization RA with beam scanning capability is designed, optimized, fabricated, and measured. We have discussed the principle of quad-polarization conversion from the unit cell level, and concluded that at least a 2-bit phase resolution is required to realize beam scanning and circular polarization conversion. An interesting phenomenon, which indicates that the far-field phase shift in the spatially-fed architecture can be achieved by changing the global reference phase only using a 1-bit phase control method, was found in this work. Based on this principle, the phase encoding arrangement for four-polarization beam scanning and the global reference phase optimization using the aperture field approach is introduced. Finally, a RA prototype is measured, and the results have verified the capability of simultaneous polarization switching and beam scanning. It is believed that the proposed low-lost multi-polarization reconfigurable RA design to be a promising candidate for mobile communication and cognitive radar applications.

Acknowledgments

This work was supported by the National Natural Science Foundation of China (U2141233, 62171416, and 62071436). (H. Yu and Z. Z. Zhang contributed equally to this work.)

Author Information

H. Yu, Z. Y. Zhang, J. X. Su, M. J. Qu, and Z. R. Li are with the State Key Laboratory of Media Convergence and Communication, Communication University of China, Beijing 100024, China.

S. H. Xu and F. Yang are with the Department of Electronic Engineering, Tsinghua University, Beijing 100084, China.

References

[1] HUM S V, PERRUISSEAU-CARRIER J. Reconfigurable reflectarrays and array lenses for dynamic antenna beam control: a review[J]. IEEE Trans. Antennas Propag., 2014, 62(1): 183-198.

[2] NAYERI P, YANG F, ELSHERBENI A Z. Beam–scanning reflectarray antennas: a technical overview and state of the art[J]. IEEE Antennas Propag. Mag., 2015, 57(4): 32-47.

[3] GAO S, GUO Y J, SAFAVI-NAEINI S A, et al. Guest editorial low-cost wide-angle beam-scanning antennas[J]. IEEE Trans. Antennas Propag., 2022, 70(9): 7378-7383.

[4] KAMODA H, IWASAKI T, TSUMOCHI J, et al. 60-GHz electronically reconfigurable large reflectarray using single bit phase shifters[J]. IEEE Trans. Antennas Propag., 2011, 59(7): 2524-2531.

[5] YANG H, et al. A 1-bit 10 × 10 reconfigurable reflectarray antenna: design, optimization, and experiment[J]. IEEE Trans. Antennas Propag., 2016, 64(6): 2246-2254.

[6] TRAMPLER M E, GONG X. Phase-agile dual-resonance single linearly polarized antenna element for reconfigurable reflectarray applications[J]. IEEE Trans. Antennas Propag., 2019, 67(6): 3752-3761.

[7] RIEL M, LAURIN J J. Design of an electronically beam scanning reflectarray using aperture-coupled elements[J]. IEEE Trans. Antennas Propag., 2007, 55(5): 1260-1266.

[8] XU H, XU S, YANG F, et al. Design and experiment of a dual-band 1-bit reconfigurable reflectarray antenna with independent large-angle beam scanning capability[J]. IEEE Antennas Wireless Propag. Lett, 2020, 19(11): 1896-1900.

[9] PEREZ-PALOMINO G, ENCINAR J A, BARBA M, et al. Design and evaluation of multi-resonant unit cells based on liquid crystals for reconfigurable reflectarrays[J]. IET Microw. Antennas Propag., 2012, 6(3): 348-354.

[10] PEREZ-PALOMINO G, et al. Accurate and efficient modeling to calculate the voltage dependence of liquid crystal-based reflectarray cells[J]. IEEE Trans. Antennas Propag., 2014, 62(5): 2559-2668.

[11] HAN J, LI L, LIU G, et al. A wideband 1 bit 12 × 12 reconfigurable beam-scanning reflectarray: design, fabrication, and measurement[J]. IEEE Antennas Wireless Propag. Lett., 2019, 18(6): 1268-1272.

[12] WANG Z, et al. 1 bit electronically reconfigurable folded reflectarray antenna based on p-i-n diodes for wide-angle beam-scanning applications[J]. IEEE Trans. Antennas Propag., 2020, 68(9): 6806-6810.

[13] CLEMENTE A, DUSSOPT L, SAULEAU R, et al. Wideband 400-element electronically reconfigurable transmitarray in X-band[J]. IEEE Trans. Antennas Propag., 2013, 61(10): 5017-5027.

[14] XIAO Y, YANG F, XU S, et al. Design and implementation of a wideband 1-bit transmitarray based on a Yagi–Vivaldi unit cell[J]. IEEE Trans. Antennas Propag., 2021, 69(7): 4229-4234.

[15] WANG M, XU S, YANG F, et al. A 1-bit bidirectional reconfigurable transmit-reflect-array using a single-layer slot element with PIN diodes[J]. IEEE Trans. Antennas Propag 2019, 67(9): 6205-6210.

[16] HALL P S, GARDNER P, KELLY J, et al. Antenna challenges in cognitive radio[C]. Taipei, Taiwan: ISAP, 2008.

[17] SHARMA S K, CHIEH J C S. Multifunctional antennas and arrays for wireless communication systems[M]. Piscataway, NJ, USA: IEEE Press, 2021.

[18] YANG H, et al. A 1600-element dual-frequency electronically reconfigurable reflectarray at X/Ku-band[J]. IEEE Trans. Antennas Propag, 2017, 65(6): 3024-3032.

[19] XIANG B J, DAI X, LUK K M. A wideband low-cost reconfigurable reflectarray antenna

with 1-bit resolution[J]. IEEE Trans. Antennas Propag., 2022, 70(9): 7439-7447.

[20] LUYEN H, ZHANG Z, BOOSKE J H, et al. Wideband, beam-steerable reflectarray antennas exploiting electronically reconfigurable polarization-rotating phase shifters[J]. IEEE Trans. Antennas Propag, 2022, 70(6): 4414-4425.

[21] ZHOU S G, et al. A wideband 1-bit reconfigurable reflectarray antenna at Ku-band[J]. IEEE Antennas Wireless Propag. Lett., 2022, 21(3): 566-570.

[22] SERUP D E, PEDERSEN G F, ZHANG S. Electromagnetically controlled beam-steerable reflectarray antenna[J]. IEEE Trans. Antennas Propag, 2023, 71(5): 4570-4575.

[23] CAO Z, LI Y, ZHANG Z, et al. Single motor-controlled mechanically reconfigurable reflectarray[J]. IEEE Trans. Antennas Propag., 2023, 71(1): 190-199.

[24] ZHANG H, WU W, CHENG Q, et al. Reconfigurable reflectarray antenna based on hyperuniform disordered distribution[J]. IEEE Trans. Antennas Propag., 2022, 70(9): 7513-7523.

[25] ZHENG K, XU K, CHEN S, et al. One-bit wideband reconfigurable reflectarray with stable beam-scanning gain for X-band application[J]. Microw. Opt. Technol. Lett., 2023, 65(8): 2323-2330.

[26] WU F, LU R, WANG J, et al. Circularly polarized one-bit reconfigurable ME-dipole reflectarray at X-band[J]. IEEE Antennas Wireless Propag. Lett., 2022, 21(3): 296-500.

[27] ZHANG M T, et al. Design of novel reconfigurable reflectarrays with single-bit phase resolution for Ku-band satellite antenna applications[J]. IEEE Trans. Antennas Propag., 2016, 64(5): 1634-1641.

[28] WU F, LU R, WANG J, et al. A circularly polarized 1 bit electronically reconfigurable reflectarray based on electromagnetic element rotation[J]. IEEE Trans. Antennas Propag., 2021, 69(9): 5585-5595.

[29] WANG M, MO Y, XIE W, et al. A 1-bit all-metal wide-angle and multi-polarization beam-scanning reconfigurable reflectarray antenna[J]. IEEE Antennas Wireless Propag. Lett., 2023, 22(5): 1015-1019.

[30] HWANG M, KIM G, KIM J, et al. A simultaneous beam steering and polarization converting S-band transmitarray antenna[J]. IEEE Access, 2022, 10: 105111-105119.

[31] ZHANG N, et al. A dual-polarized reconfigurable reflectarray antenna based on dual-channel programmable metasurface[J]. IEEE Trans. Antennas Propag., 2022, 70(9): 7403-7412.

[32] YANG H, YANG F, XU S, et al. A 1-bit multipolarization reflectarray element for reconfigurable large-aperture antennas[J]. IEEE Antennas Wireless Propag. Lett., 2017, 16:581-584.

[33] DI PALMA L, CLEMENTE A, DUSSOPT L, et al. Circularly-polarized reconfigurable transmitarray in Ka-band with beam scanning and polarization switching capabilities[J]. IEEE Trans. Antennas Propag., 2017, 65(2): 529-540.

[34] PEREIRA R, GILLARD R, SAULEAU R, et al. Dual linearly-polarized unit-cells with nearly 2-bit resolution for reflectarray applications in X-band[J]. IEEE Trans. Antennas Propag., 2012, 60(12): 6042-6048.

[35] MEI P, ZHANG S, PEDERSEN G F. A wideband 3-D printed reflectarray antenna with mechanically reconfigurable polarization[J]. IEEE Antennas Wireless Propag. Lett., 2020, 19(10):

1798-1802.

[36] CHANG D C, HUANG M C. Multiple-polarization microstrip reflectarray antenna with high efficiency and low cross-polarization[J]. IEEE Trans. Antennas Propag., 1995, 43(8): 829-834.

[37] PERRUISSEAU-CARRIER J. Dual-polarized and polarization-flexible reflective cells with dynamic phase control[J]. IEEE Trans. Antennas Propag., 2010, 58(5): 1494-1502.

[38] YANG H, et al. A study of phase quantization effects for reconfigurable reflectarray antennas[J]. IEEE Antennas Wireless Propag. Lett., 2017, 16:302-305.

[39] NAYERI P, YANG F, ELSHERBENI A Z. Reflectarray Antennas: Theory Designs and Applications[M]. Hoboken, NJ:Wiley, 2018.

[40] LO Y T, LEE S W. Antenna Handbook[M]. New York:Van Nostrand Reinhold, 1993.

[41] PRADO D R, et al. Efficient crosspolar optimization of shaped-beam dual-polarized reflectarrays using full-wave Analysis for the antenna element characterization[J]. IEEE Trans. Antennas Propag., 2017, 65(2): 623-635.

Analysis of Scattering Characteristics of Height-Adjustable Phased Array with Ultra-Wideband Dual-Linear Polarized RCS Amplitude Regulation[*]

1. Introduction

Radar cross section (RCS) is used to measure the ability of a target to scatter radar signals and is generally associated with military platforms. The low RCS characteristics of the equipment can reduce the detectability of the platform. Phased array antennas are widely used in modern combat radar systems due to their flexible beam scanning characteristics. However, the large antenna aperture increases the RCS and improves the platform's visibility. Early antenna RCS reduction (RCSR) technology mainly works on a single antenna. However, those low RCS antennas have difficulty forming an array with flexible beam scanning properties. Recently, the RCSR of array antennas has attracted significant attention. The artificial magnetic conductor (AMC) ground, multifunctional metasurface, and absorbers have been employed to realize the antenna array's low RCS property. In [13], by injecting and extracting ethanol, a slot antenna array with a tunable RCS was obtained. Recently, some researchers have relied on structural mode scattering and antenna mode scattering to analyze the antenna's in-band RCSR. In [19], an antenna was connected to a phase shifter and an amplifier in its circulation path; thus, the phase and amplitude of the antenna mode scattering field were controlled for the cancellation of structural mode scattering. Moreover, each element was connected to a delay line to provide an additional phase excitation. When the co-polarized wave was impinging on the surface, the in-band RCSR based on the phase cancellation method was achieved. Because the scattering control part is behind the antenna, the RCSR is limited by the antenna's operation frequency and polarization. Therefore, it is uneasy to realize an antenna array's in- and out-of-band RCSR for both polarizations. Moreover, due to the application of various radar systems operating at different frequency bands, enhancing the RCSR bandwidth has recently attracted considerable interest. The phased array (PA) monostatic RCS (mRCS) amplitude regulation for dual-linear polarized waves over an ultrawide frequency band has rarely been reported in the open literature.

---

[*] The paper was originally published in *IEEE Transactions on Antennas and Propagation*, 2023, 71 (11), and has since been revised with new information. It was co–authored by Pan Li, Hang Yu, Jianxun Su, Zengrui Li, Hongcheng Yin, and Zhihe Xiao.

Analysis of Scattering Characteristics of Height-Adjustable Phased Array with Ultra-Wideband Dual-Linear Polarized RCS Amplitude Regulation

In this paper, we will analyze the scattering characteristics of a height-adjustable PA (hPA) that incorporates large-angle beam scanning with ultra-wideband mRCS amplitude regulation for both polarizations. As shown in Fig. 1 (a), the hPA's in- and out-of-band scattering suppression methods are both considered. With the perfect absorption for in-band RCSR, antenna mode scattering considering the reflection coefficient with higher-order-mode circumstances is used to analyze the hPA's out-of-band monostatic RCSR (mRCSR). Different reflection phases are generated through the hPA's height control board. Based on the spatial phase cancellation method, the scattered fields will be distributed over a wide angular region instead of the main beam region to lower the monostatic RCS for normal incident waves. With the low RCS performance for x- and y-polarized waves, different amplitudes and phases loading to the hPA will bring about the controllable radiation beam and adjustable side lobe level, as shown in Fig. 1 (b). In addition, the radiation performance degradation generated by uneven heights is compensated by the phase shifter's magnitude and phase control system. Since the spatial phase cancellation process is accomplished as soon as the incident waves impinge on the hPA surface, perfect isolation between radiation and scattering performance is obtained. Thus, the limitations of the narrow band and single polarization in PA RCSR designs have been broken. Therefore, the plane phased array (PA #1), and the PA with two height control boards, namely, 5 dB (PA #2) and 15 dB mRCSR (PA #3), respectively, are designed and measured. The experimental results have certified the hPA's ultra-wide out-of-band mRCS amplitude regulation for dual-linear polarized waves.

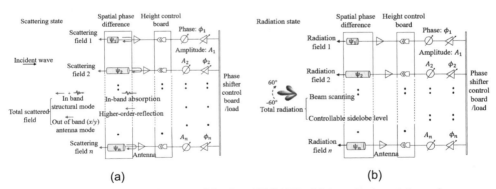

Fig. 1   Principle diagram of the low RCS hPA at (a) scattering state and (b) radiation state, respectively.

The remainder of this paper is organized as follows. In Section II, the analysis of antenna scattering characteristics and the mRCS amplitude regulation design method is proposed. In Section III, the radiation and scattering performance of hPA is experimentally verified. Finally, conclusions are offered in Section IV.

2. Scattering Methodology Upon Height-Adjustable Phase Array

2.1 Scattering Analysis of hPA

When plane waves illuminate an antenna, the antenna scattering is characterized by structural mode scattering and antenna mode scattering. The structural mode scattering is produced by the current on the antenna surface, which is related to the material, size, and shape of the antenna, and is independent of the load impedance. The antenna mode scattering represents the reradiation of reflected power from the excitation port, which is determined by the mismatched loads. When the antenna port is terminated with load Z_L, the total scattering electric field $\vec{E}(Z_L)$ for the antenna is given by [25]:

$$\vec{E}(Z_L) = \vec{E}_S(Z_L^*) + \Gamma \frac{jZ_0}{4\lambda R_a} \vec{h}_t(\vec{h}_r \cdot \vec{E}_i) \frac{e^{-jk_0 r}}{r} \quad (1)$$

where $\vec{E}_i$ is the incident electric field, $\Gamma = \dfrac{Z_a^* - Z_L}{Z_a + Z_L}$ is a modified current reflection coefficient of the antenna, $Z_0 = 377\Omega$ is the impedance of free space, $Z_a = R_a + jX_a$ is the antenna radiation impedance; k_0, λ, and r represent the wavenumber in the free space, operation wavelength, and the distance from antenna to the observation point, respectively. $\vec{h}_r$ and $\vec{h}_t$ are the effective heights of the receiving and transmitting antennas, respectively, which are functions of aspect (θ, φ). In (1), $\vec{E}_S(Z_L^*)$ is the structural mode scattering, while the term on the right side represents the antenna mode scattering because it is determined strictly from the radiation properties of the antenna and vanishes when the antenna is conjugate matched. Therefore, for the mismatched case, the energy will be reradiated when the power is reflected at the terminal load, and then the scattering field is mainly composed of antenna mode scattering.

Based on the antenna scattering theory of (1), the structural mode scattering can be obtained as the excitation port is terminated with load impedance. For the mismatched case, the right side of (1) can be used to analyze the scattering performance, and the power of the antenna mode scattering field is given by

$$\left|\vec{E}_M(\theta, \varphi)\right|^2 = \left|\Gamma\right|^2 \left|\frac{Z_0}{4\lambda R_r}\right|^2 \left|\vec{h}_t(\theta_t, \varphi_t)\right|^2 \cdot \left|\vec{h}_r(\theta_r, \varphi_r) \cdot \vec{E}_i\right|^2 \frac{e^{-j2k_0 r}}{r^2} \quad (2)$$

$$\left|\vec{h}_t(\theta_t, \varphi_t)\right|^2 = \frac{G_t(\theta_t, \varphi_t)\lambda^2 R_r}{\pi Z_0} \quad (3)$$

$$\left|\vec{h}_r(\theta_r,\varphi_r)\right|^2 = \frac{G_r(\theta_r,\varphi_r)\lambda^2 R_r}{\pi Z_0} \tag{4}$$

where $G_t(\theta_t,\varphi_t)$ is the gain of the transmitting antenna and $G_r(\theta_r,\varphi_r)$ is the gain captured by the receiving antenna, which can induce a polarization difference between the scattering antenna and the incident wave. The polarization mismatch factor η can be introduced to quantify this difference such that

$$\left|\vec{h}_r(\theta_r,\varphi_r)\cdot\vec{E}_i\right|^2 = \left|\vec{h}_r(\theta_r,\varphi_r)\right|^2\left|\vec{E}_i\right|^2\eta \tag{5}$$

Then the RCS introduced by the antenna mode scattering is given by

$$\sigma = \lim_{r\to\infty} 4\pi r^2 \frac{\left|\vec{E}_M(\theta,\varphi)\right|^2}{\left|\vec{E}_i\right|^2} = C\left|\Gamma\right|^2 e^{-j2k_0 r} \tag{6}$$

$$C = \frac{G_r(\theta_r,\varphi_r)G_t(\theta_t,\varphi_t)\lambda^2}{4\pi}\eta \tag{7}$$

For the monostatic and polarization matched case, $\vec{h}_t(\theta_t,\varphi_t)=\vec{h}_r(\theta_r,\varphi_r); \eta=1$ thus, $G_t(\theta_t,\varphi_t)=G_r(\theta_r,\varphi_r)$. The antenna gain is determined by the effective antenna aperture area A. For a phased array (PA) consisting of M × N elements, the total monostatic RCS of the PA's antenna mode scattering with additional phase contribution in each element can be approximated by

$$\sigma = \frac{4\pi}{\lambda^2}\sum_{m=1}^{M}\sum_{n=1}^{N} A_{mn}^2\left|\Gamma_{mn}\right|^2 e^{-j2(k_0 r_{mn}+\psi_{mn})} \tag{8}$$

where A_{mn} and ψ_{mn} correspond to the area and phase contribution of the mn-th element, respectively, and r_{mn} represents the distance from the mn-th element to the observation point, as depicted in Fig. 2. Based on the above analysis, the mRCSR of the PA antenna mode scattering with respect to a perfect electric conductive (PEC) plate with the same area $A_{PEC}=MN\times A_{mn}$ is expressed as

$$\text{RCSR(dB)} = 10\log_{10}\left|\sum_{m=1}^{M}\sum_{n=1}^{N}\frac{1}{(MN)^2}\left|\Gamma_{mn}\right|^2 e^{-j2(k_0 r_{mn}+\psi_{mn})}\right| \tag{9}$$

$$r_{mn} = (md_x\cos\varphi + n\sin\varphi d_y)\sin\theta \tag{10}$$

where d_x and d_y are the element periods in the x- and y-directions, respectively. Accordingly, the Γ_{mn}, r_{mn}, and ψ_{mn} dominate the mRCSR of the PA's antenna mode scattering. In [20], a coaxial delay line at the feed port of each element is used to optimize the phase excitations of the antenna array to lower the in-band monostatic RCS. However, the RCSR part is connected behind the antenna, which introduces polarization and bandwidth restrictions on the scattering suppression. Unlike the scheme presented in [20], we rely on the perfect absorption of the loads for in-band RCSR. Considering the RCSR of PA's antenna mode scattering in (9), where an hPA is designed to realize the ultra-wide out-of-band mRCSR. As shown in Fig. 2, when electromagnetic (EM) waves illuminate the hPA, the irregular reflection paths of EM waves generate multiple spatial reflection waves with different phases. The non-uniform distributions of the reflection phases lead to the scattered fields being redirected in more directions, and the backscattering power is then decreased according to the energy conservation law. Since the spatial reflection phase in (9) is polarization-independent, the mRCSR for dual-linear polarized waves can be theoretically obtained once the PA's height is determined.

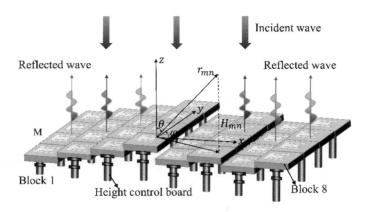

Fig. 2 Scattering diagram of an hPA with "M×N" elements.

For a 4×8-element PA, take 4×1-element as the subarray that shares the same height. We begin with the design of mRCSR for x-polarized waves. When the incident wave normally impinges on the surface, for the monostatic ($\theta = 0°$) -15 dB RCSR in the range of 3-30 GHz, (9) can be expressed as

$$\text{RCSR(dB)} = 10\log_{10}\left|\sum_{n=1}^{8}\frac{1}{64}|\Gamma_n|^2 e^{-j2\Psi_n}\right| < -15 \qquad (11)$$

In this case, $\psi_n = 2k_0 H_n$, $1/(MN)^2$ is the coefficient of the mRCSR, M=1 and N=8 could lead to $1/(MN)^2 = 1/64$; then, $\psi_1 \cdots \psi_8$ determines the number of the blocks. It is worth noting that eight randomly arranged blocks will obtain a more diffused bistatic scattering effect. Take 4.5 mm, which is a quarter of the central frequency (16.5 GHz) wavelength, as the basic holder's height to satisfy the 180° phase difference at 16.5 GHz. Based on our previous studies, the adjustable height will generate multiple spatial reflection phase cancellation consequences; thus, the wideband RCSR can be realized. However, in [27] and [28], the radiation performance was not considered.

For the design of hPA's ultra-wide out-of-band low RCS property. The antenna's inevitably higher-order-mode reflection dramatically increases the complexity of hPA's scattering analysis at an ultra-wideband frequency. For the ultra-wide out-of-band RCSR, the higher-order-mode occurs, leading to the $0<|\Gamma|\leq 1$. Then, the antenna scattering field is a mixed composition of antenna structural mode scattering and antenna mode scattering. Considering $|\Gamma|$, we rely on the right side of (1) to suppress a PA's out-of-band scattering field. For the condition of $|\Gamma|=1$, the spatial phase cancellation process is implemented to lower the monostatic RCS generated by the antenna mode scattering field. In the case of $0\leq|\Gamma|<1$, the spatial phase cancellation method combined with absorption of the antenna loads to suppress structural mode and antenna mode scattering field simultaneously. Consequently, we will propose two holders with different mRCSR amplitudes from 3 to 30 GHz. According to the spatial phase cancellation method, from (11), the spatial reflection phases corresponding to eight eligible holder heights for 15 dB mRCSR are obtained, as shown in Fig. (3). The phase difference in the range of 3-30 GHz provides the foundation for spatial phase cancellation. For 5 dB mRCSR, finding the optimal eight reflection phases is computationally complex. Thus, particle swarm optimization (PSO) is utilized to obtain the eight qualified heights. The corresponding fitness function is defined as

$$\text{fitness} = \sum_{k=0}^{K} \max\{[\sigma_R(f_k)+S_{dB}],0\} \quad (12)$$

$$f_k = f_{\min} + \frac{k(f_{\max}-f_{\min})}{K} \quad (13)$$

where f_{max} and f_{min} are the optimized minimum and maximum frequencies, respectively; K is the number of optimized frequency points; and S_{dB} is the optimized target 5 dB. The predicted mRCSR of the optimized frequency band formed the basis of the fitness function. During the optimization, the smaller the value of fitness functions is, the better the result of mRCSR. The selected holder heights for different mRCSR amplitudes are shown in Table I. The corresponding phases for eligible heights of 5 dB mRCSR are also provided in Fig. 3. With the perfect absorption for in-band RCSR, the ultra-wideband mRCS amplitude regulation of PA is obtained when placing the PA on the surface of 5 and 15 dB RCSR holders. The spatial phase cancellation process above not only suppresses the antenna mode scattering but also superimposes with absorbing to lower the antenna scattering field when the load is not perfectly absorbing. Generally, the mRCS amplitude regulation is obtained by optimizing the number of distinct heights. Based on this theoretical analysis, the location of different height blocks will not influence the mRCSR. A more diffused scattering effect can be obtained by putting the higher blocks in the middle of the array. This design method offers the advantage that the spatial phase cancellation process is accomplished when the incident wave impinges on the hPA. Thus, an integration design with perfect isolation of radiation and scattering performance is obtained.

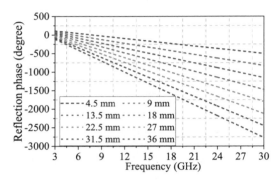

Fig. 3 The reflection phases corresponding to the eligible holder's heights "H" for 15 dB RCSR.

Table I

Distribution of different height for 0, 5, and 15 dB RCSR

Holder Height	H.1	H.2	H.3	H.4	H.5	H.6	H.7	H.8
PA #1	0	0	0	0	0	0	0	0
PA #2	4.5	9	13.5	13.5	13.5	18	13.5	13.5
PA #3	4.5	9	13.5	18	22.5	27	31.5	36

The unit of the holder's height is "mm".

2.2 Design and Analysis of Low RCS hPA

Considering the stable performance and ease of implementation, an E-shaped microstrip patch antenna is used as the basic element to construct the PA. The schematic of the patch antenna is shown in Fig. 4. The element consists of three parts: including a radiation patch, two substrates, and a ground plane. The radiation patch is connected to the ground plane through a metal via with an inner diameter of 0.9 mm. Considering the craft constraints, the element incorporates two boards of low-loss substrate (Rogers 2 mm thick AD255C, $\varepsilon r = 2.55$, $\tan \delta = 0.0013$) that are bonded using an RO4450F film ($\varepsilon r = 3.52$, $\tan \delta = 0.002$). The optimized physical geometric parameters of the patch antenna are shown in Table II. To validate the characteristics of the element, the full-wave simulation software CST Microware Studio® with periodic boundary conditions in the x- and y-directions and an open boundary in the z-direction is adopted. The simulated regular S-parameter of the patch antenna is shown in Fig. 5 (a). It can be observed that the S11 is below −25 dB at 4.5 GHz that almost no power is reflected from the load, which indicates that the incoming energy is absorbed. In this case, the scattering field is structural mode scattering. Fig. 5 (b) illustrates the reflection coefficient of the antenna under x- and y-polarized incident waves. In the higher-order-mode reflection region, $|\Gamma|$ fluctuates between 1 and 0. Higher-order-mode reflection occurs when the frequency is higher than 10 GHz, which corresponds to the wavelength of the distance between the elements.

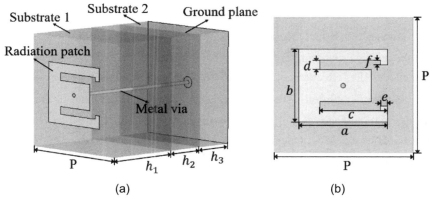

Fig. 4 (a) 3D schematic configuration of the antenna element. (b) Top view of the element.

Table II

Parameters of the antenna element

Parameter	Value (mm)	Parameter	Value (mm)
P	30	a	19
h_1	2	b	16
h_2	0.1	c	14.5
h_3	2	d	2
e	1.5	f	1

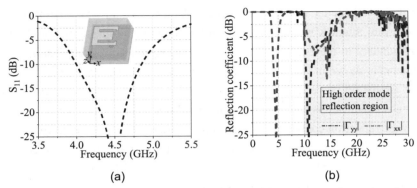

Fig. 5 Simulated (a) regular S-parameter of the antenna element and (b) its reflection coefficient for x- and y-polarized incident waves.

The uneven $|\Gamma|$ values occur in the higher frequency band, this is why we do not consider the antenna scattering as similar to that of a metasurface over an ultra-wide out-of-band frequency. The reflection coefficient of the E-shaped antenna element is obtained with periodic boundary conditions. We assume that the $|\Gamma|$ of each element in the array is the same for normal incidence. According to the parameters in Table I, the theoretical 0, 5, and 15 dB mRCSR of the 4×8-element PA (PA #1, #2, and #3) is realized in the range of 3-30 GHz with a fractional bandwidth (FBW) of 164%. To analyze the scattering characteristics of the hPA over ultra-wideband frequency, it is necessary to study the mRCSR of PA #1, #2, and #3 compared to a PEC with the same area. The simulated and theoretical results are shown in Fig. 6. Considering Fig. 6 (a), the mRCSR on the resonant frequency of the fundamental mode has almost no change for three arrays, which occurs due to the perfect absorption of the matching loads. For the higher-order-mode region, e.g., 10-15 GHz, some of the energy is absorbed while other energy is reflected from the load and reradiated through the antenna structure. Under this condition, through superimposing with the spatial cancellation progress, the reradiated scattering field is further suppressed as the mRCSR of PA #1 and PA #2 in Fig. 6 (a). Similar to the mRCSR of the fundamental mode, the absorption of the loads generates no obvious depression area for PA #3's mRCSR in the range of 10-15 GHz. Since the mRCSR design method is polarization-independent, similar results are obtained for y-polarization, as shown in Fig. 6 (b). Because the electric field on the element surface is x-polarized, according to the reciprocity theorem, no radiation or absorption occurs at the resonant frequency of the fundamental mode for y-polarized incident waves. Generally, the ideal consistency between simulation and calculation has verified the rationality of our analysis method.

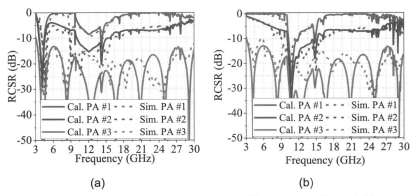

Fig. 6 Simulated and theoretically calculated mRCSR of PA #1, #2, and #3 compared to a PEC plane with the same area for (a) x-polarized waves and (b) y-polarized waves.

3. Prototype and Performance of Height-Adjustable Phased Array

The 4×8-element PA #1, #2, and #3 are fabricated and tested. The monostatic scattering performances of the three arrays are all measured. For the radiation performance, PA #1 is used as the reference antenna to illustrate the effectiveness of the radiation performance of PA #2 and PA #3. Taking a 4×1-element as the subarray, each subarray is fixed on the corresponding holder through four nylon bolts that do not affect the performance. The radiation performance is measured through the near-field testing system in an anechoic chamber, as shown in Fig. 7 (a). The mRCSRs are tested through the far-field testing system, as shown in Fig. 7 (b). The prototypes are shown in Fig. 7 (c).

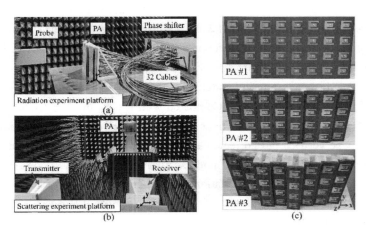

Fig. 7 The experimental platforms for (a) radiation near-field measurement, (b) scattered far-field measurement, and (c) prototype of PA #1, #2, and #3.

4. The Scattering Property of the hPA

In the measurement of scattering performance, a pair of horn antennas covering the operating band of 3-30 GHz act as receivers and transmitters. The aperture centers of the antenna, transmitter horn, and receiver horn are collinearly aligned. The radiation elements are all terminated with matching loads during the measurement. Fig. 8 (a) and (b) depict the simulated and measured mRCSR of PAs #1, #2, and #3 for both polarizations compared to a PEC plane. And the mRCSR of PA #2, and #3 compared to PA #1 are investigated as shown in Fig. 8 (c) and (d). From the figure, the 0, 5, and 15 dB mRCSRs of the phased array for x- and y-polarized waves are obtained in the range of 3-30 GHz. For the x-polarized incident wave, despite the phase cancellation introduced by holders, the perfect matching load leads to the energy being absorbed at the fundamental frequency band, and the measured structural mode scatterings of the three arrays are approximately the same, which is similar to the simulations. Besides, the antenna mode scattering dominates the scattering field for complete reflection conditions, where the mRCS amplitude regulation is clearly shown. For the higher-order-mode reflection, e.g., 10-15 GHz, where the combined absorption and phase cancellation exists, the apparent 5 dB mRCSR of PA #2 is obtained as shown in Fig. 8 (c) and (d). Additionally, for 15 dB mRCSR, the phase cancellation slightly influences the mRCSR amplitude regulation in the range of 10-15 GHz due to the absorption of loads. Considering Fig. 8, the measurement results agree well with the simulations. The effect of the reflection coefficient with higher-order-mode reflection on antenna mode scattering is experimentally verified. We further investigated the mRCSR under oblique incidence. Simulated mRCSRs of PA #1, #2, and #3 under x- and y-polarized waves with incident angles of 0°, and 30° are illustrated in Fig. 9. When the incident angle increases from 0° to 30°, larger mRCSR amplitudes for PAs #2 and #3 are obtained at a higher frequency. Owing to the development of radar netting detection systems, research on bistatic RCSR is crucial. To further verify the arrays' bistatic scattering performance for oblique *x*- and *y*-polarized incidence, the 3-D bistatic scattering patterns at 20 GHz with incident angels of 0° and 30° are also studied, as shown in Fig. 10. Compared with PA #1, the bistatic scattering fields of PA #2, and PA #3 are distributed over a wide angular region for both polarizations. Generally, we have verified our correct analytic method for in- and ultra-wide out-of-band mRCSR relying on antenna structural mode scattering and antenna mode scattering. The ultra-wideband mRCS amplitude regulation for dual-linear polarization has been realized with a stable incident angle within 30°. Furthermore, a more diffused bistatic scattering effect is obtained as well. The real-time hPA will be obtained by transferring the power of the motor beneath each element to realize the real-time control of the mRCSR amplitude.

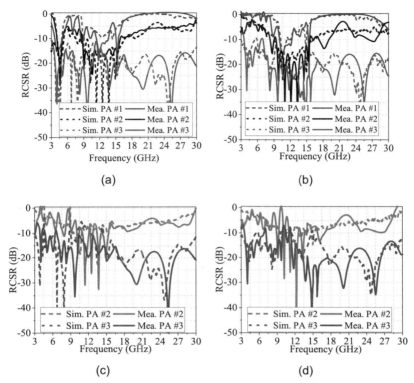

Fig. 8 Simulated and measured mRCSR of PAs #1, #2, and #3 compared to a PEC plane with the same area for (a) x- and (b) y-polarized waves. The RCSR of PA #2 and PA #3 compared to PA #1 for (c) x- and (d) y-polarized waves.

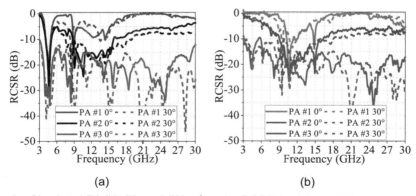

Fig. 9 Simulated PA #1, #2, and #3's specular RCSR for (a) x-polarized waves and (b) y-polarized waves with incident angles of $\theta = 0°$ and $\theta = 30°$, respectively.

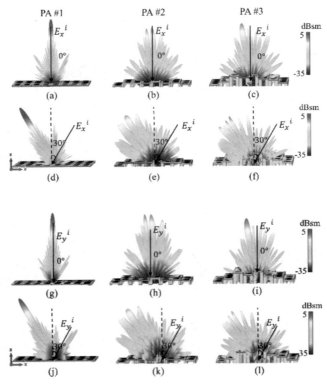

Fig. 10 Simulated 3-D bistatic RCS patterns of PA #1, #2, and #3 at 20 GHz for (a)-(f) *x*-polarized incident waves and (g)-(l) *y*-polarized incident waves with the incident angles of 0° and 30°.

5. Radiation Performance of the hPA

During the radiation performance measurement, the hPA is connected to a 32 channels phase shifter, which is used to independently control the phase and magnitude of the antenna elements. The uneven height of the antenna elements conversely introduces a non-uniform radiation wavefront that will dramatically deteriorate the radiation performance of PA #2 and PA #3. It is therefore crucial to compensate the wavefront of the low mRCS hPA into a plane wave. For an M×N-element array, according to the array theory, the array factor is given by

$$E(\theta,\varphi) = \sum_{m=1}^{M}\sum_{n=1}^{N} I_{mn} e^{j(\phi_{mn} - \phi'_{mn})} \tag{14}$$

$$\phi_{mn} = k_0[(md_x \cos\varphi + nd_y \sin\varphi)\sin\theta + H_{mn}\cos\theta] \tag{15}$$

$$\phi'_{mn} = k_0[(md_x \cos\varphi_0 + nd_y \sin\varphi_0)\sin\theta_0 + H_{mn}\cos\theta_0] \quad (16)$$

where I_{mn} is the amplitude excitation of the *mn*-th element, $\varphi_{mn'}$ is the phase of the *mn*-th element for the beam direction (θ_0, ϕ_0), and H_{mn} is the height of the *mn*-th element. Based on the above analysis, the phase and magnitude of each element are loaded to the hPA and controlled to obtain the desired radiation pattern. The measured broadside radiation patterns in the *xoz*- and *yoz*-planes at 4.5 GHz are shown in Fig. 11; the co- and cross-polarization patterns are represented by solid and dashed lines, respectively. From the figure, the main beam in the *yoz*-plane is wider than that in the *xoz*-plane, this is because the number of elements along the *x*-direction is larger than the number of elements along the *y*-direction. The good agreement between the radiation patterns of PA #1, #2, and #3 shows the influence on radiation performance introduced by uneven height holders is sufficiently compensated through the phase control system of the phase shifter. Furthermore, by assigning the amplitude of each column of the phase shifter, the Chebyshev amplitude distribution is applied to the element to obtain -20 dB side lobe levels (SLLs). Taking PA #2 as an example, the simulated and measured radiation patterns with no weighting and with a -20 dB Chebyshev amplitude weight are shown in Fig. 12. The obvious -20 dB SLL is obtained compared with the unweighted result. Furthermore, the radiation beam scanning patterns of the three arrays at 4.5 GHz are depicted in Fig. 13. The beam scanning of PA #1, #2, and #3 in the *xoz*-plane are all up to 60°. As the beam scanning angle increases, the half-power beam width increases, and the main beam suffers from scan loss and increasing SLLs. Based on the above experimental results, it can be concluded that mRCS amplitude regulation of the hPA is realized under the premise of unchangeable radiation performance.

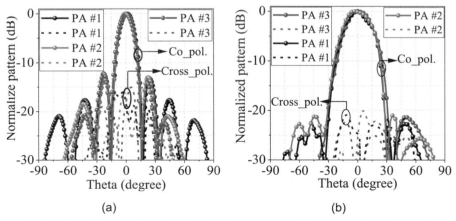

Fig. 11 Measured radiation patterns of PA #1, #2, and #3 in the (a) *xoz*- and (b) *yoz*-planes at 4.5 GHz.

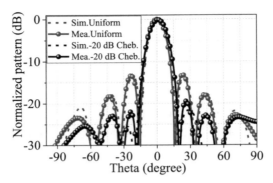

Fig. 12 Simulated and measured radiation pattern of the PA #2 with no weighting and with a -20 dB Chebyshev amplitude weight.

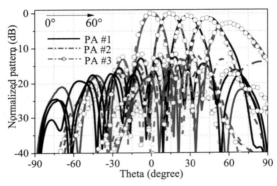

Fig. 13 Measured radiation beam scanning patterns of PA #1, #2, and #3 at 4.5 GHz in the *xoz*-plane.

For an hPA with large angle beam scanning, it is necessary to evaluate the array element's active reflection coefficient (ARC), which is related to the beam scanning direction. Since the ARC of the excitation port considers the effects of mutual coupling between other elements in the array, which can more accurately describe the impedance matching characteristics of the port. The ARC of the array's *mn*-th port is given by [33].

$$\Gamma_{mn}(\theta_0,\varphi_0) = \sum_{p=1}^{M}\sum_{q=1}^{N} S_{pq,mn} e^{-j[(p-m)d_x u + (q-n)d_y v]} \quad (17)$$

$$\text{Active VSWR}_{mn}(\theta_0,\varphi_0) = \frac{1+|\Gamma_{mn}(\theta_0,\varphi_0)|}{1-|\Gamma_{mn}(\theta_0,\varphi_0)|} \quad (18)$$

where $u = k\sin\theta_0\cos\varphi_0$, $v = k\sin\theta_0\sin\varphi_0$, $S_{pq,mn}$ is the transmission coefficient between the *pq*-th and *mn*-th element when other ports are connected to matching loads. A vector network analyzer is used to investigate the center element ARC of PA #1, #2, and #3 for

0° and 60° beam scanning. According to (18), the simulated and measured active voltage standing wave ratio (VSWR) are obtained as shown in Fig. 14. The data show that the measured and simulated active VSWRs of the three antenna arrays are less than 3 for the 0° and 60° beam directions at 4.5 GHz. Comparing the active VSWRs of the three arrays, as the mRCSR amplitude increases, the bandwidth of the active VSWR becomes narrower. It is worth pointing out that when the low RCS PAs scan to 60°, the active VSWR is less than 3.5 from 4.4-4.8 GHz, which demonstrates the good beam scanning performance of the arrays. Slight discrepancies between the measured and simulated active VSWR are observed, which are caused by the truncation effects that result from the uneven height of the antenna element in a finite antenna array. Fig. 15 gives the simulated and measured gain and aperture efficiency (AE) of PA #1, #2, and #3. The measured 1 dB gain bandwidth of the PA is from 4.3-5 GHz. The measured AE of the three PAs is approximately 80% at 4.6 GHz. The experimental results fluctuate slightly compared to the simulated results, which is caused by imperfect fabrications and measurement accuracy. The ideal measured results in section III demonstrate the independent design of scattering and radiation performance. Table III illustrates the comparison between this work and some of the newest publications related to low mRCS phased arrays, which are from the perspective of theoretical analysis. From the table, our design can realize PA's in- and ultra-wide out-of-band mRCS amplitude regulation for dual-linear polarized waves, which exceeds the performance of the existing works. Due to the independent mechanism of the radiation and scattering properties, the design method in this paper can be applied to wideband phased arrays. Furthermore, the motor-controlled mechanically reconfigurable mechanism can be adopted beneath the antenna to generate real-time control of the radiation and scattering performance.

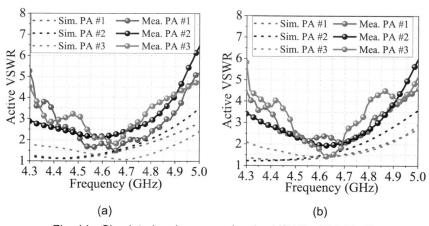

Fig. 14 Simulated and measured active VSWR of PA #1, #2, and #3 for (a) 0° and (b) 60° cases.

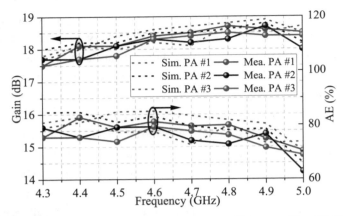

Fig. 15 Simulated and measured gains and aperture efficiency of the PA #1, #2, and #3.

Table III

Comparison of the proposed low RCS hPA with some existing works

Ref.	Radiation		Scattering (RCSR) for normal incidence				
	Operation Bandwidth (FBW)	Beam Scanning (Angle)	RCSR Bandwidth		In-Band and Out-of-Band	Dual-Linear Polarization	RCS Amplitude Regulation
			x_pol.	y_pol.			
[16]	8-12 GHz (40%)	0-60°	/	6-18 GHz (100%) 10 dB	Yes	No	No
[19]	0.96fc-1.1fc (13.6%)	No	/	0.994fc-1.027fc (3.2%) 5 dB	No	No	No
[20]	8-12 GHz (40%)	0-60°	/	8.5-12 GHz (34%) 6.4 dB	No	No	No
[22]	9.8-10.2 GHz (4%)	No	9.8-10.2 GHz (4%) 15 dB	/	No	No	No
This work	4.4-4.8 GHz (8.7%)	0-60°	3-30 GHz (164%) 5 and 15 dB	3-30 GHz (164%) 5 and 15 dB	Yes	Yes	Yes

"fc": represents the center frequency of the antenna.

6. Conclusion

In this paper, we have presented an analysis of the antenna scattering characteristics of a hPA for its mRCS amplitude regulation technology. The reflection coefficient with higher-order-mode condition is considered in antenna mode scattering to realize the hPA's 5 and 15 dB mRCSR over an ultra-wide frequency band. In the case of scattering conditions, the mRCS amplitude regulations for both polarizations are experimentally verified in the range of 3-30 GHz. For radiation performance, the tunable SLL and large angle beam steering performance have been confirmed. We have accomplished the joint design of independent radiation and scattering performance of PA, which breakthroughs the narrow band and single polarization limitations for PA's scattering suppression. Perfect isolation between radiation and scattering properties enables our design method to be employed in wideband PAs. In future works, real-time RCSR control of the hPA can be realized by introducing a motor-controlled mechanism into the system. The hPA's ultra-wideband mRCS amplitude regulation method presented in this paper has provided a new perspective for the scattering suppression of PA. And will accelerate the promotion and application of higher performance low RCS PAs in radar and low detectable systems.

Acknowledgments

This work was supported by the National Natural Science Foundation of China (U2141233, 62171416, and U2241229) and the Fundamental Research Funds for the Central Universities CUC230D039.

Author Information

P. Li, J. X. Su, and Z. R. Li are with the State Key Laboratory of Media Convergence and Communication, Communication University of China, Beijing 100024, China.

H. Yu is with the Information Materials and Intelligent Sensing Laboratory of Anhui Province, Anhui University, Hefei 230601, China.

H. C. Yin and Z. H. Xiao are with the Science and Technology on Electromagnetic Scattering Laboratory, Beijing 100854, China.

References

[1] XING Z, YANG F, YANG P, et al. A low-RCS and wideband circularly polarized array antenna co-designed with a high-performance AMC-FSS radome[J]. IEEE Antennas Wireless Propag.

Lett., 2022, 21(8): 1659-1663.

[2] ZHAO Y, et al. Broadband low-RCS metasurface and its application on antenna[J]. IEEE Trans. Antennas Propag., 2016, 64(7): 2954-2962.

[3] LIAO W J, ZHANG W Y, HOU Y C, et al. An FSS-integrated low-RCS radome design[J]. IEEE Antennas Wireless Propag. Lett., 2019, 18(10): 2076-2080.

[4] YU W, YU Y, WANG W, et al. Low-RCS and gain-enhanced antenna using absorptive/transmissive frequency selective structure[J]. IEEE Trans. Antennas Propag., 2021, 69(11): 7912-7917.

[5] LIU Z, et al. A low-RCS, high-GBP fabry–perot antenna with embedded chessboard polarization conversion metasurface[J]. IEEE Access, 2020, 8:80183-80194.

[6] LI K, LIU Y, JIA Y, et al. A circularly polarized high-gain antenna with low RCS over a wideband using chessboard polarization conversion metasurfaces[J]. IEEE Trans. Antennas Propag., 2017, 65(8): 4288-4292.

[7] MEI P, LIN X Q, YU J W, et al. Development of a low radar cross section antenna with band-notched absorber[J]. IEEE Trans. Antennas Propag., 2018, 66(2): 582-589.

[8] LIU Y, ZHAO X. Perfect absorber metamaterial for designing low-RCS patch antenna[J]. IEEE Trans. Antennas Propag., 2014, 13:1473-1476.

[9] GAO X, YIN S, WANG G, et al. Broadband low-RCS circularly polarized antenna realized by nonuniform metasurface[J]. IEEE Antennas Wireless Propag. Lett., 2022, 21(12): 2417-2421.

[10] XI Y, JIANG W, WEI K, et al. Wideband RCS reduction of microstrip antenna array using coding metasurface with low Q resonators and fast optimization method[J]. IEEE Antennas Wireless Propag. Lett., 2022, 21(4): 656-660.

[11] GAO K, CAO X Y, GAO J, et al. Design of a low-RCS circularly polarized metasurface array using characteristic mode analysis[J]. Opt. Mater. Express, 2022, 12:907-917.

[12] JIANG H, XUE Z, LI W, et al. Low-RCS high-gain partially reflecting surface antenna with metamaterial ground plane[J]. IEEE Trans. Antennas Propag., 2016, 64(9): 4127-4132.

[13] ZOU Y, KONG X, XING L, et al. A slot antenna array with reconfigurable RCS using liquid absorber[J]. IEEE Trans. Antennas Propag., 2022, 70(7): 6095-6100.

[14] CHENG Y F, FENG J, LIAO C, et al. Analysis and design of wideband low-RCS wide-scan phased array with AMC ground[J]. IEEE Antennas Wireless Propag. Lett., 2021, 20(2): 209-213.

[15] DING X, CHENG Y F, SHAO W, et al. Broadband low-RCS phased array with wide-angle scanning performance based on the switchable stacked artificial structure[J]. IEEE Trans. Antennas Propag., 2019, 67(10): 6452-6460.

[16] GOU Y, CHEN Y, YANG S. Radar cross section reduction of wideband Vivaldi antenna arrays with array-level scattering cancellation[J]. IEEE Trans. Antennas Propag., 2020, 70(8): 6740-6750.

[17] LIU Y, JIA Y, ZHANG W, et al. An integrated radiation and scattering performance design method of low-RCS patch antenna array with different antenna elements[J]. IEEE Trans. Antennas Propag., 2019, 67(9): 6199-6204.

[18] QIU L, XIAO G. A broadband metasurface antenna array with ultrawideband RCS reduction[J]. IEEE Trans. Antennas Propag., 2022, 70(9): 8620-8625.

[19] NAKAMOTO N, TAKAHASHI T, FUKASAWA T, et al. RCS synthesis of array antenna with circulators and phase shifters and measurement method for deterministic RCSR[J]. IEEE Trans. Antennas Propag., 2021, 69(1): 135-145.

[20] XIAO S, YANG S, CHEN Y, et al. In-band scattering reduction of wideband phased antenna arrays with enhanced coupling based on phase-only optimization techniques[J]. IEEE Trans. Antennas Propag., 2020, 68(7): 5297-5307.

[21] YANG P, YAN F, YANG F, et al. Microstrip phased-array in-band RCS reduction with a random element rotation technique[J]. IEEE Trans. Antennas Propag., 2016, 64(6): 2513-2518.

[22] LI Z, XIE C, YANG F, et al. A study on electromagnetic scattering characteristics of 4-D antenna arrays[J]. IEEE Trans. Antennas Propag., 2023, 71(1): 275-287.

[23] WIESBECK W, HEIDRICH E. Wide-band multiport antenna characterization by polarimetric RCS measurements[J]. IEEE Trans. Antennas Propag., 1998, 46(3): 341-350.

[24] MUNK B A. Finite Antenna Arrays and FSS[M]. New York: Wiley InterScience, 2003.

[25] LU B, GONG S X, ZHANG S, et al. A new method for determining the scattering of linear polarized element arrays[J]. Prog. Electromagn. Res., 2009, 7:87-96.

[26] BLAIR R. The general theory of antenna scattering[D]. Columbus: The Ohio State University, 1963.

[27] SU J, YU H, YIN H, et al. Breaking the high-frequency limit and bandwidth expansion for radar cross-section reduction: a low-observable technology[J]. IEEE Antennas Propag. Mag., 2021, 63(6): 75-86.

[28] LI P, YU H, SU J, et al. A low-RCS multifunctional shared aperture with wideband reconfigurable reflectarray antenna and tunable scattering characteristic[J]. IEEE Trans. Antennas Propag., 2023, 71(1): 621-630.

[29] PUES H, BOGAERS J, PIECK R, et al. Wideband quasilog-periodic microstrip antenna[J]. IEE Proceedings H Microwaves Optics and Antennas, 1981, 128(3): 159-163.

[30] OOI B L, SHEN Q. A novel E-shaped broadband microstrip patch antenna[J]. Microw Opt Techn Let., 2020: 348-352.

[31] CAO Z, LI Y, ZHANG Z, et al. Single motor-controlled mechanically reconfigurable reflectarray[J]. IEEE Trans. Antennas Propag., 2023, 71(1): 190-199.

[32] LIN J, SHEN W, YANG K. A low-sidelobe and wideband series-fed linear dielectric resonator antenna array[J]. IEEE Antennas Wireless Propag. Lett., 2016, 16: 513-516.

[33] GROSS F B. Frontiers in antennas: next generation design & engineering[M]. New York: McGraw-Hill Professional, 2011.

Design of an Active Polarizer for Wideband Quad-Polarization Conversion[*]

1. Introduction

Polarization, defined as the instantaneous orientation of the electric field vector, is one of the fundamental characteristics of electromagnetic (EM) waves. In wireless communication, the regulation of EM wave polarization mode is not negligible as communication equipment is highly sensitive to it. To enhance the anti-interference capability of wireless communication equipment, the utilization of a polarization converter, also known as "polarizer", is essential.

There are several types of frequently-employed polarization modes in daily communication, including horizontal linear polarization (HP), vertical linear polarization (VP), left-hand circular polarization (LHCP), and right-hand circular polarization (RHCP). The primary objective of the reflective polarizers is to enable switching the HP wave into the one of the above four polarization modes. In recent years, there has been a growing demand for the ability to freely switch between multiple polarization modes, driven by advancements in reconfigurable technology. As a result, active polarizers have garnered considerable attention, primarily due to their tunable capabilities.

Most of the active polarizer design focuses on the switching between two polarization modes within a shared frequency band. For instance, in [12], a proposed polarizer featuring an arrow-shaped metal patch and one PIN diode can effectively convert the HP wave into the VP and one type of circular polarized (CP) mode by switching the PIN states. Similar functional designs can be found in other researches like [13], [14], [15], [16], [17], all of which employ a single active component to achieve dual-polarization mode switching.

To enhance the functionality of polarizers, a novel active tri-polarization converter was introduced in [18]. This polarizer incorporates double-layer substrates, with a square-ring metal patch on it. The metal patch includes a pair of symmetrically positioned PIN diodes. This design enables the realization of three polarization modes (HP, VP and one kind of CP) without any shared band. But in practice, left- and right-hand circular polarization

---

[*] The paper was originally published in *IEEE Antennas and Wireless Propagation Letters*, 2023, 22 (12), and has since been revised with new information. It was co-authored by Chengxiang Xu, Jianxun Su, Meijun Qu, Zengrui Li, Kainan Qi, and Hongcheng Yin.

are orthogonal to each other, so it needs to be clearly distinguished. In [19], a symmetrical design approach is employed to achieve same frequency band switching of the three desired polarization modes (HP, LHCP and RHCP) using a 2-bit metasurface design method.

Although the previous polarizer's reconfigurability was effectively, there is still room for investigation and advancement in the low-profile design and the usage of fewer active components to possess more polarization modes. In this letter, a wideband active polarizer with quad-polarization conversion is proposed, which enables to switch of HP wave into four polarization modes (HP, VP, LHCP and RHCP) in a shared band with low-profile structure and fewer active components.

2. Principle Analysis and Polarizer Unit Design

2.1. Polarizer Unit Geometry Structure

Fig. 1(a)-(c) shows the structure of the proposed polarizer. It is composed of frontside metal patch, F4B dielectric layer (ε_r=2.65), backside metal patch, and metal ground in that order. The frontside metal patch features two pairs of butterfly-shaped structures along the x and y axes, with a PIN diode integrated in each direction. Notably, the staggered placements of the active components enable independent control of the two PIN diodes. To mitigate the impact of the direct-current control circuit on the conversion performance, a 5 nH inductor is integrated in each direction. For the PIN diodes, the SMP1340-040LF type is employed, and their equivalent circuits in both ON and OFF states, according to the datasheet, are presented in Fig. 1(d).

When the PIN diode turns ON, the equivalent circuit is a 1 Ω resistor in series with a 0.45 nH inductor, whereas when it turns OFF, the circuit is a 10 Ω resistor in series with a 0.45 nH inductor and a 0.16 pF capacitor. The control of *PIN_1* and *PIN_2* is independent. In our setup, the *PIN_1* state is displayed in the front, while the *PIN_2* state is displayed in the back. For instance, if *PIN_1* is set to ON and *PIN_2* is set to OFF, we designate this configuration as "10" based on our established rule.

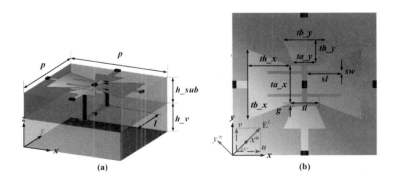

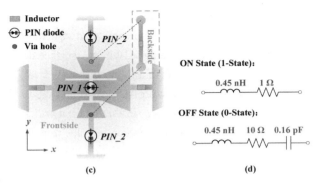

Fig. 1 The proposed active polarizer unit (a) 3D structure diagram, (b) top view and the incident wave electric field E^i (along x^* direction that points to 45° with x axis), (c) position of lumped components, and (d) equivalent circuit of PIN diodes. (Units: mm, p=10, h_sub=2.93, h_v=2, l=3.7, ta_x=2.7, tb_x=5, th_x=3, ta_y=1.5, tb_y=3, th_y=1.5, tl=2, sl=2.4, sw=0.15)

2.2. Principle Analysis

Electric field direction of the HP incident wave is along with x^*-axis as shown in Fig. 2. In the u-v coordinate system, Matrix R represents the reflection coefficient relating E^r and E^i on the periodic polarizer unit. Thus, the E^r can be expressed as,

$$\begin{bmatrix} E_u^r \\ E_v^r \end{bmatrix} = \mathbf{R} \cdot \begin{bmatrix} E_u^i \\ E_v^i \end{bmatrix} = \begin{bmatrix} R_{uu} & R_{uv} \\ R_{uv} & R_{vv} \end{bmatrix} \cdot \begin{bmatrix} E_u^i \\ E_v^i \end{bmatrix} \quad (1)$$

For clearly seeing the polarization mode of the reflected wave, the coordinate system needs to be transferred to the x^*-y^* coordinate system. The x^*-y^* coordinate system is rotated by the u-v coordinate system with the counterclockwise angle of $\theta = 45°$, and thus the relation between two coordinates is,

$$\begin{bmatrix} \widehat{x^*} \\ \widehat{y^*} \end{bmatrix} = \mathbf{O} \cdot \begin{bmatrix} \hat{u} \\ \hat{v} \end{bmatrix} = \begin{bmatrix} \cos\theta & \sin\theta \\ -\sin\theta & \cos\theta \end{bmatrix} \begin{bmatrix} \hat{u} \\ \hat{v} \end{bmatrix} \quad (2)$$

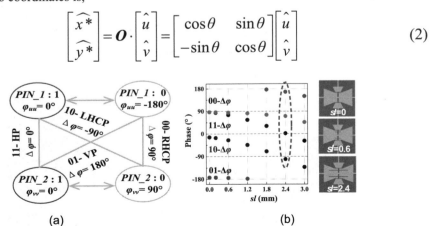

Fig. 2 (a) The principle model of the quad-polarization conversion polarizer design. (b) $\Delta\varphi$ of each state varies with sl at 10 GHz.

where **O** is the rotation matrix. Thus, the expression of **E**^{r} in the x^*-y^* coordinate system can be obtained,

$$\begin{bmatrix} E_{x*}^r \\ E_{y*}^r \end{bmatrix} = \boldsymbol{O} \cdot \boldsymbol{R} \cdot \boldsymbol{O}^T \begin{bmatrix} E^i \\ 0 \end{bmatrix} = \frac{1}{2} \begin{bmatrix} R_{uu} + R_{vv} & -R_{uu} + R_{vv} \\ -R_{uu} + R_{vv} & R_{uu} + R_{vv} \end{bmatrix} \begin{bmatrix} E^i \\ 0 \end{bmatrix} \quad (3)$$

Among them, R_{uu}, R_{vv} and φ_{uu}, φ_{vv} represent the magnitude and phase of R_{uu}, R_{vv}, respectively. The states of PIN diodes have a significant impact on R_{uu} and R_{vv}. Assuming that the reflection process is lossless, that is $R_{uu}=R_{vv}\approx 1$, the phase difference $\Delta\varphi = \varphi_{uu} - \varphi_{vv}$ becomes crucial in reconfiguring the polarization mode of the reflected wave. When $\Delta\varphi = 0 + 2n\pi, \pi + 2n\pi, \pi/2 + 2n\pi, -\pi/2 + 2n\pi$, the **E**^{r} can be expressed,

$$\begin{bmatrix} E_{x*}^r \\ E_{y*}^r \end{bmatrix} = \begin{cases} \begin{bmatrix} E^i \\ 0 \end{bmatrix}, & \Delta\varphi = 0 + 2n\pi, n \in Z \\ \begin{bmatrix} 0 \\ E^i \end{bmatrix}, & \Delta\varphi = \pi + 2n\pi, n \in Z \\ \frac{1}{2} \begin{bmatrix} (1+j)E^i \\ (-1+j)E^i \end{bmatrix}, & \Delta\varphi = \frac{\pi}{2} + 2n\pi, n \in Z \\ \frac{1}{2} \begin{bmatrix} (1-j)E^i \\ (-1-j)E^i \end{bmatrix}, & \Delta\varphi = -\frac{\pi}{2} + 2n\pi, n \in Z \end{cases} \quad (4)$$

Consequently, for freely switching the polarization modes of incident wave from HP mode to the four modes of HP, VP, RHCP and LHCP, it must be as much as feasible to actualize $\Delta\varphi = 0 + 2n\pi, \pi + 2n\pi, \pi/2 + 2n\pi, -\pi/2 + 2n\pi$, , respectively.

As the Fig. 2(a) shown, to achieve the flexible switching of four polarization modes, it is necessary that when *PIN_1* and *PIN_2* switch from state 1 to state 0, there is a reflection phase shifts of 180° in the *u*-direction and 90° in the *v*-direction, respectively. Moreover, certain operations are needed to make the "11" state the initial state so that the polarizer is equivalent to the metal plane without polarization conversion. Therefore, we employ the approach of etching slots in the butterfly structure to adjust the *u*-direction from the initial phase. This step serves two purposes: 1) it stabilizes $\Delta\varphi = 0$ under the "11" state. 2) it provides an adjustable parameter to optimize the results under each state. The results shown in Fig. 2(b) demonstrated that when *sl*=2.4 mm, $\Delta\varphi$ under the four states closely align with the desired values at 10 GHz.

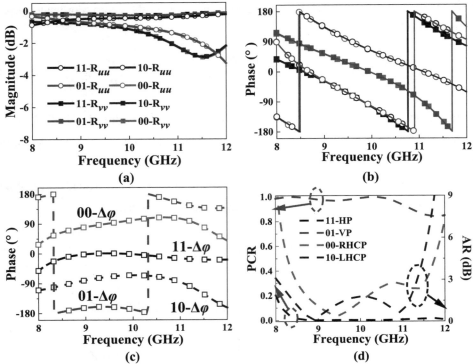

Fig. 3 The simulated reflection coefficients of proposed unit in u-v system. (a) Magnitude, (b) phase, (c) $\Delta\varphi$ and (d) PCR, circular polarization AR for each state.

It is important to note that while some reconfigurable metasurfaces can achieve 2-bit phase control in a single direction, theoretically meeting the mentioned conditions, there are limitations. The initial phase control value requires further adjustment, and the operating bandwidth is frequently restricted due to the presence of two active components in one direction. Consequently, the proposed polarizer must exhibit anisotropic behavior and be capable of producing unequal phase shifts in the u and v directions. Specifically, it should introduce 180° and 90° phase shifts in the u and v direction respectively by switching the two PIN diodes, as proposed in this design.

3. Simulation Results and Analysis

In general, polarization conversion ratio (PCR) and axial ratio (AR) are often used to describe the characteristics of linear to linear polarized and linear to circular polarized for multi-functional polarizer. The PCR and AR can be expressed as,

$$\text{PCR} = \left|\mathbf{R}_{y^*x^*}\right|^2 \Big/ \left(\left|\mathbf{R}_{y^*x^*}\right|^2 + \left|\mathbf{R}_{x^*x^*}\right|^2\right) \tag{5}$$

$$AR = \sqrt{\frac{(R_{uu}^2 + R_{vv}^2 + \sqrt{R_{uu}^4 + R_{vv}^4 + 2R_{uu}^2 R_{vv}^2 \cos(2\Delta\varphi)})}{(R_{uu}^2 + R_{vv}^2 - \sqrt{R_{uu}^4 + R_{vv}^4 + 2R_{uu}^2 R_{vv}^2 \cos(2\Delta\varphi)})}} \quad (6)$$

By analogy to the orthogonal decomposition process in (2) and rotation matric **O**, $R_y *x*$ and $R_x *x*$ can be represented by R_{uu} and R_{vv},

$$\begin{bmatrix} R_{x^*x^*} \\ R_{y^*x^*} \end{bmatrix} = \mathbf{O} \cdot \begin{bmatrix} R_{uu} \\ R_{vv} \end{bmatrix} = \frac{\sqrt{2}}{2} \begin{bmatrix} 1 & 1 \\ -1 & 1 \end{bmatrix} \begin{bmatrix} R_{uu} \\ R_{vv} \end{bmatrix} \quad (7)$$

The frequency domain solver of CST is used to simulate the proposed polarizer unit in a periodic environment. Setting the electric field direction of the incident wave along the x and y axis, the reflection coefficient results, including the magnitude and phase, in u, v direction under different PIN states, are shown in Fig. 3(a)-(b). It can be observed that the magnitude and phase at u, v direction can be independently controlled and do not affect each other.

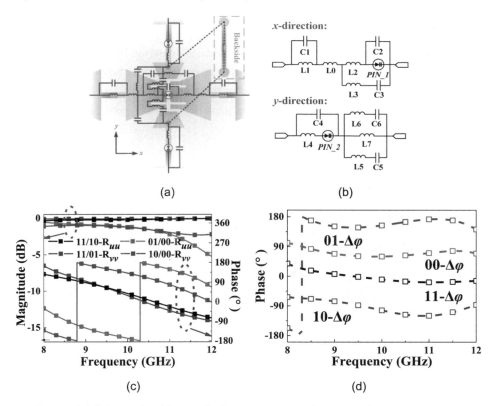

Fig. 4 (a) Schematic of the equivalent component distribution for the unit model. (b) Equivalent circuit in the x, y directions. (c) Magnitude and phase of the reflection coefficients for four states and (d) phase difference simulated by ADS. (C1=0.145 pF, C2=0.012 pF, C3=0.124 pF, C4=0.082 pF, C5=0.260 pF, C6=0.105 pF, L1=5 nH, L2= 1.8 nH, L3=0.01 nH, L4=3.05 nH, L5=5.5 nH, L6=2.6 nH, L7=5 nH)

In the wireless communication applications, 3 dB AR bandwidth is commonly used to represent the bandwidth of CP waves. Similarly, more than 90% of the PCR bandwidth was reasonable bandwidth for VP conversion while for HP conversion, it was below 10%. Ideally, we can calculate that $\Delta\varphi$ for LHCP and RHCP conversion falls within the range of (−90°−36.75°, −90°+36.75°) and (90°−36.75°, 90°+36.75°), $\Delta\varphi$ for VP and HP conversion falls within the range (180°−32.2°, 180°+32.2°) and (0°−32.2°, 0°+32.2°). As illustrated in Fig. 3(c), $\Delta\varphi$ for all the states is roughly equal to 0, -π/2, π, and π/2 in the entire X-band, respectively.

To provide a more direct representation of the control effect of the multi-functional active polarizer, the simulated AR and PCR are displayed in Fig. 3(d). The polarizer operates within 8.46–11.57 GHz (RHCP) and 8.02–11.35 GHz (LHCP) in the "00" and "10" states, respectively. In the "11" state, there is minimal PCR across the entire X-band. However, in the "01" state, a bandwidth of 8–11.2 GHz is achieved with PCR above 90%. All four polarization modes are accommodated within the X-band, and the shared band is 8.46-11.2 GHz.

Here, we employ an equivalent circuit to further validate the broadband and tunable mechanisms of the proposed polarizer. Due to the anisotropic structure of the unit patch, different equivalent circuit models are applied in the x and y directions. According to the method of equivalent circuit construction, the gaps between adjacent metal patches are considered as capacitors,

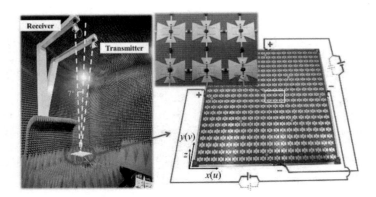

Fig. 5 Photo of fabricated polarizer and measurement environment.

whereas the metal patches as inductors. Following the approach depicted in Fig. 4(a), the equivalent circuits in the x and y directions are separately extracted, as shown in Fig. 4(b). As illustrated in Fig. 4(c)-(d), the close agreement between the simulation results of Advanced Design System (ADS) circuit and CST model strongly validates the fundamental principles of this design, including ohmic losses and phase control. This analysis provides

Design of an Active Polarizer for Wideband Quad-Polarization Conversion

valuable insights into the working principles of active polarizer and further validating its functionality of tunable control.

4. Measurement Results and Analysis

To verify the simulation results of the proposed polarizer, a prototype surface of size 210 mm× 210 mm (20×20 units) was fabricated. The model and the measuring environment are depicted in Fig. 5. The entire measurement procedure was carried out in the arch method measurement system. The bias forward voltage is set at 17 V. We always maintain the antennas on the transmit and receive sides in the same polarization mode to get accurate R_{uu} and R_{vv} measured results. By setting the different PIN diode operating states, and the measurement results are shown in Fig. 6.

The measured results show a reduction in magnitude of approximately 2 dB. The losses causing this reduction can be attributed to various factors, such as additional resistance at soldering points, substrate deformation due to thermal effects, and the inherent resistance of PIN diodes. The operating frequencies of circular polarization reflection mode with AR less than 3 dB under the state of "10" and "00" are 8.7-11.3 GHz (LHCP) and 8.61-11.17 GHz (RHCP), respectively. In the state of "11", almost no polarization conversion occurs in the whole X-band of the reflected wave. In the "01" state, the polarization conversion rate of the reflected wave is more than 90% of the bandwidth is 8-11.3 GHz (VP). The shared band is 8.7-11.17 GHz. These discrepancies between measured and simulated results could arise due to potential errors during the manufacturing and measurement processes.

As shown in Table I, when compared with the previous design, the proposed active polarizer can achieve quad-polarization conversion for the HP incident waves across a wide shared band. Notably, this polarizer effectively overcomes the limitations of a narrow shared band, intricate construction, low integration, and constrained functionality. These attributes render it a highly appealing choice for diverse applications in the field of wireless communication.

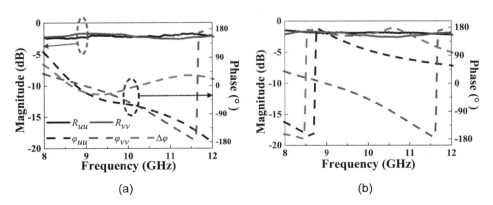

(a)   (b)

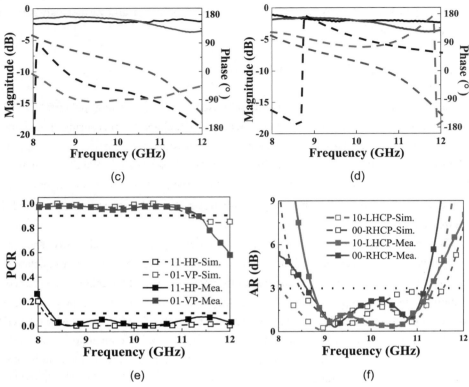

Fig. 6 The measurement results of reflection coefficients magnitude and phase and Δφ of the fabricated polarizers under (a) "11", (b) "01", (c) "10", (d) "00", respectively. And the corresponding results of (e) PCR, (f) AR for each state.

Table I

Comparison between the proposed active polarizer and previous designs

Refs.	No. of A.C.	Thick-ness	Reflection Losses	No. of Polarization Modes	Shared Band (GHz)
[14]	1	0.12 λc	2 dB	2 (VP, LHCP)	4.9-7.9
[16]	2	0.08 λc	2 dB	2 (HP, VP)	3.83-4.74
[17]	1	0.10 λc	1 dB	2 (HP, VP)	7.4-12
[18]	2	0.09 λc	1 dB	3 (HP, VP, LHCP)	N.A.
[19]	2	0.24 λc	1 dB	3 (HP, LHCP, RHCP)	3.4-8.8
This work	2	0.13 λc	2 dB	4 (HP, VP, LHCP, RHCP)	8.7-11.17

A.C.: Active components. λc: Center frequency wavelength.

5. Conclusion

To solve the reconfigurable issues of polarization regulation in wireless communication, a novel low-profile wideband active polarizer for qual-polarization conversion is proposed. In this design, utilizing the asymmetrical characteristic of structure, achieving phase shifts of 180° and 90° in the u and v direction, the proposed polarizer can convert HP waves into HP, VP, LHCP, and RHCP modes in a shared band. The active polarizer unit only contains two active components, and it resolves the previous polarizers design's issues with restricted function, poor integration, and complex structure.

Acknowledgments

This work was supported by the National Natural Science Foundation of China (U2141233 and 62171416) and Fundamental Research Funds for the Central Universities (CUC230D037).

Author Information

C. Xu, J. Su, M. Qu and Z. Li are with the School of Information and Communication Engineering, Communication University of China, Beijing 100024, China, and also with State Key Laboratory of Media Convergence and Communication, Communication University of China.

K. Qi and H. Yin are with the Science and Technology on Electromagnetic Scattering Laboratory, Beijing, 100854, China.

References

[1] PFEIFFER C, GRBIC A. Bianisotropic metasurfaces for optimal polarization control: analysis and synthesis[J]. Phys. Rev. Applied, 2014, 2: 044011.

[2] NASERI P, MATOS S A, COSTA J R, et al. Dual-band dual-linear-to-circular polarization converter in transmission mode application to K/Ka-band satellite communications[J]. IEEE Trans. Antennas Propag., 2018, 66 (12): 7128-7137.

[3] CACOCCIOLA R, RATNI B, MIELEC N, et al. Metasurface-tuning: a camouflaging technique for dielectric obstacles[J]. J. Appl. Phys., 2021, 129: 124902.

[4] LI Y, ZHANG J, QU S, et al. Achieving wide-band linear-to-circular polarization conversion using ultra-thin bi-layered metasurfaces[J]. J. Appl. Phys., 2015, 117: 044501.

[5] CHENG Y Z, FANG C, MAO X S, et al. Design of an ultrabroadband and high-efficiency reflective linear polarization convertor at optical frequency[J]. IEEE Photonics J., 2016, 8 (6): 1-9.

[6] DOUMANIS E, GOUSSETIS G, GOMEZ-TORNERO J L, et al. Anisotropic impedance surfaces for linear to circular polarization conversion[J]. IEEE Trans. Antennas Propag., 2012, 60 (1): 212-219.

[7] EULER M, FUSCO V, DICKIE R, et al. Sub-mm wet etched linear to circular polarization FSS based polarization converters[J]. IEEE Trans. Antennas Propag., 2011, 59 (8): 3103-3106.

[8] SILVEIRINHA M G. Design of linear-to-circular polarization transformers made of long densely packed metallic helices[J]. IEEE Trans. Antennas Propag., 2008, 56 (2): 390-401.

[9] GAO X, HAN X, CAO W P, et al. Ultrawideband and high-efficiency linear polarization converter based on double V-shaped metasurface[J]. IEEE Trans. Antennas Propag., 2015, 63 (8): 3522-3530.

[10] CHEN K, FENG Y, MONTICONE F, et al. A reconfigurable active Huygens' metalens[J]. Adv. Mater., 2017, 29 (17): 1606422.

[11] ZHANG X, JIANG W, JIANG H, et al. An optically driven digital metasurface for programming electromagnetic functions[J]. Nat. Electron., 2020, 3: 165-171.

[12] TIAN J, CAO X, GAO J, et al. A reconfigurable ultra-wideband polarization converter based on metasurface incorporated with PIN diodes[J]. J. Appl. Phys., 2019, 125: 135105.

[13] QIN Z, LI Y, WANG H, et al. Polarization meta-converter for dynamic polarization states shifting with broadband characteristic[J]. Opt. Express, 2022, 30 (11): 20014-20025.

[14] GAO X, YANG W L, MA H F, et al. A reconfigurable broadband polarization converter based on an active metasurface[J]. IEEE Trans. Antennas Propag., 2018, 66 (11): 6086–6095.

[15] YANG Z, KOU N, YU S, et al. Reconfigurable multifunction polarization converter integrated with PIN diode[J]. IEEE Microwave Wireless Compon. Lett., 2021, 31 (6): 557-560.

[16] SUN S, JIANG W, GONG S, et al. Reconfigurable linear-to-linear polarization conversion metasurface based on PIN diodes[J]. IEEE Antennas Wireless Propag. Lett., 2018, 17 (9): 1722-1726.

[17] LIU W, KE J C, XIAO C, et al. Broadband polarization-reconfigurable converter using active metasurfaces[J]. IEEE Trans. Antennas Propag., 2023, 71 (4): 3725–3730.

[18] PRAMANIK S, BAKSHI S C, KOLEY C, et al. Active metasurface based reconfigurable polarization converter with multiple and simultaneous functionalities[J]. IEEE Antennas Wireless Propag. Lett., 2023, 22 (3): 522-526.

[19] CUI J H, HUANG C, PAN W B, et al. Dynamical manipulation of electromagnetic polarization using anisotropic meta-mirror[J]. Sci. Rep., 2016, 6: 30771.

[20] ZHANG L, WANG Z, SHAO R, et al. Dynamically realizing arbitrary multi-bit programmable phases using a 2-bit time-domain coding metasurface[J]. IEEE Trans. Antennas Propag., 2020, 68 (4): 2984-2992.

[21] HUANG C, SUN B, PAN W B, et al. Dynamical beam manipulation based on 2-bit digitally-controlled coding metasurface[J]. Sci. Rep., 2017, 7: 42302.

[22] RODRÍGUEZ-BERRAL R, MOLERO C, MEDINA F, et al. Analytical wideband model for strip/slit gratings loaded with dielectric slabs[J]. IEEE Trans. Microw. Theory Techn., 2012, 60 (12): 3908–3918.

A Low-RCS Multifunctional Shared Aperture with Wideband Reconfigurable Reflectarray Antenna and Tunable Scattering Characteristic[*]

1. Introduction

Metasurfaces, a two-dimensional version of metamaterial, have provided unprecedented freedoms in manipulating electromagnetic (EM) waves upon interfaces. With the assistance of metasurfaces, multifunctional reflectarrays are implemented. Predominantly, reflectarray antennas have synthesized the advantages of parabolic antennas and phased arrays, such as low-cost, lightweight, and easy fabrication, are widely used in satellite communications and radars. When the reflectarray antenna is applied in military radar systems, the large aperture will contribute a noticeable scattering effect; thus, increasing the detectability of the platform. Therefore, low radar cross section (RCS) characteristics become an essential index of aperture antenna performance in military platforms.

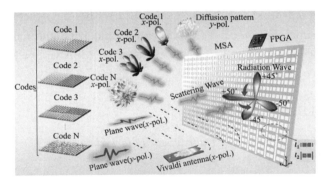

Fig. 1 Schematic of the proposed MSA.

When the methodology of metasurfaces is introduced to implement the RCS reduction (RCSR) of the antenna, a series of multifunctional low RCS antennas are accomplished. For example, the antenna loaded with functional metasurfaces, such as frequency selective surface (FSS), artificial magnetic conductor (AMC), perfect absorber metamaterial (PAM), and polarization conversion metasurfaces (PCM), which can achieve the low-

---

[*] The paper was originally published in *IEEE Transactions on Antennas and Propagation*, 2023, 71 (1), and has since been revised with new information. It was co-authored by Pan Li, Hang Yu, Jianxun Su, Liwei Song, Qingxin Guo, and Zengrui Li.

RCS property of the platform. However, the radiation pattern of the above designs cannot be flexibly controlled. Thus, the research on low RCS antenna arrays with beam scanning properties will be of great significance. According to the newest publications, most of the low RCS antenna arrays are accomplished by combing the antenna arrays with functional metasurfaces. However, despite the narrow band and different operation frequencies for radiation and scattering, most of the low RCS antenna arrays are non-reconfigurable.

It can be expected that the in-band low RCS performance of both polarizations can be applied to a wideband reconfigurable antenna array. In this paper, we put forward a single-layer multifunctional shared-aperture (MSA) with tunable control of radiation and scattering patterns for *x*-polarized waves while low RCS property for the *y*-polarized waves. As shown in Fig. 1, when MSA is illuminated by an *x*-polarized Vivaldi antenna, the required phase shifts are coded and downloaded to the MSA through a field-programmable gate array (FPGA); thus, the 2D beam scanning can be realized accompany with the low RCS property. For *x*-polarized waves, different coding sequences corresponding to the variable bistatic scattering patterns when the MSA does not perform radiation action. For *y*-polarized waves, owing to the randomly arranged 16 lattices with different sizes in *y*-direction, which are selected through the optimized multielement phase cancellation (OMEPC) method, the diffusion scattering pattern will be realized. Finally, a 16×16-element MSA is fabricated and measured, and the experimental results have verified the correctness of the proposed method.

This paper is organized as follows: Section II describes the structure and the design principle of the initial unit model 1 (Mo. 1) and the modified unit model 2 (Mo. 2). In Section III, the simulated and experimental results of the MSA will be depicted. Then we will finally give a conclusion in Section IV.

2. Unit Cell Design Principle

2.1 Initial Unit Cell Mo. 1 of Dual-Linear Polarization

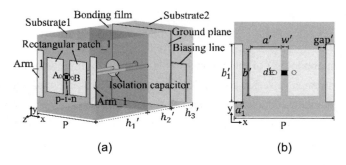

Fig. 2 (a) 3D schematic configuration of the initial unit cell Mo. 1. (b) Top view of the unit. A and B represents the positive and negative poles of the p-i-n diode.

A Low-RCS Multifunctional Shared Aperture with Wideband Reconfigurable Reflectarray Antenna and Tunable Scattering Characteristic

Table I

Parameters of the tunable unit cell Mo.1

Parameter	Value (mm)	Parameter	Value (mm)
P	18	a'	5.1
$h1'$	5	b'	10.5
$h2'$	0.1	a_1'	1.5
$h3'$	0.2	b_1'	12
w'	2	d'	0.8
gap'	2.1		

The critical step for the proposed single-layer MSA is to design a wideband unit cell with separate responses for the two perpendicular waves. The unit cell integrated with a p-i-n diode along x-direction will be introduced for the purpose of beam scanning and flexible scattering patterns. As for the RCS reduction of the y-polarization, we rely on the varying sizes of the unit to generate different reflection phases. Based on the phase cancellation method, the low RCS property could be accomplished. We start with an initial unit cell Mo. 1 as presented in Fig. 2, the unit cell consists of five parts including a phase-shift layer, two substrates, an isolation capacitor, a ground plane, and a biasing line layer. The substrates 1 (F4B, $\varepsilon_r = 2.65$, tan δ = 0.001) and 2 (FR-4, $\varepsilon_r = 4.4$, tan δ = 0.02) are bonded together through a 0.1 mm bonding film (FR-4, $\varepsilon_r = 4.4$). The phase-shift layer is composed of rectangular patch_1 and arm_1, where a low-cost p-i-n diode (Skyworks, SMP 1320-079LF) is integrated to construct the reconfigurable unit cell with 1-bit phase quantization (0/180°). The "OFF" and "ON" state of p-i-n diode is switched through changing the bias voltages applied to the diode. The isolation capacitor is deliberately designed to minimize the influence of the direct current (DC) signal on the radio frequency (RF) signal. The optimized physical geometric parameters of Mo. 1 are shown in Table I.

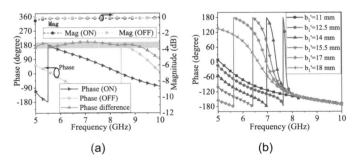

Fig. 3 Reflection responses of Mo. 1 for x- and y-polarized wave. (a) Reflection magnitude and phase for x-polarized wave under different p-i-n diode's states. (b) Reflection phases for y-polarized wave at different values of b_1'.

To validate the characteristics of Mo. 1, the full-wave simulation software CST Microware Studio® with periodic boundary conditions in the *x*- and *y*-directions and an open boundary in the *z*-direction is adopted to complete the simulation. The p-i-n diode is equivalently modeled as a resistor-inductor-capacitor (RLC) series resonant circuit with parameters R = 0.5 Ω, L = 0.7 nH for "ON" state and L = 0.5 nH, C = 0.24 pF for "OFF" state. For *x*-polarized waves, by switching the states of the diode, the reflection phase difference is $180 \pm 20°$, while the amplitude maintains greater than -0.5 dB from 5-9.4 GHz as shown in Fig. 3(a).

For *y*-polarized waves, it needs to be noted that the length of arm_1 should always be larger than that of the rectangular patch_1 (b'). Thus, the capacitance between arm_1 and rectangular patch_1 keeps unchanged, bring about the performance in *x*-direction being independent of the changeable b_1' along *y*-direction. When b_1' varies from 11 to 18 mm, the reflection phases are shown in Fig. 3(b), and no phase difference emerges when the frequency is greater than 8 GHz. Consequently, the common operation frequency of the unit cell for *x*- and *y*-polarized waves is limited to 5-8 GHz (FBW = 46.2%). Then, a method aims to break the bandwidth restraint for *y*-polarized waves will be mentioned in Section II-C.

2.2 Analysis of Unit Cell Mo. 1 Based on ECM Method

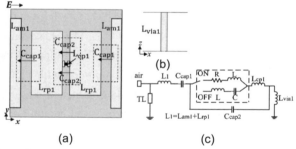

Fig. 4   (a) Capacitances and inductances on the phase-shift layer. (b) Inductance on the metalized via. (c) ECM of Mo. 1 for *x*-polarized waves.

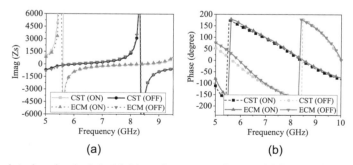

Fig. 5   Simulated and calculated (a) imaginary part of Z_s and (b) the reflection phases for *x*-polarized waves.

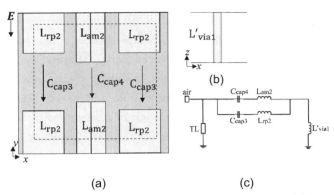

Fig. 6 (a) Capacitances and inductances on the phase-shift layer. (b) Inductance on the metalized via. (c) ECM of Mo. 1 for y-polarized waves.

In order to investigate the bandwidth constraints of the Mo. 1, the ECM of the unit cell has been analyzed. The inductor (L) is related to the current distribution in the metallic patches and the capacitor (C) results from the electric field distribution in the metallic gaps. Due to the asymmetry structure, there are different ECMs along x- and y-directions.

For the x-polarized waves, the capacitances and inductances on the phase-shift layer and metalized via are shown in Figs. 4(a) and (b). The ECM of Mo. 1 is presented in Fig. 4(c), in which the part enclosed by the dashed line is the equivalent circuit of the p-i-n diode. The F4B spacer in the design is modeled by a transmission line (TL) with the characteristic Z_0/ε_r, where Z_0 is 377 Ohms. And L_1 is the sum of L_{am1} and L_{rp1}. In addition, the surface impedances (Z_s) of the metallic patches are studied. Because of the low loss dielectric, the real part of the equivalent impedance is very small and can be ignored. From Fig. 5, at 5.5 GHz, the imaginary part of Z_s is infinite at "ON" state, and the reflection phase is close to 0°. In this case, the unit cell is analogous to perfect magnetic conductor (PMC). Meanwhile, the imaginary part of Z_s is close to 0° at "OFF" state, while the reflection phase is about 180°, then the unit cell is equivalent to a perfect electric conductor (PEC). It is the same at 8.3 GHz. The single layer unit switches between PMC and PEC at two frequencies has produced the $180\pm20°$ phase difference at a wide frequency band. We can learn the good agreement between calculated and simulated reflection phases from Fig. 5(b).

As Mo. 1 is excited by y-polarized waves, the circuit parameters are shown in Figs. 6(a) and (b). Fig. 6(c) shows the corresponding ECM of the Mo. 1. We take b_1' equals to 11 and 15 mm as examples. The coincident calculated and simulated phases under different values of b_1' are shown in Fig. 7(a). In order to investigate the resonant frequency of the rectangular patch_1, the reflection coefficient of the phase-shift layer with/without arm_1 is presented in Fig. 7(b). When simulated with arm_1, we have studied the reflection coefficient for different b_1'. From the figure, two perfect reflection points are obtained when simulated with arm_1, the lower one is generated by the arm_1 and the higher one is generated by rectangular patch_1. The common frequency of the three conditions in Fig. 7(b) shows the rectangular patch_1 is resonant at around 10.5 GHz. Then the less current is flowing on the arm_1 at the corresponding resonant frequency. Fig. 8 gives the current distribution on the unit cell at 8.5 GHz (near the resonant frequency of the rectangular patch_1) and 6 GHz when b_1' equals 11 and 15 mm. At 8.5 GHz, most of the current on the rectangular patch_1 results in the unchanged current distribution on the arm_1 regardless of any values of b_1'. At 6 GHz, different b_1' engender the varying current on the arm_1. Therefore, the resonant state of rectangular patch_1 has restrained the bandwidth of y-polarization at a higher frequency.

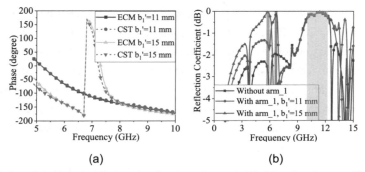

Fig. 7 (a) Simulated and calculated reflection phases and (b) reflection coefficient of the phase-shift layer with/without arm_1 for y-polarized waves under different b_1'.

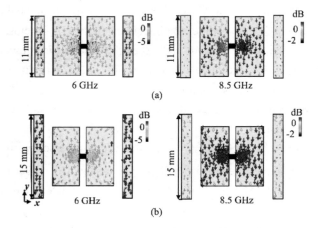

Fig. 8 Current distributions on the phase-shift layer of Mo. 1 for the y-polarized waves. (a) At 6 GHz and 8.5 GHz when b_1' =11 mm. (b) At 6 GHz and 8.5 GHz when b_1' = 15 mm.

Table II
The value of the capacitances and inductances in the ECM of Mo. 1

	x-pol.			y-pol.	
Parameter	Value		Parameter	Value	
	"ON"	"OFF"		$b_1'=11$mm	$b_1'=15$mm
L_1	1 nH	1 nH	C_{cap3}	0.05 pF	0.05 pF
C_{cap1}	0.12 pF	0.12 pF	C_{cap4}	0.014 pF	0.05 pF
L_{cp1}	1.8 nH	1.8 nH	L_{am2}	10.8 nH	14.4 nH
C_{cap2}	0.11 pF	0.11 pF	L_{rp2}	4.5 nH	4.5 nH
L_{via1}	0.6 nH	0.6 nH	$L_{via1'}$	0.2 nH	0.2 nH

The corresponding parameters in the ECM of Mo. 1 are presented in Table II. As for x-polarized waves, the capacitances and inductances keep unchanged at different states of p-i-n diode. In the case of y-polarized waves, according to the values of C_{cap3} and L_{rp2}, the rectangular patch_1 is resonant at around 10.5 GHz, which is consistent with the simulation result in Fig. 7(b). The C_{cap4} becomes larger as b_1' increases owing to the closer distance between adjacent arm_1. We found that the smaller b' can right-shift the rectangular patch_1's resonant frequency and leading to a significant phase difference for the y-polarized waves at a higher frequency. However, the smaller b' will lead to the shrinking C_{cap1} and C_{cap2}; then, the phase coverage of x-polarization will be less than $180 \pm 20°$. Therefore, it is crucial to propose a modified unit cell with smaller b' while ensuring wideband performance of x-polarization.

2.3 Modified Unit Cell Mo. 2 of Dual-Linear Polarization

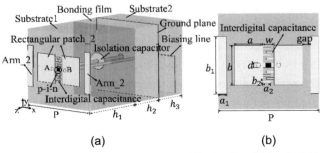

Fig. 9 (a) 3D schematic configuration of the modified unit cell Mo. 2. (b) Top view of the unit. A and B represents the positive and negative poles of the p-i-n diode.

Table III

Parameters of the modified unit cell Mo. 2

Parameter	Value (mm)	Parameter	Value (mm)
P	18	a	5.1
h_1	4.7	b	6.9
h_2	0.1	a_1	1.5
h_3	0.2	b_1	10
w	2	d	0.8
gap	2.1	a_2	1.5
b_2	0.28		

With the smaller size of rectangular patch_1 in the y-direction, the modified unit cell Mo. 2 integrated with interdigital capacitance (IC) is presented in Fig. 9(a). The IC, which is equivalently modeled as an inductor-capacitor in parallel connection, is utilized to make up for the decreased C_{cap1} and C_{cap2}. Therefore, the wideband performance for x-polarized waves can be maintained. The unit cell Mo. 2 shared a 3D structure with Mo. 1, except that the phase-shift layer is replaced by the structure in Fig. 9(b). Table III shows the optimized physical geometric parameters of the unit cell Mo. 2.

For the x-polarized incident waves, the reflection responses presented in Fig. 10(a) are the same as that of the Mo. 1. In the case of y-polarized waves, due to the right-shift resonant frequency of the rectangular patch_2, a significant phase difference is established when the frequency is higher than 8 GHz. Fig. 10(b) shows the reflection phases for different b_1 in the range of 5-10 GHz. Therefore, the common operation band of the Mo. 2 for the two perpendicular waves is from 5.2 to 9.4 GHz with the FBW of 57.5%. Compared with Mo. 1, the Mo. 2 improves the FBW from 46.2% to 57.5%.

To further investigate the independent performance of Mo. 2 for the two perpendicular polarized waves, we have studied the phase differences between "OFF" and "ON" states of the p-i-n diode under the x-polarized incident waves when b_1 varies from 6.9 to 18 mm. From Fig. 10(c), the reflection phase differences for different b_1 are approximately $180 \pm 20°$ from 5.2 to 9.4 GHz. It can be seen that the size of the arm_2 in y-direction has little impact on the performance of the x-polarization. Meanwhile, Fig. 10(d) presents the consistent reflection phases and magnitudes for y-polarized waves. It comes out that the states of diode in x-direction do not influence the performance of y-polarization. Thus, the unit cell Mo. 2 has realized the independent control of x- and y-polarized waves over a wide frequency band.

A Low-RCS Multifunctional Shared Aperture with Wideband Reconfigurable Reflectarray Antenna and Tunable Scattering Characteristic

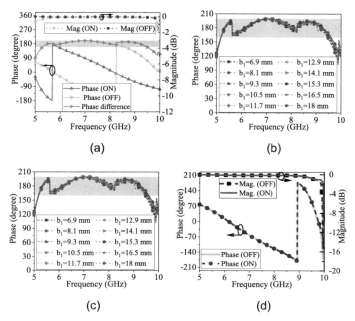

Fig. 10 (a) Reflection magnitudes and phases for x-polarized wave under different p-i-n diode's states. (b) Reflection phases for y-polarized waves. (c) Phase differences for x-polarized waves under different values of b_1. (d) Phases and magnitudes of y-polarization for "OFF" and "ON" states of p-i-n diode.

2.4 Analysis of Modified Unit Cell Mo. 2 Based on ECM Method

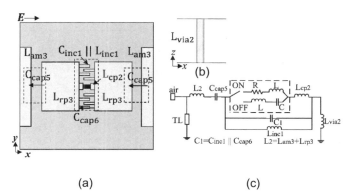

Fig. 11 (a) Capacitances and inductances on the phase-shift layer. (b) Inductance on the metalized via. (c) ECM of Mo. 2 for x-polarized waves.

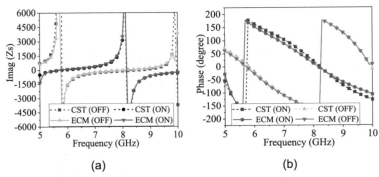

Fig. 12 Simulated and calculated (a) imaginary part of Z_s and (b) the reflection phases for x-polarized waves at two states of diode.

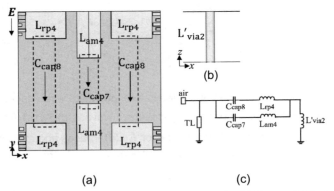

Fig. 13 (a) Capacitances and inductances on the phase-shift layer. (b) Inductance on the metalized via. (c) ECM of Mo. 2 for y-polarized waves.

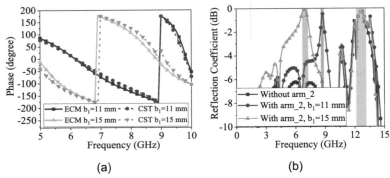

Fig. 14 Simulated and calculated (a) reflection phases and (b) reflection coefficient of the phase-shift layer with/without arm_2 for y-polarized waves under different b_1.

Table IV

The value of the capacitances and inductances in the ECM of Mo. 2

Para.	x-pol. Value		Para.	y-pol. Value	
	ON	OFF		b_1 =11mm	b_1 =15mm
L_2	1.2 nH	1.2 nH	C_{cap8}	0.023 pF	0.023 pF
C_{cap5}	0.09 pF	0.09 pF	C_{cap7}	0.022 pF	0.038 pF
L_{cp2}	2.45 nH	2.45 nH	L_{rp4}	7.5 nH	7.5 nH
C_1	0.15 pF	0.15 pF	L_{am4}	15.4 nH	18 nH
L_{inc1}	50 nH	50 nH	L'_{via2}	0.2 nH	0.2 nH
L_{via2}	1.5 nH	1.5 nH			

Similarly, when Mo. 2 is normally illuminated by x-polarized waves, the circuit parameters are shown in Figs. 11(a) and (b). The IC is modeled as a parallel connection of L_{inc1} and C_{inc1}. Combined with the equivalent model of the diode, the ECM of Mo. 2 for x-polarized waves is illustrated in Fig. 11(c). It needs to be noted that the L_2 is the sum of L_{am3} and L_{rp3}, while the C_1 is the parallel connection of C_{inc1} and C_{cap6}. Similarly, the imaginary part of Z_s in Fig. 12(a) demonstrates the wideband property of the Mo. 2. The calculated reflection phases for different p-i-n diode's states are also provided in Fig. 12(b). In the case of y-polarized incident waves, the circuit parameters and the ECM of Mo. 2 are shown in Fig. 13. The additional capacitance and inductance that introduced by IC can be ignored because of their slight influences on the performance of y-polarization. The well agreement between simulated and calculated reflection phases is shown in Fig. 14(a). Similarly, the reflection coefficient of the phase-shift layer is also studied in Fig. 14(b). And the resonant frequency of the rectangular patch_2 is around 12.5 GHz, which is right-shifted respected to that of the rectangular patch_1. And Table IV shows the corresponding parameters in the ECM of Mo. 2.

From Tables II and IV, as for the x-polarized waves, the smaller C_{cap5} than C_{cap1} is obtained by the decreased size of rectangular patch_2 in y-direction. In addition, the capacitance C_1 is greater than C_{cap2} to compensate for the decreased C_{cap5}, so the wideband performance for x-polarized waves is maintained. In the case of y-polarization, similarly, the smaller b than b' produces the smaller C_{cap8} compared to the C_{cap3}. According to C_{cap8} and L_{rp4}, the resonant frequency of rectangular patch_2 is approximately 12.5 GHz, which is consistent with the simulation result in Fig. 14(b). Coincident with that presented in Fig. 10(b), the right-shift resonant frequency of rectangular patch_2 produces the significant phase differences of y-polarization when the frequency is higher than 8 GHz. The above analysis shows that the IC integrated into Mo. 2 has realized the independent control of the two perpendicular waves over a wide frequency band.

3. Simulation and Measurement

3.1 Scattering Performance of MSA

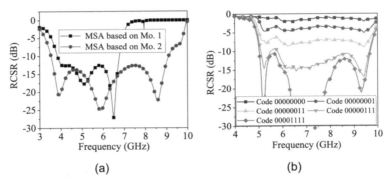

Fig. 15 (a) Theoretically calculated RCSR for *y*-polarized wave based on units Mo. 1 and Mo. 2 through OMEPC method. (b) Simulated RCSR for *x*-polarized wave at different codes based on unit cell Mo. 2.

In the design of low RCS property, the OMEPC method is considered to take as much phase information as possible to maximize the RCSR bandwidth for the *y*-polarized incident waves. According to the superposition principle, when plane waves normally impinge on the MSA, the RCS of the surface is the vector superposition of N scattering lattices. Take a PEC of the same area as the reference, the RCSR is computed by

$$\text{RCSR(dB)} = 10\log_{10}\left(\left|\frac{\sqrt{\sigma_{MSA}}}{\sqrt{\sigma_{PEC}}}\right|^2\right) = 10\log_{10}\left(\left|\frac{1}{N}\sum_{n=1}^{N}|\Gamma_n|e^{j\varphi_n}\right|^2\right) \quad (1)$$

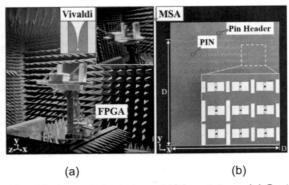

Fig. 16 1-bit 16 × 16-element single-layer MSA prototype. (a) System assembly. (b) Top view of the MSA.

Table V
Selected values of b_1 for 16 lattices based on Mo. 2 and their distribution location on the aperture

y- \ x-	lattice_1	lattice_2	lattice_3	lattice_4
lattice_1	16.9 mm	14.4 mm	10.4 mm	15.4 mm
lattice_2	16.9 mm	10.9 mm	10.4 mm	7.9 mm
lattice_3	10.4 mm	16.9 mm	16.9 mm	16.4 mm
lattice_4	9.4 mm	7.9 mm	16.9 mm	10.4 mm

where the Γ_n and φ_n are the reflection coefficient and reflection phase of the nth element. The reflection coefficient is approximate 1, then the RCSR of the MSA is only related to the reflection phase of the lattices. A 4×4-element sub-array constructs a lattice that shares the same value of b_1; thus, 16 lattices are needed to build up the MSA. In order to make a contrast, 16 lattices based on Mo. 1 and Mo. 2 are selected to calculate the RCSR for the y-polarized waves, respectively. In Fig. 15(a), when y-polarized wave normally illuminates the MSA, the theoretically calculated 10-dB RCSR bandwidth of Mo. 2 is 4-9.2 GHz, and the RCSR bandwidth of Mo. 1 is 4-6.8 GHz. Therefore, the modified unit cell Mo. 2 has effectively expanded the RCSR bandwidth for the y-polarized waves. According to [24], the single-layer structure will face a short circuit point for the low RCS design, which limits the RCS reduction bandwidth, but the multi-layer structure will avoid this problem. However, the multi-layer structure in our design will bring about the deterioration of radiation performance. So, the MSA in this paper is an integration of radiation and scattering characteristics, and is also a joint optimization process of radiation and scattering structures.

Based on the selected lattices, a 16×16-element MSA is fabricated and measured as shown in Fig. 16. The b_1 of the lattices and their distribution locations on the aperture are shown in Table V. The proposed array has an effective size of 288 mm × 288 mm, and 256 p-i-n diodes are used to dynamically tune the phase shift of all reflection elements independently. A wideband x-polarized Vivaldi antenna is used to illuminate the MSA to reduce the feed blockage.

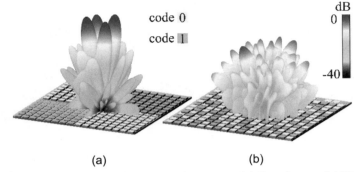

Fig. 17 Scattering patterns for *x*-polarized plane wave. (a) Four-beam. (b) Diffusion-beam.

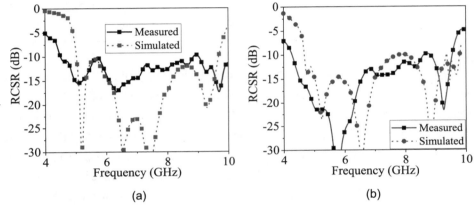

Fig. 18 Simulated and measured RCSR for (a) *x*-polarized wave, based on the code 00001111, and (b) *y*-polarized wave.

The radiation and scattering performance are measured in the far-field anechoic chamber. For the measurement of radiation pattern, the Vivaldi antenna is connected to the vector network analyzer (VNA). When measuring the backward RCSR, the feeding antenna is connected with a matched load. The p-i-n diode at "OFF" and "ON" states, encoded as "0" or "1", respectively. By tuning different coding sequences loaded to the MSA via FPGA, various bistatic scattering patterns and tunable backward RCSRs for the *x*-polarized wave can be obtained. If the two columns on the surface are grouped, the backward RCSRs for five coding sequences are shown in Fig. 15(b); the RCSR amplitudes are flexibly regulated from 5.2-9.6 GHz. According to the chessboard and random distribution coding sequences loaded to the MSA, the four- and diffusion-beam scattering patterns for the *x*-polarized wave at 8 GHz are shown in Fig. 17. Fig. 18 shows the simulated and measured backward RCSRs for the *x*- and *y*-polarized waves. The measured 10-dB RCSR bandwidth for the *x*-polarized wave based on the code 00001111 is in the range of 4.5-10 GHz, while the 10-

dB RCSR for the *y*-polarized wave is from 4.2-9.6 GHz. The RCSR is greater than 15 dB at some frequency points, which is result from the reflection phase difference being almost 180°. Then the common RCSR frequency for the *x*- and *y*-polarized waves is from 4.5-9.6 GHz (FBW=72.3%). The slight deviation compared to the simulated results is mainly caused by imperfect fabrications and measurements.

3.2 Radiation Performance of MSA

The primary mechanism of the single-layer MSA is to adjust the phase distribution of the array aperture to realize the excepted radiation patterns.

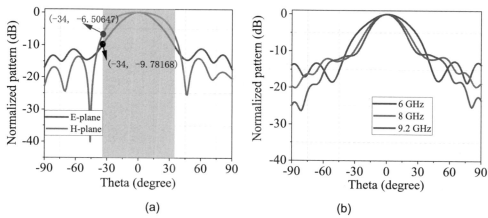

Fig. 19 Simulated radiation patterns (a) in the E- and H- planes at 7 GHz and (b) in the E-plane at 6, 8, and 9.2 GHz.

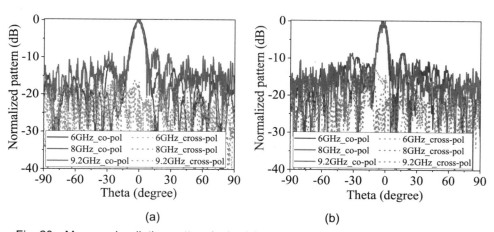

Fig. 20 Measured radiation pattern in the (a) E- and (b) H- plane at 6, 8, and 9.2 GHz.

Table VI

Comparison of the proposed single-layer MSA with some existing low RCS reflectarrays

Ref.	Radiation Performance		Scattering Performance				Wideband reconfigurable reflectarray with low RCS for both polarizations
	Radiation bandwidth	Beam scanning property	x-pol.		y-pol.		
			RCSR bandwidth	Tunability	RCSR bandwidth	Tunability	
[19]	6.8-7.8 GHz (13.7%) +	No	No	No	10 dB from 2-4.5 GHz (76.9%) 8 dB from 9.1-10.8 GHz (17.1%)	No	No
[20]	8.55-9.2 GHz (7%) +	No	10 dB from 4.3-10 GHz (79.7%) and 11-13 GHz (16.7%)	No	10 dB from 4.3-10 GHz (79.7%) and 11-13 GHz (16.7%)	No	No
[21]	9-10.5 GHz (15.4%) +	No	10 dB from 7-9.2 GHz (27.16%) 5 dB from 10.5-13 GHz (21.27%)	No	7 dB from 7-13 GHz (60%)	No	No
[22]	9.2-10.6 GHz (14%) +	No	10 dB from 8.4-14.9 GHz (55.8%)	No	10 dB from 4.6-8.8 GHz (62.6%) and 9.6-13.8 GHz (36%)	No	No
[23]	5.5 GHz*	30° - 60°	No	No	0-8 dB at 5.5 GHz	Yes	No
This work	5.5-9.5 GHz (53.3%) +	0 – 50°	≥ 10 dB from 4.5-10 GHz (75.9%)	Yes	10 dB from 4.2-9.6 GHz (78.3%)	No	Yes

*: Impendence bandwidth of the antenna; +: -3 dB gain bandwidth of the array.

Table VII

Measured half-power beam width in E-plane and H-plane at three frequencies

Frequency (GHz)	HPBWs in E-plane	HPBWs in H-plane
6	10.5°	9.8°
8	8.2°	7.2°
9.2	7.8°	6.2°

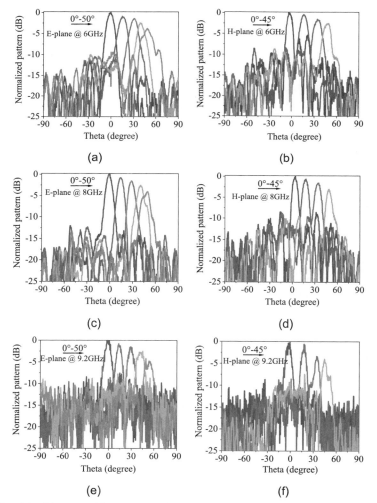

Fig. 21 Measured beam scanning patterns of the MSA at (a) (b) 6 GHz, (c) (d) 8 GHz, and (e) (f) 9.2 GHz. The left column is the scanning patterns in the E-plane, while the right column is in the H-plane.

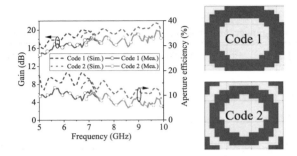

Fig. 22 Simulated and measured gain and aperture efficiency versus frequency for code 1 and code 2.

According to the approach used in microstrip reflectarray design, the phase shift (φ_r) of each unit cell for the steerable performance can be determined as:

$$\varphi_r = \varphi_f - k \cdot \hat{u}_0 \cdot \vec{r}_{mn} + \Delta\varphi \tag{2}$$

where φ_f is the phase delay from the feed to the $(m,n)^{th}$ element, k is the free space wavenumber, $\hat{u}_0$ is the beam direction, and $\vec{r}_{mn}$ is the position vector of the $(m,n)^{th}$ element, $\Delta\varphi$ is the reference phase that provides additional freedom for the design. This is because that different $\Delta\varphi$ does not alter the phase difference between two arbitrary units in the aperture. Meanwhile, in the case of 1-bit MSA, when $\Delta\varphi$ changes φ_r changes as well, leading to different phases obtained by each unit. Especially when being quantized, it will result in different p-i-n diode's states on the aperture; the relative phase of two arbitrary elements may change. The reference phase for each scanning angle is optimized during the measurement to improve the radiation performance of the MSA.

Considering the phase shift quantization, a 1-bit MSA with phase shift 180° is chosen in our design. Then we quantized the phase shift as follows:

$$\varphi_r^q = \begin{cases} 0°, \text{if } -90° \leqslant \varphi_r \leqslant 90° \\ 180°, \text{other cases} \end{cases} \tag{3}$$

The required phase shifts are coded and downloaded to FPGA. And a wideband x-polarized Vivaldi antenna is used to excite the MSA, the edge illumination taper of around -10 dB in the E-plane at 7 GHz is finally selected for the design of the MSA, while the fringe radiation level in H-plane is -6.5 dB [see Fig. 19(a)]. In addition, the beam width of Vivaldi antenna in E-plane becomes narrower with the increasing frequency [see Fig. 19(b)]. The asymmetric radiation pattern of the primary feed, together with the inconsistent beam width over the wide frequency band, may cause the performance degradation of the MSA. For an MSA with wideband performance, the feed location selection becomes more challenging. In order to improve the gain as well as beam-steering performance over a wide frequency band, a thorough investigation of the feed location is carried out. An optimal feed location is determined when all the factors (near-field effect, oblique incidence, feed blockage, and wideband condition) are fully considered. Therefore, after a comprehensive parametric study, a tradeoff is made by placing the feed at F (0, 0, 210) mm, with approximately F/D=0.7, where D is the effective side length of the MSA. The radiation patterns in the E- and H- planes at 6, 8, and 9.2 GHz are shown in Fig. 20; the co- and cross-polarization patterns are represented by solid and dashed lines, respectively. The average cross-polarization level is 15 dB less than co-polarization. In Fig. 20, different measured cross-polarized level at E- and H-planes for theta = 0° is mainly caused by misalignment errors between the receiving horn and the MSA in the far-field antenna test system. Besides, the high environmental noise of the microwave anechoic chamber has also affected the cross-polarization levels. The Table VII shows the measured half-power beam width (HPBWs) at 6, 8, and 9.2 GHz. From the Table, the corresponding beam width of the array will become narrower with the increasing frequency. Further, the beam scanning patterns at three frequencies are depicted in Fig. 21. From the figure, the beam-scanning within ±50° in E-plane while ±45° in H-plane at three frequencies are obtained. As the beam scanning angle becomes larger, the HPBW increases, and the main beam suffers from scan loss and increasing side band levels (SLLs). To further demonstrate the wideband performance, the gains versus frequency are simulated and measured when MSA is loaded with coding sequences at 5.4 and 7 GHz, which generates the widest 3-dB gain band width, namely code 1 and 2, respectively. The gain junction for the two codes is studied to maximize the 3-dB gain bandwidth of the MSA. Fig. 22 shows the gain and aperture efficiency (AE) versus frequency for code 1 and code 2. From the figure, the measured maximum AE is 13.2% at 5.6 GHz, while the 3-dB gain bandwidth of the MSA is from 5.5-9.5 GHz (FBW = 53.3%), which confirms the wideband radiation performance of the MSA. The measured gain loss is approximately 2 dB over the entire frequency band, which is mainly caused by the malfunctioning p-i-n diodes in the fabricated prototype, the influence of the room for p-i-n diode direct current lines and digital commands, and the inaccurate characterization of the equivalent parameters of p-i-n diodes. Besides, the experimental accuracy has also resulted in the performance degradation. Moreover, the AE decline is mainly due to 1-bit quantization loss, asymmetric radiation patterns of the feed antenna, and fixed F/D. The radiation pattern of the feeding source over a wide frequency band and F/D can be optimized to realize a higher AE. Furthermore, the design of wideband reconfigurable unit cell with more phase bits is also a significant way to improve AE. The measured radiation and scattering properties shown in the above figures demonstrate the wideband characteristics of MSA. A comparison of the proposed 1-bit MSA with some previous low RCS reflectarrays is listed in Table VI. From the table, the main contribution of the proposed work includes simple structure, wideband, low-cost, beam scanning, tunable scattering patterns, and low RCS properties for both x- and y-polarized waves at the same wide frequency band.

4. Conclusion

A single layer wideband low RCS MSA with tunable control of radiation and scattering characteristics has been proposed and experimentally verified. The interdigital capacitance (IC) integrated into the unit cell is used to break the bandwidth limitation for the y-polarized waves. Combined with the OMEPC method, the RCS reduction for both polarizations is realized at the same wide frequency band. Furthermore, the radiation beams and bistatic scattering patterns for x-polarized waves can be real-time switched in a digital manner by using an FPGA. The measured gains and backward RCSRs for the x- and y-polarized waves confirm the wideband performance of the multifunctional MSA. The MSA in our paper is a joint optimization process of radiation and scattering characteristics over a wide frequency band. Considering the superior performances, such as wideband real-time beam steering and tunable scattering patterns with low RCS for both polarizations over a wide frequency band, the proposed single-layer MSA is potentially useful for radar and low detectable platforms.

Acknowledgments

This work was supported by the National Natural Science Foundation of China (U2141233, 62171416, and 62071436) and the Fundamental Research Funds for the Central Universities (CUC220D013). *(Corresponding author: Jianxun Su)*

Author Information

P. Li, H. Yu, J. X. Su, Q. X. Guo, and Z. R. Li are with the State Key Laboratory of Media Convergence and Communication, Communication University of China, Beijing 100024, China.

L. W. Song is with the School of Electromechanical Engineering, Xidian University, Xi'an, Shaanxi 710071, China.

References

[1] WANG J, LI Y, JIANG Z, et al. Metantenna: when metasurface meets antenna again[J]. IEEE Trans. Antennas Propag., 2020, 68 (3): 1332-1347.

[2] YANG G, ZHANG Y, ZHANG S. Wide-band and wide-angle scanning phased array antenna for mobile communication system[J]. IEEE Open Journal of Antennas and Propag., 2021, 2: 203-212.

[3] HAN J, LI L, LIU G, et al. A wideband 1-bit 12× 12 reconfigurable beam-scanning reflectarray: design, fabrication, and measurement[J]. IEEE Antennas Wireless Propag. Lett., 2019, 18 (6): 1268-1272.

[4] LIU Y, HAO Y, WANG H, et al. Low RCS microstrip patch antenna using frequency-selective surface and microstrip resonator[J]. IEEE Antennas Wireless Propag. Lett., 2015, 14: 1290-1293.

[5] LIAO W J, ZHANG W Y, HOU Y C, et al. An FSS-integrated low-RCS radome design[J]. IEEE Antennas Wireless Propag. Lett., 2019, 18 (10): 2076-2080.

[6] ZHANG C, GAO J, CAO X, et al. Low scattering microstrip antenna array using coding artificial magnetic conductor ground[J]. IEEE Antennas Wireless Propag. Lett., 2018, 17 (5): 869-872.

[7] LIU T, CAO X, GAO J, et al. RCS reduction of waveguide slot antenna with metamaterial absorber[J]. IEEE Trans. Antennas Propag., 2013, 61 (3): 1479–1484.

[8] LIU Y, ZHAO X. Perfect absorber metamaterial for designing low RCS patch antenna[J]. IEEE Antennas Wireless Propag. Lett., 2014, 13: 1473–1476.

[9] LIU Y, LI K, JIA Y, et al. Wideband RCS reduction of a slot array antenna using polarization conversion metasurfaces[J]. IEEE Antennas Wireless Propag. Lett., 2016, 64 (1): 326-331.

[10] ZHAO Y, CAO X, GAO J, et al. Broadband low-RCS metasurface and its application on antenna[J]. IEEE Trans. Antennas Propag., 2016, 64 (7): 2954-2962.

[11] ZHENG Q, GUO C, DING J, et al. A broadband low-RCS metasurface for CP patch antennas[J]. IEEE Trans. Antennas Propag., 2020, 69 (6): 3529-3534.

[12] QIU L, XIAO G. A broadband metasurface antenna array with ultra-wideband RCS reduction[J]. IEEE Trans. Antennas Propag., 2022, 70 (9): 8620-8625.

[13] CHENG Y F, FENG J, LIAO C, et al. Analysis and design of wideband low-RCS wide-scan phased array with AMC ground[J]. IEEE Antennas Wireless Propag. Lett., 2021, 20 (2): 209-213.

[14] DING X, CHENG Y F, SHAO W, et al. Broadband low-RCS phased array with wide-angle scanning performance based on the switchable stacked artificial structure[J]. IEEE Trans. Antennas Propag., 2019, 67 (10): 6452-6460.

[15] GOU Y, CHEN Y, YANG S. Radar cross section reduction of wideband vivaldi antenna arrays with array-level scattering cancellation[J]. IEEE Trans. Antennas Propag., 2022, 70 (8): 6740-6750.

[16] LIU Y, JIA Y, ZHANG W, et al. An integrated radiation and scattering performance design method of low-RCS patch antenna array with different antenna elements[J]. IEEE Trans. Antennas Propag., 2019, 67 (9): 6199–6204.

[17] HUANG H, SHEN A, OMAR Z. Low-RCS and beam-steerable dipole array using absorptive frequency-selective reflection structures[J]. IEEE Trans. Antennas Propag., 2019, 68 (3): 2457-2462.

[18] YIN L, YANG P, GAN Y, et al. A low cost, low in-band RCS microstrip phased-array antenna with integrated 2-bit phase shifter[J]. IEEE Trans. Antennas Propag., 2021, 69 (8): 4517-4526.

[19] HUANG H, SHEN Z. Low-RCS reflectarray with phase controllable absorptive frequency-selective reflector[J]. IEEE Trans. Antennas Propag., 2019, 67 (1): 190-198.

[20] SUN Z, CHEN Q, GUO M, et al. Low-RCS reflectarray antenna based on frequency selective rasorber[J]. IEEE Antennas Wireless Propag. Lett., 2019, 18 (4) :693-697.

[21] MEI P, ZHANG S, CAI Y, et al. A reflectarray antenna designed with gain filtering and low-RCS properties[J]. IEEE Trans. Antennas Propag., 2019, 67 (8): 5362-5371.

[22] XU J, XU H, LUO H, et al. A low-RCS folded reflectarray combining dual-metasurface and rasorber[J]. IEEE Antennas Wireless Propag. Lett., 2022, 21 (12): 2462-2466.

[23] CALLAGHAN P, GIANNAKOU P, KING S G, et al. Linearly polarized reconfigurable reflectarray surface[J]. IEEE Trans. Antennas Propag., 2021, 69 (10): 6480-6488.

[24] SU J, YU H, YIN H, et al. Breaking the high-frequency limit and bandwidth expansion for radar cross-section reduction: a low-observable technology[J]. IEEE Antennas Propag Mag., 2021, 63 (6): 75-86.

[25] SU J, LU Y, LIU J, et al. A novel checkerboard metasurface based on optimized multielement phase cancellation for superwideband RCS reduction[J]. IEEE Trans. Antennas Propag., 2018, 66 (12): 7091-7099.

[26] KAMODA H, IWASAKI T, TSUMOCHI J, et al. 60-GHz electronically reconfigurable large reflectarray using single-bit phase shifters[J]. IEEE Trans. Antennas Propag., 2011, 59 (7): 2524-2531.

[27] ZHANG N, CHEN K, ZHENG Y, et al. Programmable coding metasurface for dual-band independent real-time beam control[J]. IEEE Journal on Emerging and Selected Topics in Circuits and Systems, 2020, 10 (1): 20-28.

[28] YANG H, YANG F, XU S, et al. A 1-bit 10×10 reconfigurable reflectarray antenna: design, optimization, and experiment[J]. IEEE Trans. Antennas Propag., 2016, 64 (6): 2246-2254.

[29] HAN J, LI L, LIU G, et al. A wideband 1 bit 12×12 reconfigurable beam-scanning reflectarray: design, fabrication, and measurement[J]. IEEE Antennas Wireless Propag. Lett., 2019, 18 (6): 1268–1272.

[30] YANG H, YANG F, XU S, et al. A study of phase quantization effects for reconfigurable reflectarray antennas[J]. IEEE Antennas Wireless Propag. Lett., 2016, 16: 302-305.

[31] YU A, YANG F, ELSHERBENI A Z, et al. Aperture efficiency analysis of reflectarray antennas[J]. Microw. Opt. Technol. Lett., 2010, 52 (2): 364-372.

Reconfigurable Bidirectional Beam-Steering Aperture with Transmitarray, Reflectarray, and Transmit-Reflect-Array Modes Switching[*]

1. Introduction

The demand for a low-cost, high-gain antenna with electronically beam scanning capability is widespread in modern wireless communication systems. These antennas are used in a wide range of applications including satellite communications, point-to-point terrestrial links, deep-space communication links, and radars. The most traditional manner of implementing electronic beam steering is by using a phased array. However, their disadvantage is each element needs to be connected to transmit/receive (T/R) module, which leads to very high implementation costs.

Spatially-fed array antennas have attracted much attention in recent years, which remove the layout of complex feed networks and mitigate much of the loss associated with transmission line feeds, due to their attractive qualities, namely their low cost, ease of manufacturing, low weight. These kinds of novel antennas are promised as low-cost, high-gain, and beam-scanning antenna alternatives. There are mainly three kinds of space-fed array antennas, namely reflectarray (RA), transmitarray (TA), and transmit-reflect array (TRA), which usually include a feed antenna and a phase-modulated aperture. By making the aperture electronically reconfigurable through integrating tunable devices such as liquid crystal, micro-electro-mechanical system (MEMS) switches, varactor diodes, and positive-intrinsic-negative (PIN) diodes, within the aperture element. The surface with a reconfigurable phase-modulated function can be synthesized to realize beam-forming. A reconfigurable RA aperture can reflect electromagnetic (EM) waves from the feed antenna and produce a reconfigurable backward pencil beam over the lower hemisphere, as shown in Fig.1 (a). For the reconfigurable TA, its aperture can transmit the EM energy to the other side and generate a reconfigurable forward beam over the upper hemisphere, as shown in Fig.1 (b). A lot of works has been proposed to extend their functionalities, such as wideband

---

[*] The paper was originally published in *IEEE Transactions on Antennas and Propagation*, 2023, 71 (1), and has since been revised with new information. It was co-authored by Hang Yu, Pan Li, Jianxun Su, Zengrui Li, Shenheng Xu, and Fan Yang.

or multiband operation, high aperture efficiency, wide-angle beam scanning, and multi-polarization. However, these antennas can only achieve beamforming and beam scanning in one hemisphere, either in TA or RA functions, while the other hemisphere is not utilized. Multiple antennas are needed to combine to realize beamforming, such as Russian radar 64N6E Big Bird has a double-sided antenna aperture and two feed horn antennas to realize 3-D electronically beam scanning performance, which increases the space of the device and the fabrication cost.

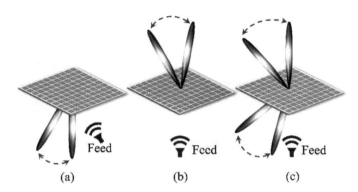

Fig. 1　Spatially-fed array architectures. (a) RA. (b) TA. (c) TRA.

The concept of the TRA, which includes a feed antenna and a phase-modulated aperture with the transmitted and reflected beams, was introduced to realize low-cost full-space EM control, as shown in Fig.1 (c). Bidirectional coverage can be divided into simultaneous and time division operating modes. The main difference between two operating modes is that, in simultaneous operating mode, the transmitted and reflected beams can be generated at the same time and achieve independent beam scanning. While in time division mode, the transmitted or reflected beam is switched within a FPGA response time. In [23], a 1-bit reconfigurable TRA with beam-reconfigurable capability was achieved for simultaneous bidirectional coverage. However, the transmitted and reflected beams are always symmetrical with respect to the radiation aperture, which limits its applications. Since the transmitted and reflected beams are generated at the same time, 3 dB gain loss is invited. A multilayer receive-transmit type element, which can electronically switch between the transmitted and reflected modes, was proposed in [24]. The main shortcoming of this design is the high loss in the varactor diodes and it cannot provide bidirectional coverages simultaneously.

Due to the aforementioned limitations of previous works, the value and the novelty of this work is proposing a reconfigurable multifunctional transmit-reflect array as the name implies is a multifunction integration containing RA, TA, and TRA function for simultaneous and time division bidirectional beam-steering, which is not reported to the best of the authors' knowledge. The proposed reconfigurable unit cell can achieve the 1-bit phase resolution both in transmission mode and reflection mode by controlling the four states of two PIN diodes. Then, a 16 × 16-element multifunctional aperture with 512 PIN diodes fed by a pyramidal horn antenna is fabricated. The measured results show that the aperture can realizes pencil-beam scanning of transmitted or reflected beams demonstrating its 3-D beam steering ability. To realize bidirectional beam independent steering simultaneously, the superposition of the aperture fields (SAF) and the particle swarm optimization (PSO) algorithm are used to obtain unit cell mode arrangement and phase distribution, then realize beam direction, low sidelobe levels (SLLs), and peak gain for each bidirectional beam. In contrast to the work in [23], the transmitted and reflected beams can be controlled independently, which greatly expands its application scenario. A shared aperture with electronically reconfigurable design, which is based on a field programmable gate array (FPGA)-based digital control circuit, enables switching in RA, TA, and TRA modes. The proposed method can manipulate the directions of single beam and the bidirectional beams, opening a new avenue for the design of high-performance multifunctional reconfigurable antenna systems and providing a promising candidate in the next-generation wireless communication applications.

This work is organized as follows. Section II describes the operating principle and structure of the proposed unit cell. Section III presents the implementation and measurements of the reconfigurable bidirectional spatially-fed array working on simultaneous and time division operating modes. Finally, Section IV concludes this work.

2. Unit Cell Design

2.1 Operating Principle

To realize the proposed multifunctional shared aperture shown in Fig. 1, the key step is to create a novel unit cell with the capacity of the backward reflection and forward transmission mode switching and at least 1-bit tunable phase shift. The traditional reconfigurable RA unit cell usually consists of one layer of metal ground plane, which can ensure the incident EM wave is totally reflected, and at least one layer of phase-shift layer. For the reconfigurable TA with 1-bit phase resolution, the current works are usually based on the receive-and-radiate method, which introduces receiving and radiating layers that are connected through metallization via holes or coupling slots. If the TA structure is not well matched to the free-space, the incident EM wave will be reflected, resulting in a high insertion loss. Therefore, the transmission mode is desirable to be the most "transparent" as possible, introducing very low loss so the EM field of the propagating wave is not severely attenuated, whereas the reflection mode is desirable to be a perfect reflecting surface so the incident wave can be entirely reflected.

To simultaneously control of the reflected and transmitted wavefronts, parallel narrow metal strips with uniform spacing are employed to replace the ground plane, which can deflect incident waves with the polarization parallel to the strips into a designed direction and transmit the orthogonal polarization component. In our previous work , a low-loss reconfigurable TA unit cell consisting of a tunable polarization converter between a pair of orthogonally oriented subwavelength metal gratings was presented to realize beam steering. By switching the states of the two mirror arranged diodes (ON/OFF or OFF/ON), due to the interference between the multiple polarization couplings in the Fabry-Pérot-like cavity, the incident EM wave is efficiently transformed to its cross-polarized wave with 1-bit phase resolution. Now, if we let the tunable metallic resonator works at a co-polarized response state, namely that this resonator is without polarization conversion function in this situation, the upper grating and the bottom grating can be considered as a "transparent" plane and a ground plane, respectively. This structural framework can be regarded as the traditional RA design framework. Manipulating the phase of the co-polarized reflected field can be realized by turning co-polarized response characteristics. In this category, tunable metallic resonator should have co-polarized and cross-polarized tunable capability to realize transmission mode and reflection mode switching.

2.2 Unit Cell Structure and Simulation

Following this idea, the proposed metasurface-based unit cell configuration is shown in Fig. 2. It consists of a reconfigurable metallic resonator between a pair of orthogonally oriented subwavelength metal gratings and a direct current (DC) bias line layer. The dielectric layers of substrates use the commonly utilized F4B substrate with a relative permittivity of 2.65 and a loss tangent of 0.001. Substrate 2 and substrate 3 are bonded together by a 0.1 mm Rogers RO4450F bonding film with a relative permittivity of 3.52. The metallic resonator consists of a rectangular patch and outer ring, where two PIN diodes are integrated. Two cut slots and a vertical line are added to achieve independent control of two diodes. The vertical line is connected to the DC ground. Substrate 1 is drilled to reserve space for the placement of diodes. The MACOM MADP-000907-14020 PIN diodes are selected for this wok, which is modeled as a series of lumped RLC elements: R = 7.8 Ω and L = 30 pH for the ON state and C = 25 fF and L = 30 pH for the OFF state. To control the DC bias voltages separately applied to the PIN diodes on the metallic layers, meandering lines are designed as blocking components to isolate radio frequency (RF) and DC signals and placed at the weakest points of the electric field. The DC via hole is used to connect the metallic resonator layer and bias line layer. The direction of the bias line is parallel to the x-direction grating to reduce effects on its transmission properties. The key parameters are in mm as follows: h1 = 1.8, h2 = 1.8, h3 = 0.5, P = 15, t = 0.5, g = 2.3, w = 7.1, wk = 0.4, wg = 1, ws1 = 2.6, r2 = 1, and ws2 = 2.3.

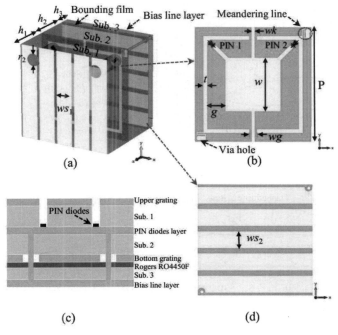

Fig. 2 Configuration of the proposed unit cell structure. (a) Schematic. (b) Center patch layer. (c) Stack up. (d) Bottom grating layer.

Since two PIN diodes are integrated into the unit cell, the proposed unit cell should have four states: "00" (OFF/OFF), "01" (OFF/ON), "10" (ON/OFF), and "11" (ON/ON). The EM performance of the unit cell for four states is simulated by the full-wave simulation software CST Microware Studio® with periodic boundary conditions in the x and y directions, and open (add space) boundary in the -z and +z directions. An x-polarized incident EM wave with -z direction propagation is imping on the unit cell. The reflection and transmission coefficients of four states under normal incidents are shown in Fig. 3. Here, R and T are used to evaluate reflection and transmission coefficients, respectively. Fig. 3(a) illustrates that the unit cell converts an x-polarized incident wave into a y-polarized transmitted wave with an insertion loss of less than -0.8 dB within 8.5-10.3 GHz for both state "10" and state "01". The co-polarization and cross-polarization reflection coefficients in the operating frequency band are < -10 dB and -40 dB, respectively. The co-polarization transmission coefficients are < -30 dB, which reveals a high cross-polarization purity of the transmitted wave. Since the two states of "10" and "01" can be equivalent to mirror unit cells, an inherent phase difference of 180° between the two states over the operating band is generated, which can be observed in Fig. 3(b). The slight fluctuation is due to the asymmetry of the bias line.

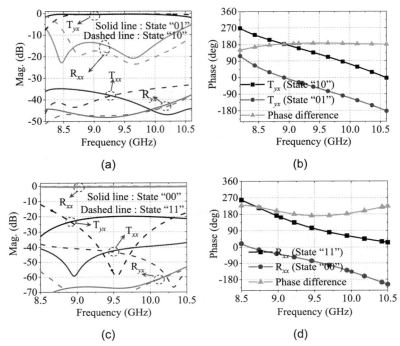

Fig. 3 Simulated performance of unit cell under normal incidence. (a) Magnitude and (b) phase for states "10" and "01". (c) Magnitude and (d) phase for states "11" and "00".

Figs. 3(c) and (d) illustrate the simulated reflection and transmission coefficients for states "11 and state "00", where the metallic resonator works at co-polarized response states without polarization conversion function. From Fig. 3(c), it can be observed that the co-polarization reflection coefficient loss is < 0.4 dB regardless of state "11 and state "00", the cross-polarization reflection coefficients are < -50 dB, and the transmission coefficients are < -20 dB in the band of 8.8-10.3 GHz, indicating that x-polarized incident wave is totally nearly reflected without polarization conversion for both states. Besides, as shown in Fig. 3(d), the x-polarized reflection phase difference between states "00" and states "11" is 180°±20° at 8.9-10.3 GHz. The conceptual illustration of EM wave propagation when the *x*-polarized incident wave illuminates the proposed reconfigurable unit cell at four states is shown in Fig. 4, which indicates that this novel unit cell is capable of the backward reflection and forward transmission mode switching and 1-bit tunable phase shift.

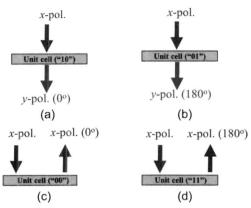

Fig. 4 Conceptual illustration of EM wave propagation when the x-polarized incident wave illuminates the proposed reconfigurable unit cell at four states: (a) "10", (b) "01", (c) "00", and (d) "11".

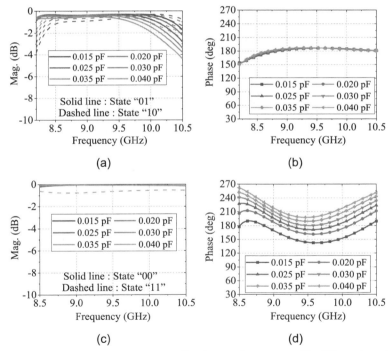

Fig. 5 Simulated transmission and reflection performance with different equivalent capacitance C of the PIN diode. (a) Magnitude and (b) phase difference of Tyx for states "10" and "01". (c) Magnitude and (d) phase difference of Rxx for states "00" and "11".

Since the equivalent parameters of the PIN diode are related to the operating frequency and assembly accuracy, the difference in the equivalent parameters (especially the equivalent capacitance C) dramatically influences the frequency response. Simulated transmission and reflection with different equivalent capacitance C of the PIN diode are shown in Fig. 5. Figs. 5 (a) and (b) give the magnitude and phase difference of Tyx for states "10" and "01". It is observed that the operating frequency slightly shifts to high frequency when the capacitance increases, while the phase difference is almost unchanged due to the mirror structure-property. Figs. 5 (c) and (d) give the magnitude and phase difference of Rxx for states "00" and "11". The reflection magnitudes are almost unchanged, while the phase differences have a large variation range. It can be concluded that the reflection performance is more sensitive than transmission performance. This phenomenon can cause the measured results of the reflection beam to deviate from the simulation ones, especially the SLLs.

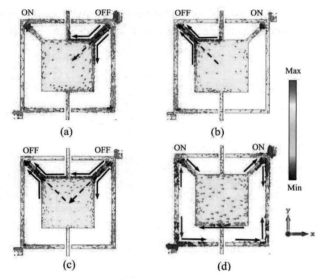

Fig. 6 Surface current distributions of reconfigurable metallic patch layer when the x-polarized incident wave illuminates the proposed unit cell at four states: (a) "10", (b) "01", (c) "00", and (d) "11".

2.3 Surface Current Analysis

To understand the physical mechanism of the proposed unit cell at four states in further detail, we investigated the surface currents at 9.5 GHz on the reconfigurable metallic resonator at four states, as shown in Fig. 6. When the unit cell is working at states "10" and "01", as shown in Figs. 6(a) and (b), strong surface currents (black arrows) are mainly located on one side of the OFF state and flow to or away from the edge of the center patch. Two black arrows are combined into a blue synthetic arrow, which can be considered as an electric dipole resonance generated by a metal cut-wire structure. Due to the existence of two orthogonally oriented metal gratings, which allow the newly generated y-polarized component to pass through while blocking the x-polarized component for forward scattering, the overall result is that the x-polarized incident waves are converted to their orthogonal polarization in transmission mode. Since the surface current distributions in the two cases in the y-direction show opposite directions, indicating a 180° phase difference in two newly generated transmission waves. When the unit cell is working at state "00" [see Fig. 6 (c)], two blue synthetic arrows are included which can be combined into an x-direction arrow, indicating that no polarization conversion is generated. Since x-direction grating can be considered as a ground plane, the x-polarized incident EM wave are totally reflected after a phase modulation. While for the state "11" [see Fig. 6 (d)], four purple arrows in the y-direction are induced in the out ring, no effective current is generated due to the direction being opposite to each other. The remaining currents represented by black arrows can be synthesized into an x-direction current. Similar to state "00", the x-polarized incident EM wave is reflected after another phase modulation. Additionally, because the resonant modes of the states "11" and "00" are different, this ultimately causes the x-polarized reflected wave to have a 180° phase difference.

3. Implementation and Measurements

3.1 Antenna System Configuration

To experimentally validate the performance of the proposed reconfigurable strategy, a 16 ×16-unit-cell shared array aperture with dimensions of D along the x- and y-axes is adopted. The proposed array has an effective size of 240 mm × 240 mm, and 512 PIN diodes are used to control the individual element by bias line and FPGA. The bias line arrangement at the array level of the proposed 16 ×16-unit-cell shared array aperture is shown in Fig. 7. The antenna prototype was fabricated using a printed circuit board (PCB) fabrication technique with a minimum feature size of approximately 0.1 mm. Therefore, the minimum width of the DC bias line and clearance is 0.1 mm. The period of bottom grating layers is 3 mm, the metallic strip width is 2.3 mm, and the gap width is 0.7 mm. In order to reduce the influence of the bias line on the transmission performance, the direction of the bias line is parallel to the grating and the total bias line width should be smaller than the metallic strip width. Thus, the number of bias lines over a metallic strip is limited to 12. When the size of the array and the number of the bias lines increase, the designer needs to increase the width of the metallic strip or use a multilayer PCB for the bias line network instead.

In the manufacturing process, substrate 2 and substrate 3 were processed first and bonded together by a 0.1 mm Rogers RO4450F bonding film. Then, the two-layer board is drilled, including metalized vias (for the DC bias layer) and air vias (for connecting the DC bias line and fixing substrate 1). Next step is the PIN diode soldering process using surface mount technology (SMT), which is a method that attaches electronic components onto the surface of the PCB. The SMT manufacturing process is broadly grouped into 3 stages, namely: solder paste printing, components placement, and reflow soldering. After this, substrate 1 is processed and drilled, including air vias for the reserved space for the PIN diodes and the holes for the fixing structure. Finally, 3-mm-diameter plastic screws were used to align and further hold together the different layers. A feed horn antenna with a pattern power factor of qf = 3.5 at 9.5 GHz is placed on the z-axis at the point of z = -140 mm, which is optimized by balancing the spillover efficiency and illumination efficiency, to produce an x-polarized incident spherical wave. And 3D printing technology was used to print the mechanical support bracket for realizing the assembly of the horn antenna and the reconfigurable aperture. The fabricated antenna is depicted in Fig. 8.

The measurement was carried out using the antenna near-field testing system in an anechoic chamber. As shown in Fig. 8 (d), the electric field is measured using a standard X-band near-field waveguide probe, which is in front of the aperture. It is worth noting that when testing reflected beams, the entire antenna should be rotated 180° compared to testing transmitted beams. The scanning step is 13.6 mm ($\lambda/2$ at the highest test frequency), and the near E-field value at each pixel can be obtained by scanning the probe along the x- and y-axis. The far-field radiation pattern can be readily obtained from the measured near-field data using the fast-Fourier transformation (FFT).

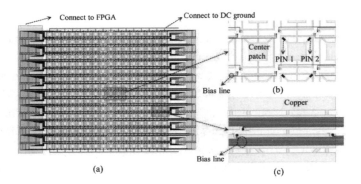

Fig. 7 Details on bias line arrangement at array level.
(a) Overall schematic (grating layers are hided).
(b) Zoom view in the center (grating layers are hided).
(c) Zoom view on the edge.

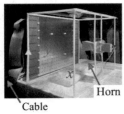

(a)

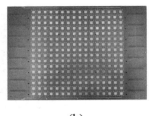

(b)

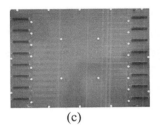
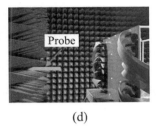

(c)   (d)

Fig. 8 Fabricated antenna. (a) System assembly. (b) PIN layer. (c) DC bias line layer. (d) Measurement of the transmitted beam in the near-field system.

3.2 Coding Arrangements of TA and RA Modes

For a space-fed planar array antenna, to generate a pencil-shaped beam in the desired direction $\hat{u}0$ ($\theta 0$, $\varphi 0$), the phase of each unit cell (m, n) should be set as

$$\varphi_{m,n} = k(|\vec{r}_{m,n} - \vec{r}_f| - \vec{r}_{m,n} \cdot \hat{u}_0) + \varphi_0 \quad (1)$$

where $\vec{r}_f$ and $\vec{r}_{m,n}$ are the position vector of feed antenna and the (m,n)-th unit cell, respectively. k is the free space wavenumber and $\varphi 0$ is a constant phase, which is the optimized value for a 1-bit space-fed planar array. $\varphi m,n$ is converted into a value between -180° and 180°. In this design, a 1-bit unit cell is adopted as the element, and the quantization phase ($\varphi_{m,n}^q$) is defined as

$$\varphi_{m,n}^q = \begin{cases} 0°, & \text{if } -90° \leq \varphi_{m,n} \leq 90° \\ 180°, & \text{other cases} \end{cases} \quad (2)$$

The codes for a pencil-shaped beam in the upper hemisphere (TA function) should be set as

$$\text{Code}_{m,n}^{TA} = \begin{cases} \text{"10", if } \varphi_{m,n}^q = 0° \\ \text{"01", if } \varphi_{m,n}^q = 180° \end{cases} \quad (3)$$

And the codes for a pencil-shaped beam in the lower hemisphere (RA function) can be written as

$$\text{Code}_{m,n}^{RA} = \begin{cases} \text{"00", if } \varphi_{m,n}^q = 0° \\ \text{"11", if } \varphi_{m,n}^q = 180° \end{cases} \quad (4)$$

3.3 TA Mode Operation

To verify the capability of the TA function for producing forward beams. In this case, the unit cell only works as transmission mode with two codes of "01" and "10". First, we investigate the broadside radiation performance with ($\theta T0$, $\varphi T0$) = (0°, 0°) at transmission mode. According to the quantized phase distribution calculated using (1)–(3), the coding distribution over the aperture plane is determined.

The full-wave simulation of the overall antenna model is performed in CST Microwave Studio using the transient solver. The simulated and measured radiation patterns in xoz-, yoz-, and diagonal (D)-plane at 9.0, 9.5, and 10 GHz are plotted in Fig. 9. Since the reconfigurable aperture generates polarization conversion in transmission mode, the y-polarization components are dominant in the upper hemisphere. Measured half power beamwidths (HPBWs) of y-polarization components at 9.5 GHz in xoz- and yoz-planes are 8.5o and 8.8o, respectively. Measured SLLs at xoz- and yoz-planes are -15.5 and -15.0 dB, and x-polarization levels are below -26 dB. Fig. 10 presents the simulated and the measured radiation gains within the frequency band of 8.5-10.5 GHz. The measured gain at 9.5 GHz is 21.4 dBi for the transmitted beam, corresponding to an aperture efficiency of 19.2%, and the 3-dB gain bandwidth is 19.5% (8.8 to 10.7 GHz). To show the transmission beam scanning ability, the measured y-polarized radiation patterns within ±60° scan range in xoz-plane and ±45° scan range in yoz (D)-planes are plotted in Fig. 11, where they are normalized to the gain of the transmitted broadside beam. The different scanning angle ranges are due to the different element radiation patterns caused by the structural asymmetry.

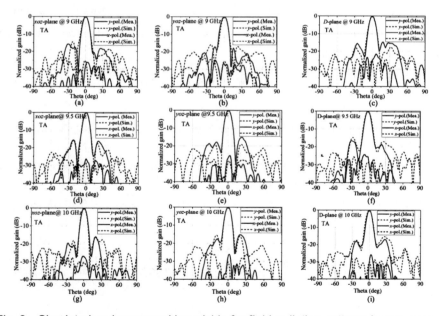

Fig. 9 Simulated and measured broadside far-field radiation patterns in xoz-, yoz-, and D-plane. (a), (b), and (c) are the patterns at 9.0 GHz. (d), (e), and (f) are the patterns at 9.5 GHz. (g), (h), and (i) are the patterns at 10 GHz.

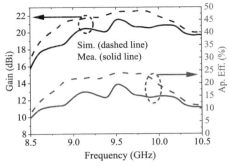

Fig. 10 Gain and aperture efficiency of broadside beam of TA modes.

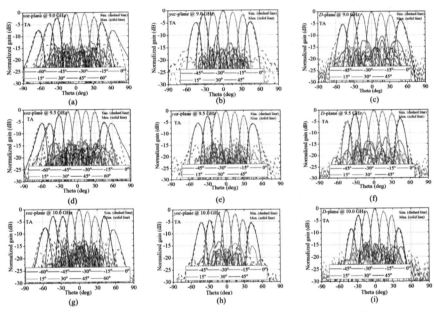

Fig. 11 Radiation patterns of y-polarization for beam scanned every fifteen degrees in xoz-, yoz-, and D-plane. (a), (b), and (c) are the patterns at 9.0 GHz. (d), (e), and (f) are the patterns at 9.5 GHz. (g), (h), and (i) are the patterns at 10 GHz.

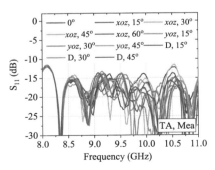

Fig. 12 Measured reflection coefficient of the focal source in TA modes.

In the TA case, the measured main beam directions well coincide with the target angles. When the beam scanning angle is increasing, the main lobe beamwidth is wider, the SLLs increase slightly, and the gain decreases. Fig. 12 gives the measured reflection coefficient of the focal source in TA modes. It can be seen that the different beams have a small effect on the reflection coefficient at the input of the horn antenna. The experimental results have demonstrated its 2-D beam steering capability over the upper hemisphere.

3.4 RA Mode Operation

In the RA case, the unit cell only works in reflection mode with two codes of "00" and "11". We first investigate the backward beam's radiation performance with ($\theta R°$, $\varphi R°$) = (180°, 0°). The RA coding distribution over the aperture plane can be calculated according to the quantized phase distribution using (4). Fig. 13 presents the simulated and the measured radiation gains within the frequency band of 8.5-10.5 GHz. The measured gain at 9.5 GHz is 21.0 dBi, corresponding to an aperture efficiency of 17.3%, and the 3 dB gain bandwidth is 19.9% (8.6 to 10.5 GHz). The differences between simulated and measured gain are mainly due to the additional loss caused by the malfunctioning PIN diodes in the fabricated prototype, the influence of the room for PIN diode DC lines and digital commands, and the inaccuractely characterization of the equivalent parameters of PIN diodes.

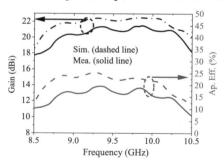

Fig. 13 Gain and aperture efficiency of reflected broadside beam.

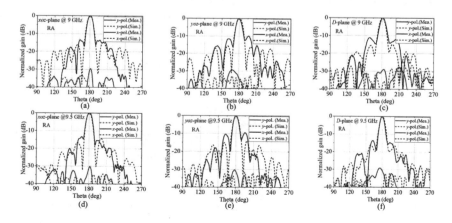

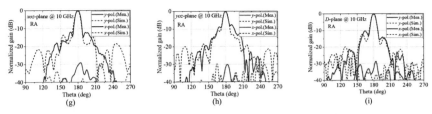

Fig. 14 Simulated and measured broadside far-field radiation patterns in xoz-, yoz-, and D-plane. (a), (b), and (c) are the patterns at 9.0 GHz. (d), (e), and (f) are the patterns at 9.5 GHz. (g), (h), and (i) are the patterns at 10 GHz.

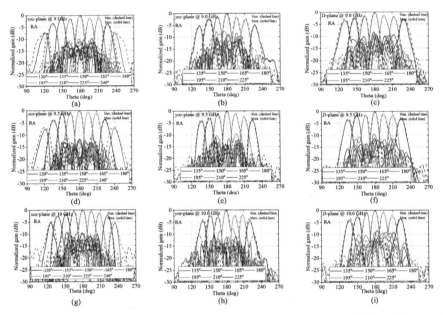

Fig. 15 Radiation patterns of x-polarization for beam scanned every fifteen degrees in xoz-, yoz-, and D-plane. (a), (b), and (c) are the patterns at 9.0 GHz. (d), (e), and (f) are the patterns at 9.5 GHz. (g), (h), and (i) are the patterns at 10 GHz.

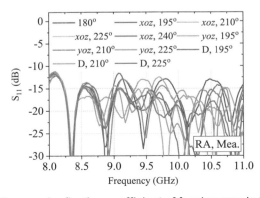

Fig. 16 Measured reflection coefficient of focal source in RA modes.

Table I

Gain loss analysis for broadside beams at 9.5 GHz

Ideal Directivity	28.6 dB
Quantization Loss	3 dB
Spillover Loss	0.4 dB
Taper Loss	1.4 dB
Horn Miss-Matching Loss	0.2 dB
Element Loss	0.8 dB (T) / 0.4 dB (R)
Feed Blockage Loss	0 dB (T) / 0.6 dB (R)
Reflection Loss and Cross-Polarization Loss	0.4 dB (T) / 0.5 dB (R)
Measured Loss	1 dB (T) / 1.1 dB (R)
Measured Gain	21.4 dBi (T) / 21.0 dBi (R)
Aperture Efficiency	19.2% (T) / 17.3% (R)

Fig. 14 presents the measured radiation patterns in xoz-, yoz and D-plane at 9.0, 9.5, and 10 GHz. Since the reconfigurable aperture are in a co-polarized response, the x-polarization components are dominant in the lower hemisphere. Measured HPBWs at 9.5 GHz of x-polarization components in xoz- and yoz-planes are 8.1o and 9.3o, respectively. SLLs in xoz- and yoz-planes are -13.3 and -9.4 dB, the SLLs increase is mainly attributed to the feed blockage for the reflected beam, which can be reduced for a larger array. Besides, in order to reduce the feed blockage, the schemes of offset feed, and low-profile feed antenna such as the Vivaldi antenna usually are adopted. The y-polarization levels are below -30 dB.

The measured x-polarization radiation patterns within ±60° scan range in xoz-plane and ±45° scan range in yoz (D)-planes are plotted in Fig. 15, where they are normalized to the gain of the reflected broadside beam, to show the reflection beam scanning ability. It's worth noting that the SLLs decrease as the main beam angles deviate the feed direction in the yoz-plane scanning. In the RA case, the measured main beam directions well coincide with the target angles. Fig. 16 gives the measured reflection coefficient of focal source in RA modes. The experimental results have demonstrated its 2-D beam steering capability over the lower hemisphere.

The above results prove the 3-D beam steering ability of the proposed aperture that can realize time division beam scanning of transmitted and reflected beams. The gain loss analysis for the reflection or transmission broadside beams is systematically analyzed and summarized in Table I. For more in-depth discussions on gain loss calculations for arbitrary RA and TA, we refer the reader to [31]-[33]. Since the phase is limited to 1-bit, the quantization loss is approximately 3 dB. It indicates that the maximum aperture efficiencies for TA and RA modes are proximately 50%.

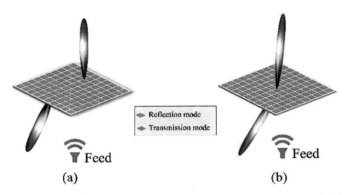

Fig. 17 Two design methods for TRA function. (a) GAD design and (b) SAF design.

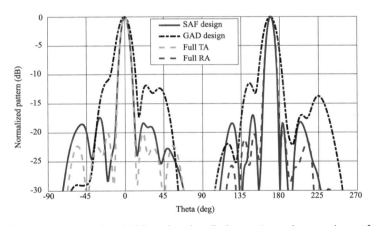

Fig. 18 Calculated normalized bidirectional radiation patterns in yoz-plane of the GAD design and SAF design.

Table II

Calculated radiation characteristics of GAD and SAF methods

Method	Directivity (dB)	SLLs (dB)
GAD design	19.1 (T) / 18.9 (R)	-10.5 (T) / -11.5 (R)
SAF design	21.2 (T) / 21.0 (R)	-17.4 (T) / -17.0 (R)

3.5 TRA Function Verification

TRA function integration with the simultaneous bidirectional beam switching can be considered as a multi-beam optimization problem. Two direct design methods are available for realizing a reconfigurable aperture with a reflection beam and a transmission beam at the same time. The first approach is geometrical aperture division (GAD) design, the aperture surface is divided into two sub-arrays where each sub-array can then radiate a reflection or transmission beam in the required direction, as shown in Fig. 17(a). Another approach for bidirectional beam designs is by using the SAF associated with reflection or transmission modes on the aperture, as shown in Fig. 17(b). The entire array is divided into the transmission and reflection modes optimization distribution. Namely, only part of elements in the TRA aperture work in transmission mode, and the remaining elements work in the reflection mode, the TRA array can be regarded as two sparse configurations with respect to the full TA/RA.

To demonstrate the bidirectional design capabilities of two approaches, we will study a bidirectional radiation performance which the transmitted beam direction is set to $(\theta T°, \varphi T°) = (0°, 90°)$ and the reflected one is $(\theta R°, \varphi R°) = (170°, 90°)$. The radiation pattern of space-fed array antennas can be fast analyzed by array theory; however, the polarization of the array model is not considered in the analysis here. Due to only one kind of polarization being dominated in each hemisphere, the impact of polarization can be ignored for simplicity.

The calculated normalized radiation pattern of two methods in yoz-plane using array synthesized are plotted in Fig. 18. It can be seen that both design methods can realize a bidirectional radiation pattern in the required directions. The beamwidths of SAF design are identical to the full TA and RA, while the GAD design has wider beamwidths. The radiation patterns also show that the side-lobe level is -11.5 dB for the GAD design and -17.0 dB for the SAF design. Calculated bidirectional radiation characteristics of GAD design and SAF design are summarized in Table II.

For the bidirectional beam designed with the GAD approach, one second of the power from the feed horn to generate each beam and the amplitude distribution in each zone is maximum at the outer edge, which results in a significant increase in the SLLs and beamwidth. For the bidirectional beam designed using the SAF method, the high side-lobes are due to the amplitude errors, which alter the required illumination taper on the aperture.

Next, we will give the optimization process of SAF design. The optimized design of two sparse antenna arrays should be performed to minimum SLLs and peak gain for each bidirectional beam. Since the location for each element is fixed, the element codes only need to be optimized. When the EM wave impinges on the coding particle with the state "00" and "11", it will be totally reflected to the backward space; and in contrast, it will be totally transmitted to the forward space when the states are "10" and "01". Once the encoding arrangement of the array is determined, the far-field radiation pattern of a space-fed array antenna can be fast and efficiently calculated using the array synthesized theory. A powerful global search algorithm, namely the PSO algorithm, is used for TRA array synthesis problems to find the optimal codes of each bidirectional beam.

Since the array aperture contains 256 elements and each element has four states, the solution space consists of 4256 cases. In order to reduce computational complexity, the following optimization procedures are employed as follows:

1) Generate 256 binary variables (A) using a random initialization with an equal 0/1 probability for each element. If Am,n = 1 means that (m, n)-th unit cell is working in the reflection mode, while Am,n = 0 is in transmission mode. Therefore, the excitations ($I_{m,n}^{RA}$) of the unit cell in reflection mode are Am,n, the excitations ($I_{m,n}^{TA}$) of the unit cell in transmission mode are |1-Am, n|.

2) Determine the beam directions of transmission and reflection waves. Then, generate 2 variables (φ_0^{RA} and φ_0^{TA}) in the range of [0 2π] for two constant phases. Using (1) and (2), to obtain the RA and TA needed quantization phases $\varphi_{m,n}^{RA}$ and $\varphi_{m,n}^{TA}$. The quantization phase of each element for TRA is

$$\varphi_{m,n}^{TRA} = I_{m,n}^{RA}\varphi_{m,n}^{RA} + I_{m,n}^{TA}\varphi_{m,n}^{TA} \tag{5}$$

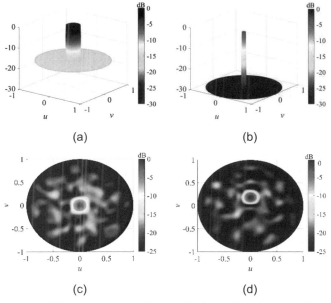

Fig. 19 (a) Upper and (b) lower bounds of the radiation pattern mask in the top hemisphere. Array synthesized asymmetric bidirectional radiation patterns of (c) transmitted beam with the direction of (θT°, φT°) = (0°, 90°) in the top hemisphere, and (d) reflected beam with the direction of (θR°, φR°) = (170°, 90°) in the bottom hemisphere.

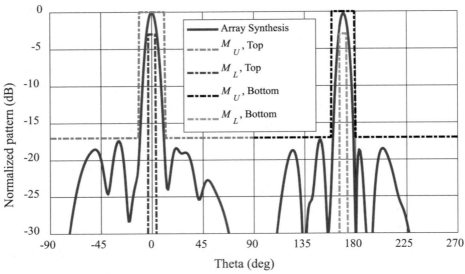

Fig. 20 Array synthesized asymmetric bidirectional radiation patterns in yoz-plane at 9.5 GHz.

3) Compute the far-field radiation pattern in the lower hemisphere from $I_{m,n}^{RA}$ and $\varphi_{m,n}^{TRA}$. And compute the far-field radiation pattern in the upper hemisphere from $I_{m,n}^{TA}$ and $\varphi_{m,n}^{TRA}$. It's noted that the impact of polarization is ignored for simplicity, due to only one kind of polarization being dominated in each hemisphere. To implement spectral transforms in the array theory formulation, a direct relationship between the array elements (m, n) and angular coordinates ($u = \sin\theta\cos\varphi$, $v = \sin\theta\sin\varphi$) is needed.

4) To achieve bidirectional patterns, far-field masks are defined in full space for the optimization to minimum SLLs and peak gain. The mask definition usually requires defining upper and lower bound values of the desired pattern in the entire angular range, as shown in Figs. 19(a)-(b) and Fig. 20. The required masks for radiation patterns are typically circular contours defined in the direction of each beam. The fitness function defined for this optimization is

$$\text{Fitness}_{TRA} = \text{Fitness}_{TA} + \text{Fitness}_{RA} \tag{6-1}$$

$$\text{Fitness}_{TA,RA} = w_1 \sum_{\substack{(u,v) \in \text{mainbeam} \\ \text{and} |E(u,v)| < M_L(u,v)}} (|E(u,v)| - M_L(u,v))^2$$

$$+ w_2 \sum_{\substack{(u,v) \notin \text{mainbeam} \\ \text{and} |E(u,v)| > M_L(u,v)}} (|E(u,v)| - M_U(u,v))^2. \tag{6-2}$$

where MU and ML are the upper and lower bounds of the radiation pattern mask, respectively, and E is the far-field radiation pattern of the antenna. The first term in the fitness function (6-2) calculates the absolute difference between the radiation pattern and the lower bound for evaluating the main beam performance. The second term evaluates the sidelobe performance the entire angular range that does not belong to the main beam area. w1 and w2 are the weights of two terms, in this case, w1 = 2 and w1 = 1. This function considers the performance of every point in the visible space.

5) PSO optimization is employed to optimize A, φ_0^{RA}, and φ_0^{TA} to minimize the fitness function in the optimization loop. The coefficient of constriction is used in the equations of particle motion to accelerate the algorithm convergence.

6) When the algorithm meets the convergence conditions, the optimized A, φ_0^{RA}, and φ_0^{TA} values are obtained. Using the principles in (3) and (4), we can calculate the optimized codes which correspond to ON/OFF states of 512 diodes. The flowchart of bidirectional beams optimization is shown in Fig. 21.

In our first example, the transmitted beam direction is set to $(\theta T°, \varphi T°) = (0°, 90°)$ and the reflected one is $(\theta R°, \varphi T°) = (170°, 90°)$. The upper and lower bounds of the radiation pattern mask in the top hemisphere are shown in Figs. 19 (a) and (b). The pattern masks in the lower hemisphere are the same, which are not shown here. The 2-D radiation pattern mask in the yoz-plane is also shown in Fig. 20. The SLLs are set at -17 dB which is the minimum value according to our optimization experience, due to 1-bit phase quantization and sparse configurations all resulting in an increase of SLLs. The computation time for pattern optimization depends on the far-field resolution and the population of particles.

In the optimization, a far-field pattern with 361× 361 points spaced in the full space is used here. A swarm population of 200 particles is selected. Array synthesized asymmetric bidirectional radiation patterns of the transmitted beam with the direction of $(\theta T°, \varphi T°) = (0°, 90°)$ in the top hemisphere, and reflected beam with the direction of (170o, 90o) in the bottom hemisphere are illustrated in Figs. 19(c) and (d). Fig. 20 shows the 2-D synthesized asymmetric bidirectional radiation patterns, which can be seen the synthesized patterns are limited in upper and lower bounds. Optimized transmission and reflection mode distribution, where white represents the reflection mode and black represents the transmission mode, is shown in Fig. 22(a). Fig. 22(b) gives optimized codes for the total RTA with the transmission beam toward $\theta T° = 0°$ and the reflection beam toward $\theta R° = 170°$ in yoz-plane. The measured reflection coefficient of focal source is plotted in Fig. 23. The measured far-field patterns are plotted in Fig. 24(a). The measured SLLs for transmitted and reflected beams are -15.9 and -12.4 dB, respectively.

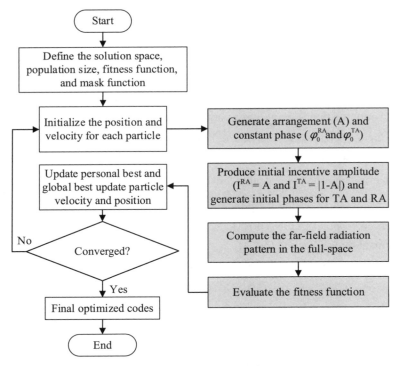

Fig. 21 Flowchart of bidirectional beam optimization.

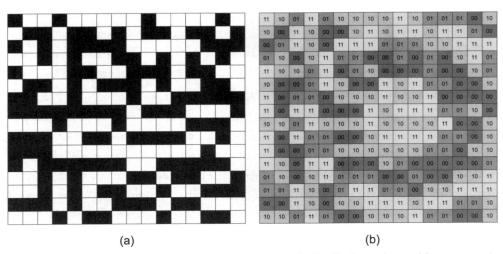

Fig. 22 (a) Optimized transmission and reflection mode distribution, where white represents the reflection mode and black represent the transmission mode. (b) Optimized codes for the total RTA with the transmission beam toward θT0= 0o and the reflection beam toward θR0 = 170o in yoz-plane.

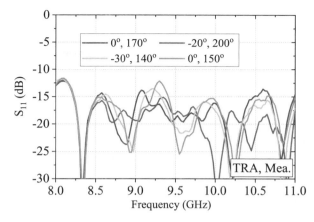

Fig. 23 Measured reflection coefficient of focal source in TRA modes.

The gain loss analysis for the TRA mode is systematically analyzed and summarized in Table III. It is worth mentioning that the input energy is divided between reflection and transmission, resulting in at least 3 dB bidirectional beam loss; the maximum aperture efficiencies for TRA modes are proximately 25%.

Table III

Gain loss analysis for bidirectional beams at 9.5 GHz

Ideal Directivity	28.6 dB
Quantization Loss	3 dB
Bidirectional Beams Loss	4.3 dB (T) / 4.8 dB (R)
Spillover Loss	0.4 dB
Taper Loss	1.4 dB
Horn Miss-Matching Loss	0.2 dB
Element Loss	0.8 dB (T) / 0.4 dB (R)
Feed Blockage Loss	0 dB (T) / 0.6 dB (R)
Reflection Loss and Cross- Polarization Loss	0.4 dB (T) / 0.5 dB (R)
Measured Loss	1 dB (T) / 1.1 dB (R)

Table III Continued

Measured Gain	17.1 dBi (T) / 16.2 dBi (R)
Aperture Efficiency	7.1% (T) / 5.7% (R)

Since the optimized codes are related to the bias voltage controlled by FPGA, the functionalities of bidirectional beams can be dynamically adjusted as desired. To show the bidirectional beam scanning ability, the measured patterns at 9.5 GHz with other different beam directions in yoz-plane are plotted in Figs. 24(b)-(d). And the corresponding measured reflection coefficients of focal source in TRA modes are plotted in Fig. 23. It's clear that the measured results prove the reconfigurability of bidirectional beams.

Due to this kind of planar antenna requires an installation platform with struts for mechanical stiffness with room for PIN diode DC lines and digital commands. Since a small amount of energy from the feed is incident on the edge installation platform and reflected, resulting in higher SLLs of the reflection beam. Microwave absorbing materials can be used to cover the edge installation platform to suppress SLLs.

In this design, each element integrates two PIN diodes, 512 PIN diodes are connected to two 256-way FPGA-based voltage control boards. In the TA mode, since 256 PIN diodes work in the ON state with a forward voltage of 1.33 V and a biasing current of 10 mA, while it is negligible for the OFF-state PIN diodes, the total DC power consumption is estimated to be 1.33 V×10 mA×256=3.4 W. In RA and TRA modes, the number of diodes working in the "ON" state is related to the scanning angle. According to coding arrangements of the RA and TRA modes. It can be concluded that the number of diodes working in the "ON" state, and the power consumption of the whole antenna panel is approximately 3.4-3.5 W. The extra control board will also have some power consumption, such as the voltage bias circuit, voltage conversion circuit, and the FPGA; this will take approximately 10-20 W.

The above results prove the 3-D beam steering ability of the proposed aperture that can realize simultaneous beam scanning of transmitted and reflected beams. Table IV summarizes the working modes and radiation polarization characteristics of the proposed bidirectional aperture to achieve simultaneous and time division bidirectional beam scanning. It is worth mentioning that due to the transmission and reflection beams are polarization dependent, which may be limited in some application scenarios. To solve above issue, a linear transmission polarization converter can be placed in front of this antenna.

A performance comparison of reconfigurable spatially-fed array between this work and previous research is presented in Table V. The advantage and original contribution of this work proposes novel functionalities, i.e., the integration of reconfigurable TA, RA, and TRA for simultaneous and time division full-space beam scanning.

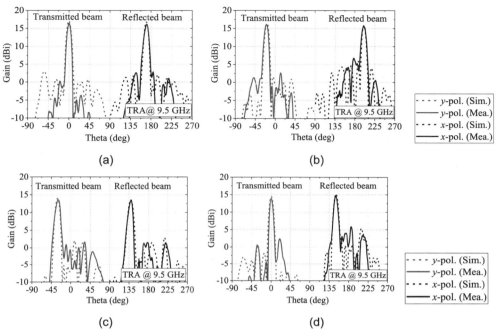

Fig. 24 Simulated and measured far-field patterns of TRA mode in yoz plane at 9.5 GHz. (a) $\theta T° = 0°$, $\theta R° = 170°$, (b) $\theta T° = -20°$, $\theta R° = 200°$, (c) $\theta T° = -30°$, $\theta R° = 140°$, and (d) $\theta T° = 0°$, $\theta R° = 150°$.

Table IV

Working modes and radiation polarization characteristics of the proposed bidirectional aperture

Feed polarization	Working modes		Two PIN diodes working states	Main beam direction	Array dominant polarization
x-polarization	Time division bidirectional beam scanning	TA	"10" and "01"	Upper hemisphere	y-polarization
		RA	"00" and "11"	Bottom hemisphere	x-polarization
	Simultaneous bidirectional beam scanning	TRA	"00", "01", "10", and "11"	Two hemispheres	y-polarization in the upper hemisphere x-polarization in the bottom hemisphere

Table V

Performance comparison of reconfigurable spatially-fed array between this work and previous research

Ref.	Freq. (GHz)	Number of elements	Tunable devices	Mode number	Operating modes	Gain (dBi)	Aperture efficiency (%)
[13]	20.4	16×16	PIN diodes	1	RA	20.4	24
[14]	5.0	12×12	PIN diodes	1	RA	19.2	15.3
[18]	13.6	16×16	PIN diodes	1	TA	22.3	25.6
[19]	10	10×10	PIN diodes	1	TA	19.1	25.8
[24]	5.7	6×6	Varactor diodes	2	TA, RA	12.4, ~12.5	11.8, ~12.1
[23]	5.35	16×16	PIN diodes	1	TRA*	17.2 (T) / 15.4	8.2 / 5.4
This work	9.5	16×16	PIN diodes	3	TA, RA, TRA	21.4, 21.0, 17.1 / 16.2	19.2, 17.3, 7.1 / 5.7

*The transmitted and reflected beams of the proposed antenna are always symmetrical with respect to the radiation aperture.

4. Conclusion and Discussion

In this paper, we present a multifunctional integrated reconfigurable aperture, which can manipulate reflected or transmitted waves both in simultaneous and time division operating modes. A 256-element aperture fed by a pyramidal horn antenna is designed to produce a pencil or bidirectional beam in desired directions at two half-spaces. Different beam states (RA, TA, and TRA modes) can be switched dynamically by controlling the bias voltages using an FPGA-based digital control circuit. The measured results are consistent with the numerical simulations, which prove the bidirectional functionalities of a reconfigurable antenna. Besides, by using the superposition optimization of the aperture fields, which is associated with reflection or transmission modes on a shared aperture, reconfigurable bidirectional beams are generated with the SLLs suppression. This method paves the way for attaining a bidirectional reconfigurable spatially-fed array.

For its value in practical applications in the specific system, unlike traditional TA or RA antennas, which can only achieve beamforming and beam scanning in one hemisphere while the other hemisphere is not utilized. Due to their low-cost electronic bidirectional beam coverage forming capabilities required by advanced communication systems, reconfigurable bidirectional aperture antennas have a great potential in some special scenarios, including radar systems, indoor wireless communications, tunnel relay communications, and broadcasting base stations.

It is worth stressing that some challenges are remained and need to be further investigated, such as continuous or multi-bit phase element design for improving efficiency, wide-angle beam scanning, feed blockage, the polarization dependent of transmission and reflection beams, and the complex layout of biasing lines in large-scale aperture.

Acknowledgments

This work was supported by the National Natural Science Foundation of China (U2141233, 62171416, and 62071436), and in part by the Fundamental Research Funds for the Central Universities (CUC220A001, CUC210A001, and CUC220D013).

Author Information

Hang Yu is with the Information Materials and Intelligent Sensing Laboratory of Anhui Province, Anhui University, Hefei 230601, China, and also with the State Key Laboratory of Media Convergence and Communication, Communication University of China, Beijing 100024, China (e-mail: yuhang232425@163.com).

Pan Li, Jianxun Su, and Zengrui Li are with the State Key Laboratory of Media Convergence and Communication, Communication University of China, Beijing 100024, China (e-mail: lp@cuc.edu.cn, sujianxun_jlgx@163.com, zrli@cuc.edu.cn).

Shenheng Xu and Fan Yang are with the Department of Electronic Engineering, Tsinghua University, Beijing 100084, China (e-mail: shxu@tsinghua.edu.cn, fanyang@tsinghua.edu.cn).

References

[1] SOLEIMANI M, ELLIOTT R C, KRZYMIEŃ W A, et al. Hybrid beamforming for mmWave massive MIMO systems employing DFT-assisted user clustering[J]. IEEE Trans. Veh. Technol., 2020, 69 (10): 11646-11658.

[2] XU Y, GAO Z, WANG Z, et al. RIS-enhanced WPCNs: joint radio resource allocation and passive beamforming optimization[J]. IEEE Trans. Veh. Technol., 2021, 70(8): 7980-7991.

[3] ZHENG B, YOU C, ZHANG R. Double-IRS assisted multi-user MIMO: cooperative passive beamforming design[J]. IEEE Trans. Wirel. Commun., 2021, 20(7): 4513-4526.

[4] KOPP B A, BORKOWSKI M, JERINIC G. Transmit/receive modules[J]. IEEE Trans. Microw. Theory Tech., 2002, 50(3): 827-834.

[5] CHEN C N, et al. 38-GHz phased array transmitter and receiver based on scalable phased array modules with endfire antenna arrays for 5G MMW data links[J]. IEEE Trans. Microw. Theory Tech., 2021, 69(1): 980-999.

[6] HUM S V, OKONIEWSKI M, DAVIES R J. Realizing an electronically tunable reflectarray using varactor diode-tuned elements[J]. IEEE Microwave Wireless Compon. Lett., 2005, 15(6): 422-424.

[7] RIEL M, LAURIN J J. Design of an electronically beam scanning reflectarray using aperture-coupled elements[J]. IEEE Trans. Antennas Propag., 2007, 55(5): 1260-1266.

[8] KAMODA H, IWASAKI T, TSUMOCHI J, et al. 60-GHz electronically reconfigurable large reflectarray using single-bit phase shifters[J]. IEEE Trans. Antennas Propag., 2011, 59(7): 2524-2531.

[9] CARRASCO E, BARBA M, ENCINAR J A. X-band reflectarray antenna with switching-beam using PIN diodes and gathered elements[J]. IEEE Trans. Antennas Propag., 2012, 60(12): 5700-5708.

[10] BILDIK S, DIETER S, FRITZSCH C, et al. Reconfigurable folded reflectarray antenna based upon liquid crystal technology[J]. IEEE Trans. Antennas Propag., 2015, 63(1): 122-132.

[11] HUM S V, PERRUISSEAU-CARRIER J. Reconfigurable reflectarrays and array lenses for dynamic antenna beam control: a review[J]. IEEE Trans. Antennas Propag., 2014, 62(1): 183–198.

[12] PERRUISSEAU-CARRIER J, SKRIVERVIK A K. Monolithic MEMS-based reflectarray cell digitally reconfigurable over a 360 phase range[J]. IEEE Antennas Wireless Propag. Lett., 2008, 7: 138–141.

[13] WU F, LU R, WANG J, et al. Circularly polarized one-bit reconfigurable ME-dipole reflectarray at X-band[J]. IEEE Antennas Wireless Propag. Lett., 2022, 21(3): 496-500.

[14] HAN J, LI L, LIU G, et al. A wideband 1 bit 12 × 12 reconfigurable beam-scanning reflectarray: design, fabrication, and measurement[J]. IEEE Antennas Wireless Propag. Lett., 2019, 18(6): 1268-1272.

[15] LAU J Y, HUM S V. Reconfigurable transmitarray design approaches for beamforming applications[J]. IEEE Trans. Antennas Propag., 2012, 60(12): 5679–5689.

[16] REIS J R, CALDEIRINHA R F S, HAMMOUDEH A, et al. Electronically reconfigurable FSS-inspired transmitarray for 2-D beamsteering[J]. IEEE Trans. Antennas Propag., 2017, 65(9): 4880-4885.

[17] CLEMENTE A, DUSSOPT L, SAULEAU R, et al. Wideband 400-element electronically reconfigurable transmitarray in X-band[J]. IEEE Trans. Antennas Propag., 2013, 61(10): 5017–5027.

[18] XIAO Y, YANG F, XU S, et al. Design and implementation of a wideband 1-bit transmitarray based on a Yagi–Vivaldi unit cell[J]. IEEE Trans. Antennas Propag., 2021, 69(7): 4229-4234.

[19] YU H, SU J, LI Z, et al. A novel wideband and high-efficiency electronically scanning transmitarray using transmission Metasurface polarizer[J]. IEEE Trans. Antennas Propag., 2022, 70(4): 3088-3093.

[20] REIS J R, VALA M, CALDEIRINHA R F S. Review paper on transmitarray antennas[J]. IEEE Access, 2019, 7: 94171-94188.

[21] LIU S, CHEN Q. A wideband, multifunctional reflect-transmit-array antenna with polarization-dependent operation[J]. IEEE Trans. Antennas Propag., 2021, 69(3): 1383-1392.

[22] CHEN Q, SAIFULLAH Y, YANG G, et al. Electronically reconfigurable unit cell for transmit-reflect-arrays in the X-band[J].Opt. Express, 2021, 29(2): 1470-1480.

[23] WANG M, XU S, YANG F, et al. 1-bit bidirectional reconfigurable transmit-reflect-array using a single-layer slot element with PIN diodes[J]. IEEE Trans. Antennas Propag., 2019, 67(9): 6205-6210.

[24] LAU J Y, HUM S V. A planar reconfigurable aperture with lens and reflectarray modes of operation[J]. IEEE Trans. Microw. Theory Tech., 2010, 58(12): 3547-3555.

[25] WU L X, CHEN K, JIANG T, et al. Circular-polarization-selective Metasurface and its applications to transmit-reflect-array antenna and bidirectional antenna[J]. IEEE Trans. Antennas Propag., 2022, 70(11): 10207-10217.

[26] CHERNYAK V S, IMMOREEV I Y. A brief history of radar[J]. IEEE Aero. El. Sys. Mag., 2009, 24(9): B1-B32.

[27] NIU T, et al. Polarization-dependent thin-film wire-grid reflectarray for terahertz waves[J]. Appl. Phys. Lett., 2015, 107(3): no. 031111.

[28] YI H, QU S W, NG K B, et al. Terahertz wavefront control on both sides of the cascaded Metasurfaces[J]. IEEE Trans. Antennas Propag., 2018, 66(1): 209-216.

[29] http://www.wang-ling.com.cn/EN/products_view.asp? id=483

[30] GRADY N K, HEYES J E, CHOWDHURY D R, et al. Terahertz metamaterials for linear polarization conversion and anomalous refraction[J]. Science, 2013, 340(6138): 1304–1307.

[31] ABDELRAHMAN A H, YANG F, ELSHERBENI A Z, et al. Analysis and design of transmitarray antennas[M]. San Rafael, USA: M&C Publishers, 2017.

[32] NAYERI P, YANG F, ELSHERBENI A Z. Reflectarray Antennas: Theory Designs and Applications[M]. Hoboken, NJ, USA: Wiley, 2018.

[33] YU A, YANG F, ELSHERBENI A Z, et al. Aperture efficiency analysis of reflectarray antennas[J]. Microw. Optical Techno. Lett., 2010, 52(2): 364–372.

[34] YANG H, et al. A study of phase quantization effects for reconfigurable reflectarray antennas[J]. IEEE Antennas Wireless Propag. Lett., 2017, 16: 302–305.

[35] NAYERI P, YANG F, ELSHERBENI A Z. Design of a single-feed quad-beam reflectarray antenna[J]. IEEE Trans. Antennas Propag., 2012, 60(2): 1166–1171.

[36] NAYERI P, YANG F, ELSHERBENI A Z. Design of single-feed reflectarray antennas with asymmetric multiple beams using the particle swarm optimization method[J]. IEEE Trans. Antennas Propag., 2013, 61(9): 4598–4605.

[37] CLERC M. Particle Swarm Optimization[M]. Amsterdam: ISTE Publishing, 2006.

On the Enforcement of Electric Field Boundary Condition in the Moment Method Solution of Volume-Surface Integral Equation*

1. Introduction

In the analysis of electromagnetic (EM) scattering or radiation properties, the volume-surface integral equation (VSIE) is widely used to model the composite inhomogeneous dielectric-perfect electric conductor (PEC) objects because of its generality and accuracy. However, during the method of moments (MoM) solution process, the peak memory usage consumed by the VSIE is quite large due to the three-dimensional volumetric unknowns of dielectrics, even if fast algorithms such as the multilevel fast multipole algorithm (MLFMA) are adopted. To reduce the memory usage, some articles discussed how to explicitly enforce conditions into the VSIE. For example, the continuity condition (CC) of the electric flux density can be explicitly enforced on the dielectric-PEC interfaces. The CC provides the normal component of electric flux density to eliminate the volumetric unknowns, no matter what material the dielectric is, isotropic or anisotropic. Nevertheless, the CC can only be typically enforced on the dielectric-closed PEC interfaces. As stated in [3], in order to rigorously enforce the CC for open PEC structures, separate expressions for surface currents on both sides of PEC should be considered, which needs to appropriately modify the surface integral equation (SIE) part of the VSIE. Therefore, the situation that can be enforced the CC is limited because of the complicated implementation. In [8] and [9], another condition, the electric field boundary condition (BC), was used to reduce the volumetric knowns and to improve the solution accuracy, by vanishing the tangential component (TC) of the electric field in the adjacent tetrahedrons of dielectric-PEC interfaces. Moreover, unlike the CC, the use of BC is independent of the type of PEC structures, that is, the BC can be enforced on either the dielectric-closed or -open PEC interfaces. However, the existing implementations were executed based on the piecewise-constant dielectric basis functions or the quasi-orthogonal characteristic ones, leading to poor versatility.

Because of the convenience of discretizing arbitrarily shaped objects and the quality

---

* The paper was originally published in *IEEE Antennas and Wireless Propagation Letters*, 2022, 21 (9), and has since been revised with new information. It was co–authored by Jinbo Liu, Xumeng Men, Hui Zhang, Zengrui Li, and Jiming Song.

of naturally avoiding pseudo line or surface charges, the divergence-conforming RWG and SWG basis functions based on triangles and tetrahedrons are being widely used. In this letter, based on the RWGs and SWGs, it is presented how to explicitly enforce the BC on the dielectric-PEC interfaces to reduce the volumetric unknowns. In the actual calculation, because of the modeling and calculation errors, the BC may not be rigorously satisfied. To be more specific, at an observed point located on the dielectric-PEC interfaces, the TC of electric field obtained by superimposing SWG basis functions may not be strictly zero. Nevertheless, we can always find a relationship among the four SWG basis functions belonging to a same tetrahedron terminated by a PEC triangle to minimize the TC. Under this relationship, the four coefficients of SWG basis functions are not independent, while one of them can be expressed by the other three, which is no longer an unknown.

2. Formulations

2.1 BC-Based VSIE

Assume that an arbitrarily shaped PEC surface S is partly covered by a dielectric region V, illuminated by an EM wave $(\vec{E}^i, \vec{H}^i)$. The equivalent volume current $\vec{J}_V$ and surface currents $\vec{J}_S$ respectively exist in V and on S, generating the scattered fields $(\vec{E}^s, \vec{H}^s)$ as

$$\begin{cases} \vec{E}^s(\vec{r}) = -j\omega\left[\vec{A}_S(\vec{r}) + \vec{A}_V(\vec{r})\right] - \left[\nabla\varphi_S(\vec{r}) + \nabla\varphi_V(\vec{r})\right] \\ \vec{H}^s(\vec{r}) = \frac{1}{\mu_0}\nabla\times\left[\vec{A}_S(\vec{r}) + \vec{A}_V(\vec{r})\right] \end{cases} \quad (1)$$

Besides, the vector and scalar potentials are expressed as

$$\begin{cases} \vec{A}_T(\vec{r}) = \mu_0 \int_T \vec{J}_T(\vec{r}')\frac{e^{-jk_0|\vec{r}-\vec{r}'|}}{4p|\vec{r}-\vec{r}'|}dT' \\ \varphi_T(\vec{r}) = \frac{j}{\omega\varepsilon_0}\int_T \nabla'\cdot\vec{J}_T(\vec{r}')\frac{e^{-jk_0|\vec{r}-\vec{r}'|}}{4p|\vec{r}-\vec{r}'|}dT' \end{cases} \quad T = S \text{ or } V \quad (2)$$

with the free space wavenumber k0 and the time-harmonic factor ejωt.

In Region V, according to the principle of volume equivalence, the volume integral equation (VIE) is written as

$$\frac{\vec{D}(\vec{r})}{\varepsilon(\vec{r})} - \vec{E}^s(\vec{r}) = \vec{E}^i(\vec{r}) \quad \vec{r} \in V \quad (3)$$

where $\vec{D}$ is the electric flux density in V, and ε is the $\vec{r}$-dependent permittivity. On the PEC surface S, by vanishing the TC of electric field, the electric field integral equation (EFiE) is formed as

$$\hat{n}_S(\vec{r}) \times \vec{E}^s(\vec{r}) = -\hat{n}_S(\vec{r}) \times \vec{E}^i(\vec{r}) \quad \vec{r} \in S \tag{4}$$

where $\hat{n}_S$ denotes the outward unit normal of S. The EFiE can be combined with the VIE to form the EFiE-VIE, the commonly used VSIE type. If S is a closed surface, according to the magnetic field boundary condition, the magnetic field integral equation (MFiE)

$$\vec{J}_S(\vec{r}) - \hat{n}_S(\vec{r}) \times \vec{H}^s(\vec{r}) = \hat{n}_S(\vec{r}) \times \vec{H}^i(\vec{r}) \quad \vec{r} \in S \tag{5}$$

can also be established. To avoid the internal resonance problems, the combined field integral equation (CFiE) can be applied on the closed PEC surface as

$$\text{CFIE} = a\,\text{EFIE} + (1-a)\eta_0\,\text{MFIE} \tag{6}$$

where α is real with 0<α<1, η0 is the intrinsic impedance of free space. Furthermore, the CFiE can be in conjunction with VIE to form the CFiE-VIE, another generalized VSIE type.

In the MoM solution, $\vec{J}_S$ and $\vec{D}$ are respectively expanded using the RWG basis functions $\vec{f}_i^S$ and the SWG basis functions $\vec{f}_i^V$ as

$$\begin{cases} \vec{J}_S = \sum_{i=1}^{N_S} I_i^S \vec{f}_i^S \\ \vec{D} = \sum_{i=1}^{N_V} I_i^V \vec{f}_i^V \end{cases} \tag{7}$$

where NS and NV are the numbers of RWG and SWG basis functions, while the total number of unknowns is N = NS + NV. I_i^S and I_i^V are the corresponding surface and volumetric unknown expansion coefficients, respectively.

On the dielectric-PEC interfaces S', the BC of electric field is satisfied, i.e., the TC of electric field is zero as

$$\hat{n}_S(\vec{r}) \times \vec{E}(\vec{r}) = \hat{n}_S(\vec{r}) \times \vec{D}(\vec{r})/\varepsilon(\vec{r}) = 0 \quad \vec{r} \in S' \tag{8}$$

On the Enforcement of Electric Field Boundary Condition in the Moment Method Solution of Volume-Surface Integral Equation

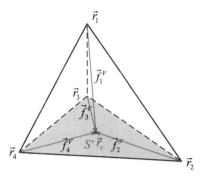

Fig. 1 Four SWGs contained by a same tetrahedron terminated by PEC triangle.

The VSIE usually requires conformal meshes on the dielectric-PEC interfaces for accuracy. In other words, the faces of tetrahedrons on S' should be completely terminated by PEC triangles. Assume a face of a tetrahedron of volume v is terminated by a PEC triangle, as shown in Fig. 1. The four SWG basis functions contained by the same tetrahedron are $\vec{f}_k^V$ (k=1, 2, 3, 4), which are being on the faces of area sk. At the center of the PEC triangle $\vec{r}_c$, we can define a vector function as

$$\vec{F}(\vec{r}_c) = \hat{n}_S(\vec{r}_c) \times \vec{D}(\vec{r}_c)$$
$$= \hat{n}_S(\vec{r}_c) \times \sum_{k=1}^{4} I_k \vec{f}_k^V(\vec{r}_c) = \hat{n}_S(\vec{r}_c) \times \sum_{k=1}^{4} I_k \frac{s_k}{3v}(\vec{r}_c - \vec{r}_k) \quad (9)$$

where Ik is the expansion coefficient of $\vec{f}_k^V$. According to (8), $\vec{F}$ should be zero. In this situation, the four coefficients I1, I2, I3 and I4 are not independent, while one of them can be expressed by the other three. Nevertheless, in the actual calculation, due to the lack of freedom degrees of the SWG basis functions, $\vec{F}$ may not be strictly zero. However, we can always find a relationship between the four coefficients to let $|\vec{F}|$ take the minimum value. Under this relationship, any one of the four coefficients, e.g., I4, can be expressed by the other three.

In what follows, the approach how to minimize $|\vec{F}|$ will be introduced. Define a real tensor $\bar{\bar{P}}$ and a real vector $\vec{C}$ as

$$\bar{\bar{P}} = \left[\frac{s_1}{3v} \hat{n}_S \times (\vec{r}_1 - \vec{r}_c) \quad \frac{s_2}{3v} \hat{n}_S \times (\vec{r}_2 - \vec{r}_c) \quad \frac{s_3}{3v} \hat{n}_S \times (\vec{r}_3 - \vec{r}_c) \right]$$
$$\vec{C} = \frac{s_4}{3v} \hat{n}_S \times (\vec{r}_4 - \vec{r}_c) \quad (10)$$

Accordingly, (9) can be rewritten as

$$\vec{F} = \bar{\bar{P}} \cdot \vec{I}' + I_4 \vec{C} \tag{11}$$

where $\vec{I}' = \{I_1 \ \ I_2 \ \ I_3\}^T$. Using the least-square (LS) method [14, (2.0.5)], if $|\vec{F}|$ to get the minimum value, we have

$$I_4 = -\vec{C} \cdot \bar{\bar{P}} \cdot \vec{I}' \Big/ |\vec{C}|^2 \tag{12}$$

Therefore, when $|\vec{F}|$ takes the minimum value, I4 can be directly calculated by I1, I2, and I3, which is no longer a volumetric unknown. Actually, besides I4, anyone of I1, I2 or I3 can also be chosen as the reduced unknown. However, if $\vec{n}$ is exact on the top of $\vec{r}_c$, $\vec{f}_1^V$ will be strictly perpendicular to the PEC triangular face. In this case, the vector $\vec{C}$ for $\vec{f}_1^V$ is zero, whereas Equation to calculate the coefficient of $\vec{f}_1^V$ is invalid. Consequently, there is no way to represent I1 by I2, I3, and I4. To avoid this situation, the unknowns defined on the faces touched PEC triangles such as I1 in Fig. 1 are not recommended to be reduced.

2.2 MoM Implementation

Let {IS} and {IV} denote the unknown coefficient subvectors of the RWG and SWG basis functions defined on PEC surfaces and in dielectrics, respectively. {IV} can be further classified into two categories: 1) $\{I_{V_2}\}$ which is the subvector related to the picked SWG basis functions to be reduced defined on the tetrahedrons with at least one face terminated by a PEC triangle, and 2) $\{I_{V_1}\}$ related to the other three SWGs on same tetrahedrons mentioned in 1) and the rest of irrelevant SWGs. The numbers of elements of $\{I_{V_1}\}$ and $\{I_{V_2}\}$ are NV1 and NV2, respectively, while NV1 + NV2 = NV. Then using the Galerkin's testing, the VSIE is converted into a non-singular matrix equation as

$$\begin{bmatrix} Z_{SS} & Z_{SV_1} & Z_{SV_2} \\ Z_{V_1 S} & Z_{V_1 V_1} & Z_{V_1 V_2} \\ Z_{V_2 S} & Z_{V_1 V_2} & Z_{V_2 V_2} \end{bmatrix} \begin{Bmatrix} I_S \\ I_{V_1} \\ I_{V_2} \end{Bmatrix} = \begin{Bmatrix} b_S \\ b_{V_1} \\ b_{V_2} \end{Bmatrix} \tag{13}$$

where the submatrix [ZPQ] (P/Q is S, V1 or V2) represents the interactions between various types of basis and test functions defined above, while {bP} is the subvector of excitation.

Due to the enforcement of the BC, according to (12), $\{I_{V_2}\}$ can be represented as a function of $\{I_{V_1}\}$ as

$$\{I_{V_2}\} = [D]\{I_{V_1}\} \tag{14}$$

where [D] is a sparse matrix. Substituting (14) into (13) yields

$$\begin{bmatrix} Z_{SS} & Z_{SV_1} + Z_{SV_2}D \\ Z_{V_1S} & Z_{V_1V_1} + Z_{V_1V_2}D \end{bmatrix} \begin{Bmatrix} I_S \\ I_{V_1} \end{Bmatrix} = \begin{Bmatrix} b_S \\ b_{V_1} \end{Bmatrix} \quad (15)$$

As is seen, the unknown subvector $\{I_{V_2}\}$ is eliminated from the matrix equation, and the dimension of the matrix is reduced from N × N to N × (NS + NV1). The number of reduced unknowns NV2 is equal to that of triangles discretized from dielectric-PEC interfaces. Therefore, the time cost and memory requirement will be saved, and the larger the proportion NV2 is in N, the more significant the savings. Furthermore, the reduced matrix-vector product (MVP) implementation requested in the iterative solution is easy to be accelerated by the MLFMA.

3. Numerical Results

All the following calculations serially run on a computer with 3.2 GHz CPU and 32 GB RAM in single precision. Restarted GMRES with a restart number 100 is used as the iterative solver to reach the target residual relative error of 0.001. The MLFMA is used to accelerate the solution, while the size of the leaf box is 0.25 λ (λ is wavelength in the free space). Table I lists the computational details for all cases such as the time costs by the total solution process and by one MVP, the peak memory usage and the number of iterations.

The first case is a coated PEC sphere illuminated by an x-polarized plane wave from z-axis. The radius of the PEC sphere is 3 λ, and the thickness of the coating material is 0.05 λ with the relative permittivity εr = 2.2 − j0.00198. After discretization with an average mesh size 0.059 λ, the numbers of triangles and tetrahedrons are 76,114 and 246,942, respectively. The bistatic radar cross sections (RCSs) at xoz and yoz planes are calculated by four types of formulations, i.e., the EFIE-VIE, the CFIE-VIE, and those enforced the BC (denoted by BC-EFIE-VIE and BC-CFIE-VIE). For comparison, another approach enforcing the CC of the electric flux density directly, is also implemented in this case. Table II lists the root-mean-square (RMS) errors of different results compared with the analytical result from the Mie series. It is observed that the numerical results agree well with the analytical one, no matter whether they are with or without the enforcement of the BC or CC. That is to say, the BC-based approaches have high accuracy, while the CC-based ones have been verified in the existing works. Additionally, from Table I, it is found that compared with the EFIE-VIE or CFIE-VIE, the BC- and CC- based approaches have similar advantages, such as the reduction of peak memory, iteration numbers and total solution time. This is because although these two approaches are based on different conditions, both of them can reduce the volumetric unknowns in the tetrahedrons located on the PEC-dielectric interfaces.

Table I

Computational details for different cases and approaches

Cases	Approaches	Total unknowns	Reduced unknowns	Peak memory (GB)	T_{tot}(min)	T_{MVP}(sec)	Iteration numbers
Coated PEC sphere	EFIE-VIE	685,598	—	22.6	109	6.06	189
	CFIE-VIE				97.1	6.18	48
	BC-EFIE-VIE	609,484	76,114	20.0	95.6	5.85	182
	BC-CFIE-VIE				85.9	5.99	45
	CC-EFIE-VIE	609,484	76,114	20.1	95.5	5.86	178
	CC-CFIE-VIE				85.9	6.01	43
Coated PEC almond	EFIE-VIE	34,629	—	2.65	325	0.334	305
	CFIE-VIE				60.3	0.339	41
	BC-EFIE-VIE	32,225	2,404	2.38	307	0.322	299
	BC-CFIE-VIE				56.9	0.328	40
Double-face coveredPEC square plate	EFIE-VIE	118,219	—	5.38	45.5	1.16	578
	BC-EFIE-VIE	103,327	14,892	4.71	31.4	1.01	550

Note: T_{tot} and T_{MVP} denote the time costs by the total solution process and by one MVP in the iterative solution, respectively. The number of iterations for the second case denotes average number per observation angle.

The second case is a coated PEC almond containing sharp tips, as shown inside Fig. 2. The length of the PEC part is 250 mm, and the coating thickness with εr =2.2 − j0.00198 is 10 mm. Discretized with an average mesh size 0.063 λ, the numbers of triangles and tetrahedrons are 2,404 and 14,147, respectively. This case is a typical low RCS target, while the numerical results are usually more sensitive to computational accuracy. Illuminated by a θ-polarized plane wave at 3 GHz, the monostatic RCS shown in Fig. 2 is calculated by the EFIE-VIE and CFIE-VIE with or without the BC, and the observation range is 0°⩽φ⩽180° with 1° interval at xoy plane. For comparison, the result from commercial software Altair FEKO based on the SIE method is also given. Excellent agreement is observed among all curves over almost all directions. It is indicated that when the sharp structures are contained, the BC-based VSIE is also valid.

Table II

RMS errors between numerical results from different codes and analytical result from mie series for coated pec sphere (unit: dB)

Codes	EV	CV	BC-EV	BC-CV	CC-EV	CC-CV
xoz	0.067	0.094	0.060	0.091	0.058	0.090
yoz	0.021	0.023	0.019	0.022	0.018	0.021

Note: EV and CV are the abbreviations of EFIE-VIE and CFIE-VIE.

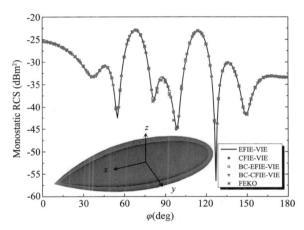

Fig. 2 Monostatic RCS at *xoy* plane for a PEC almond of 250 mm length coated with 10 mm thickness dielectric with ε_r=2.2 − j0.00198, illuminated by a θ-polarized normalized plane wave.

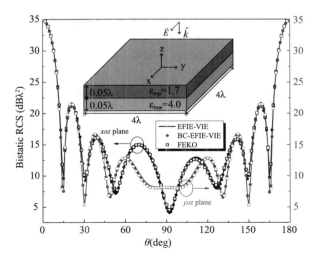

Fig. 3 Bistatic RCS for a double-face covered PEC square plate, illuminated by a *x*-polarized plane wave from *z*-axis.

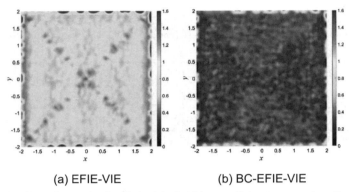

(a) EFIE-VIE (b) BC-EFIE-VIE

Fig. 4 Distribution of magnitude of electric field tangential component on the top face of PEC square, normalized by the magnitude of incident plane wave.

In the third case, a PEC square plate with 4λ side length is sandwiched by two dielectric slabs of the same lateral sizes, as shown inside Fig. 3. The relative permittivity of the top and the bottom slabs are εtop = 1.7 and εbtm = 4.0, respectively, and their thicknesses are both 0.05λ. Each slab is modeled by average mesh size 0.071λ, while the number of tetrahedrons is 45,908 with respect to 7,446 triangles. The incident plane wave is x-polarized from z-axis, and the EFIE-VIE and BC-EFIE-VIE are used to calculate the bistatic RCS at xoz and yoz planes. In this case, when the BC is enforced, the volumetric unknowns related to both sides of the PEC can be reduced. Therefore, the number of reduced volumetric unknowns in the BC-EFIE-VIE implementation is two times of the number of triangles. The numerical results as well as the FEKO result are shown in Fig. 3, and they consist well with each other. Fig. 4 shows the distribution of the magnitude of electric field TC on the top face of PEC square, which is normalized by the magnitude of incident wave. It can be found that the magnitude of the TC from the EFIE-VIE is about two times larger than that from the BC-EFIE-VIE. Because the TC is theoretically zero, the smaller magnitude obtained from numerical solution means the higher accuracy. Therefore, the enforcement of BC can also improve the solution accuracy.

From Table I, it is seen that for all cases, when the BC is enforced in either EFIE-VIE or CFIE-VIE, both the memory usage and the time cost will be reduced. Moreover, due to the reduction of the total unknowns and the numerical errors from non-zero TC, the BC-based approaches converge faster than the conventional ones. That is to say, under the premise of ensuring accuracy, the application of BC can always improve the solution efficiency.

4. Conclusion

A novel BC-based VSIE is proposed to improve the performance in the calculation of

EM properties for composite objects containing dielectric-PEC interfaces. Different from the existing methods, the common RWG and SWG basis functions based on triangular and tetrahedral meshes are used to discretize the currents, which are more versatile. Using the LS method, the relationship among four SWGs in one tetrahedron terminated by a PEC triangle is established to minimize the TC of electric field on dielectric-PEC interfaces. Typical numerical experiments indicate that the BC-based VSIE can always be safely used with high accuracy and efficiency.

Acknowlegments

This work was supported in part by the National Natural Science Foundation of China under Grant 61971384 and Grant 62071436, and in part by the Fundamental Research Funds for the Central Universities under Grant CUC220B007.

References

[1] CHEW W C, JIN J M, MICHIELSSEN E, et al. Fast and efficient algorithms in computational electromagnetics[M]. Houston: Artech House, 2001.

[2] LU C C, CHEW W C. A coupled surface-volume integral equation approach for the calculation of electromagnetic scattering from composite metallic and material targets[J]. IEEE Transactions on Antennas & Propagation, 2002, 48(12): 1866-1868.

[3] HE M, LIU J, WANG B, et al. On the Use of Continuity Condition in the Fast Solution of Volume-Surface Integral Equation[J]. IEEE Antennas and Wireless Propagation Letters, 2017, 16(99): 625-628.

[4] LIU J, YUAN J, LI Z, et al. A well-conditioned integral equation for electromagnetic scattering from composite inhomogeneous bi-anisotropic material and closed perfect electric conductor objects[J]. IET Microwaves, Antennas & Propagation, 2021,15(4): 404-414.

[5] HARRINGTON R F. Field Computation by Moment Methods[M]. New York: Macmillan, 1968.

[6] SONG J M, LU C C, CHEW W C. Multilevel fast multipole algorithm for electromagnetic scattering by large complex objects[J]. IEEE Transactions on Antennas and Propagation, 1997, 45(10): 1488-1493.

[7] LIU J, SONG J M, ZHANG H, et al. Solving the surface current distribution for open PEC-dielectric objects using the volume surface integral equation[J]. IEEE Antennas Wireless Propagat. Lett., 2022,21(1): 89-93.

[8] MAKAROV S N, KULKARNI S D, MARUT A G, et al. Method of moments solution for a printed patch/slot antenna on a thin finite dielectric substrate using the volume Integral equation[J]. IEEE Transactions on Antennas & Propagation, 2006, 54(4): 1174-1184.

[9] ZHANG T, HE S. Accuracy Improvement of the VSIE Solution in Frequency Response Simulation[C]. Singapore: IEEE Asia-Pacific Microwave Conference (APMC), 2019.

[10] RAO S M, WILTON D R, GLISSON A W. Electromagnetic scattering by surfaces of arbitrary shape[J]. IEEE Trans. Antennas Propagat., 1982,30(5): 409-418.

[11] SCHAUBERT D H, WILTON D R, GLISSON A W. A tetrahedral modeling method for electromagnetic scattering by arbitrarily shaped inhomogeneous dielectric bodies[J]. IEEE Transactions on Antennas and Propagation, 1984, 32(1): 77-85.

[12] WANG J, WEBB J P. Hierarchal vector boundary elements and p-adaption for 3-D electromagnetic scattering[J]. Antennas & Propagation IEEE Transactions on, 1997, 45(12): 1869-1879.

[13] GRAGLIA R D, WILTON D R. Higher order interpolatory vector bases for computational electromagnetics[J]. IEEE Transactions on Antennas and Propagation, 1997, 45(3): 329-342.

[14] PRESS W H, FLANNERY B P, TEUKOLSKY S A, et al. Numerical Recipes: The Art of Scientific Computing [M]. 3rd ed. Cambridge: Cambridge University Press, 2007.

[15] SAAD Y A. Generalized minimum residual algorithm for solving nonsymmetric linear systems[J]. SIAM J. Stat. Comput, 1986, 7(3): 856-869.

[16] WOO A C, WANG H T G, SCHUH M J, et al. EM programmer's notebook-benchmark radar targets for the validation of computational electromagnetics programs[J]. IEEE Trans. Antennas Propagat., 1993,35(1): 84-89.

Using the Interpolative Decomposition to Accelerate the Evaluation of Radome Boresight Error and Transmissivity[*]

1. Introduction

Radome is a type of equipment protecting antennas from the influence of external environment. However, the existence of radome also has negative impacts on the electromagnetic performance of antenna, such as main lobe broadening, side lobe elevation, beam pointing deviation, etc. Meanwhile, boresight error (BSE) and transmissivity are very important and frequently concerned. In the design of radome and antenna-radome system (ARS), various numerical methods can be used to evaluate the ARS performance and the effect of the radome. The high-frequency methods, such as physical optics and the shooting and bouncing ray method, have clear physical concepts with fast speed. However, because they are based on the principle of local field approximation, it is not considered the electromagnetic coupling between antenna and radome as well as different parts of the radome, resulting in large numerical error. On the other hand, the integral equation (IE)-based method of moments (MoM), as a famous full-wave method, has a generally acknowledged high accuracy. Moreover, because the MoM naturally satisfies the radiation boundary condition, it is ideally suitable for the solution of scattering or radiation problems, including the evaluation of radome BSE and transmissivity. With the rapid development of fast algorithms such as the multilevel fast multipole algorithm (MLFMA), the MoM can be used to calculate the radome and ARS of extremely large electric size.

Using the MoM, radome BSE and transmissivity can be calculated by comparing the radiation patterns from the antenna and the ARS. Compared with the excitation sources of the aperture electric field, using the physical antenna directly is better able to simulate the real distribution of radiated field, and to precisely evaluate the BSE and transmissivity. During the evaluation, a simple comparison is executed between the sum patterns of the antenna with and without the radome to determine transmissivity , while the difference patterns are used to obtain the BSE. For antenna arrays, they can realize the beam scanning by changing the phase of feeding at each port of the element, then the BSE and transmissivity

---

[*] The paper was originally published in *IEEE Antennas and Wireless Propagation Letters*, 2022, 21 (9), and has since been revised with new information. It was co-authored by Jinbo Liu, Chen Yang, Mang He, Jin Yuan, and Zengrui Li.

will change accordingly. Therefore, repeated computations are inevitable for different scan angles. On the one hand, every scan angle relates to a separate right-hand side (RHS) vector in the MoM solution, while the matrix equation corresponding to each RHS vector should be independently solved. To obtain detailed radome results, the number of scan angles is large, leading to huge calculation time, even if fast algorithms are adopted. On the other hand, because the interval between adjacent scan angles is usually narrow, the RHS matrix formed by multiple RHS vectors is rank-deficient. Utilizing this property, the calculation process can be improved. In [15-19], the interpolative decomposition (ID) techniques were successfully adopted to accelerate the MoM solution of scattering problems with the rank-deficient RHS matrix. In these methods, the skeleton RHS vectors are extracted by the ID first, and then only these unknown coefficients related to the skeleton RHSs are solved, followed by the restoration process to obtain the full solutions from the skeleton ones. To alleviate vast core memory requested by the ID procedure for the problems with large numbers of RHSs or/and unknowns, multi interval strategies and multilevel algorithm were proposed. However, the existing articles focused on the scattering problems such as the solution of monostatic radar cross sections. To the best of the authors' knowledge, the ID technique has never been used to solve radiation problems, which is the main work of this letter.

There is a natural difference between the RHS matrices formed by scattering problems and by radiation problems. That is, the scattering RHS matrix has no zero elements, whereas most rows of the radiation RHS matrix are null. Consequently, in our proposed ID method, the radiation RHS matrix can be specifically compressed first, which is then decomposed by the ID procedure. Compared with the existing ID methods which are designed for scattering problems, the implementation of the proposed method is much easier, while the solution process is more efficient.

2. IE-Based MoM Combined with the ID Technique

In a variety of IEs, the volume-surface integral equation (VSIE) is widely used due to its versatility and accuracy. The VSIE is completely competent to model the antenna and ARS which can be classified as perfect electric conductor (PEC) or composite dielectric-PEC objects. Using the Galerkin' MoM combined with the commonly used RWG and SWG basis functions, the matrix equation with RHS matrix generated from the VSIE is written as

$$[Z]_{N \times N}[I]_{N \times M} = [V]_{N \times M} \qquad (1)$$

where $[Z]$ is the impedance matrix, $[V]$ is the RHS matrix, and $[I]$ is the unknown coefficients matrix. Besides, N is the number of unknowns, while M is the columns number of $[V]$, i.e., the number of RHS vectors. Traditionally, each unknown column vector in $[I]$ should be solved independently.

2.1 Conventional ID Technique

Thanks to the ID technique, $[V]$ can be approximately decomposed into

$$[V]_{N\times M} \approx [V^s]_{N\times M_2}[R]_{M_2\times M} \tag{2}$$

where $[V^s]$ is the matrix composed of the so-called skeleton column vectors of $[V]$ of Ms number, and $[R]$ is the column projection matrix. The decomposition in is restricted by a threshold ε_{ID} for accuracy control to ensure

$$\left\|[V^s][R]-[V]\right\|_2 \approx \varepsilon_{ID} \tag{3}$$

Deeming $[V^s]$ as the RHS matrix forms a new matrix equation

$$[Z]_{N\times N}[I^s]_{N\times M_2} = [V^s]_{N\times M_2} \tag{4}$$

from which can obtain the skeleton coefficients matrix $[I^s]$. Substituting (2) and (4) into (1), the full unknown matrix $[I]$ can be restored by

$$[I]=[Z]^{-1}[V]\approx [Z]^{-1}[V^s][R]=[Z]^{-1}[Z][I^s][R]=[I^s][R] \tag{5}$$

It is quite clear that the solution time for (4) is about Ms/M of that for (1). If $[V]$ is rank-deficient, i.e., the $(Ms+1)$th greatest singular value of $[V]$ is tiny, Ms should be much smaller than M. Hence, if the ID procedure to find $[V^s]$ and $[R]$ is fast and not large-memory needed, the saving for solving $[I]$ due to the ID adoption will be considerable. However, when the N rows or/and M columns numbers of $[V]$ are large, the ID procedure will consume vast core memory, which is a serious drawback. To reduce the memory usage, a two-step strategy can be used. The core idea is that, first figure out the skeleton basis functions in the rows space of $[V]$ as

$$[V]_{N\times M} \approx [L]_{N\times M_s}[V^l]_{N_s\times M} \tag{6}$$

then use $[V^l]$ instead of the original $[V]$ to find the skeleton RHSs as

$$[V^l]_{N_s\times M} \approx [V^{ls}]_{N_s\times M}[R]_{M_s\times M} \tag{7}$$

combined with

$$[V^s]_{N\times M_s} \approx [L]_{N\times M_s}[V^{ls}]_{N_s\times M_s} \tag{8}$$

In (6)-(8), N_s is the number of skeleton basis functions, $[L]$ is the row projection matrix. Because $N^s \ll N$, the memory usage needed by the ID is significantly reduced. Furthermore, hierarchical or multilevel algorithms could be adopted to execute both the left (6) and right (7) decompositions [15] - [19], further reducing the memory requirement.

2.2 Modification of ID for Radiation Problems

For radiation problems, the most common types of antenna ports are electric edge port and magnetic fluid port. In the MoM implementation, delta-gap voltage sources are imposed on the RWG basis functions belonging to these ports, while each scan angle relates to an RHS vector, i.e., a column of $[V]$. Assume that jth RWG basis function defined on a common triangle side of length l_j belongs to nth antenna port located in (x_n, y_n, z_n) coordinates. For the mth scan angle $(\theta_m^3, \varphi_m^3)$, the entries of $[V]$ are calculated as

$$V_{j,m} = l_j w_n e^{-jk_0(x_n \sin\theta_m^3 \cos\varphi_m^3 + y_n \sin\theta_m^3 \sin\varphi_m^3 + z_n \cos\theta_m^3)} \tag{9}$$

where w_n is the weighted value of nth port, and k_0 is the wavenumber in the free space. When the interval between adjacent scan angles is narrow, the difference between corresponding column vectors of $[V]$ should be small. Therefore, $[V]$ is usually rank-deficient, which is one of the prerequisites to ensure the ID technique work.

Different from scattering problems that the incident electromagnetic wave excites the whole objects, the radiation source is only loaded on the antenna ports as (9). Therefore, there are many rows of $[V]$ being null, not affecting the ID procedure. Let J denote the set of nonzero-row indices of $[V]$, a compressed matrix can be extracted as

$$[\overline{V}]_{\widehat{N} \times M} = [V(J,:)]_{\widehat{N} \times M} \tag{10}$$

where $\overline{N}$ is the number of elements in the set J and $\overline{N} \ll N$. Thus, the ID procedure can be reduced to

$$[\overline{V}]_{\widehat{N} \times M} \approx [\overline{V}^s]_{\widehat{N} \times M_s}[R]_{M_s \times M} \tag{11}$$

which yields the exact same column projection matrix $[R]$ as done in (2), while $[\overline{V}^s]$ is formed by the non-zero rows of $[V^s]$. Obviously, (11) has a much lower computational complexity than (2). Furthermore, a multi-interval strategy can also be applied for a large M to obtain $[R]$, as done in [15] and [16].

Compared with the existing ID methods, in addition to the low computational complexity, other advantages of the proposed method are stated as follows.

1) Easy implementation. As stated in 1), the decomposition in rows space is discarded. As a result, the corresponding procedure is avoided, which is quite complicated. Actually, for the proposed method, the ID procedures for antenna and for ARS are identical, because they have the same compressed matrix $[\overline{V}]$.

2) User friendly. In the worst situation that the interval among scan angles is too big, $[V]$ is not rank-deficient, leading to $Ms = M$. The ID technique is then out of action, while the time and memory consumed by the ID procedure are totally wasted. However, because of the low computational complexity, this waste can be completely ignored. More importantly, the proposed method can still yield reliable results in this situation. Therefore, the proposed

method can always be used without concerning whether it brings additional burden.

3. Numerical Experiments

The following calculations are carried out on a workstation configured by 3.2 GHz CPU with 16 cores and 32 GB DDR4 memory, while per core has one thread. The GMRES with a restart number 50 and zero initial guesses is used as the iterative solver to reach 0.001 residual error. The preconditioner based on sparse approximate inverse (SAI) is used to accelerate the convergence speed. The MLFMA is implemented to accelerate matrix-vector product computations requested by iterative solution with fixed 0.25λ leaf boxes. When the ID technique is used, the threshold ε_{ID} in (3) is 0.0001. The voltage magnitude loaded on each port of the antenna array is equal.

The BSE and transmissivity of a single-layer dielectric radome with super-spheroidal shape are calculated, enclosing a planar antenna array working at 6 GHz, as shown in Fig. 1. The antenna array is composed of 116 half-wave dipole elements of PEC strip, while each element is x-polarized. The array plane parallels with xoy plane. The distances between adjacent elements along x- and y-axis are both 0.55λ. The antenna is placed at a distance of 0.25λ from the base where exists a circular PEC ground plane of 0.2 m radius. The governing equation of the radome axisymmetric surface is

$$x^2 + y^2 = (R/L)^2 (L^v - Z^v)^{2/v}$$

In this case, the radius R of the base is 0.2 m, the height L is 0.6 m, and the curve parameter v is 1.38. The wall thickness of the radome is 5 mm, and the relative permittivity is $\varepsilon_r = 2.2 - j0.00198$. Discretized by moderate mesh size, the PEC antenna array and ground totally generate 12,276 triangles concerning 16,985 RWG basis functions, while the dielectric radome is with 447,075 tetrahedrons and 968,657 SWG basis functions. When the main beam scans to θ^s angle at a fixed φ^s plane, the BSE can be calculated from the difference radiation patterns as

$$BSE(\theta^S) = Ang_{ARS}(\theta^s) - Ang_{ant}(\theta^s)$$

where Ang_{ant} and Ang_{ARS} mean the null angular locations of difference radiation patterns from the antenna and the ARS, respectively. The transmissivity is calculated as

$$T(\theta^s) = \left[\left|\vec{E}_{ARS}(\theta^s)\right|^2 \bigg/ \left|\vec{E}_{ant}(\theta^s)\right|^2 \right]\%$$

where $\vec{E}_{ant}$ and $\vec{E}_{ARS}$ denote the far radiated fields from the antenna and the ARS, respectively. Please note that since the direction of the main beam is θ^s, $\vec{E}_{ant}(\theta^s)$ is actually the peak of radiated field for the antenna. On the contrary, because of the existence of BSE, $\vec{E}_{ARS}(\theta^s)$

may not catch the peak value of the ARS radiated field. Obviously, accurate results of BSE and transmissivity depend upon the accurate calculation of the radiation pattern or field.

The BSE and transmissivity of the radome are calculated at *xoz* and *yoz* planes, respectively, while the scan angle θ^s is varied from 0° to 30° in 0.5° intervals. Therefore, it is a typical multiple RHS radiation problem with 61 RHS vectors. Due to the use of the SAI preconditioner, the calculation for antenna and ARS at each scan angle can always be converged in dozens of iterations. Using the direct solving method, the existing ID method presented in [19] (denoted by *exist.*) and the proposed ID method (*prop.*), the numerical results of BSE and transmissivity are calculated and shown in Fig. 2 and Fig. 3, respectively. Besides, the results excited by the aperture electric field but not the real antenna are also given for comparison. Meanwhile, the resolution for BSE is 6′, under which the BSE results from different methods excited by the real antenna are identical. For easy reading, only one curve of the results from the real antenna is drawn for each plane in Fig. 2. As is seen, both the existing and the proposed ID methods can recover the complete results faithfully, demonstrating the high accuracy of the ID technique. On the other hand, compared with the real antenna results, the results excited by the aperture electric field show a distinct difference. This is because the aperture field cannot correctly simulate the real distribution of the radiated field, which leads to rough results. This phenomenon exhibits the necessity of the use of real antennas during the calculation.

Table I gives the comparison of the peak memory (Mem_{peak}), the total solution time (T_{tot}), and the memory (Mem_{ID}, including $[I^s]$) and time (T_{ID}) consumed in the ID procedure, during the calculation of the sum patterns of antenna and ARS at *xoz* plane. Please note *Mem*ID is proportional to the number of threads. Compared with the direct method, the ID methods speeded up several times, because the ID solution is only executed on the skeleton RHS matrix $[V^s]$ with a much smaller columns number, rather than on the original RHS matrix $[V]$. Another finding is that the ID consumption can be totally ignored in contrast to the total consumption, although the memory and time consumed in the proposed ID method are a few times less than the existing one. This phenomenon is very different from that in the scattering problems, in which situation $[V]$ is a full matrix and the ID consumption cannot be neglected. The third finding is that in the calculation of antenna or ARS, the *Ms* value for the existing method is a little larger than the proposed method (9 vs. 8), inevitably leading to more solution time. The reason is likely that, for the existing method, the decomposition (6) in rows space may suffer from the limit of the computer machinery accuracy due to the involvement of zero elements, leading to a slightly larger *Ms*. In contrast, the similar procedure (10) in the proposed method is error-free, because the zero elements in the RHS matrix are directly removed.

Using the Interpolative Decomposition to Accelerate the Evaluation of Radome Boresight Error and Transmissivity

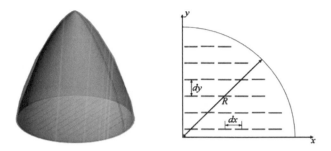

(a) Model of ARS (b) Model of a quarter of antenna array

Fig. 1 Structure of a single-layer radome enclosed antenna array, which is composed of 116 half-wave dipole elements.

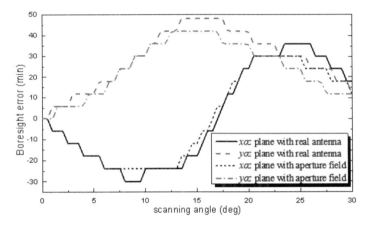

Fig. 2 Radome BSE at xoz and yoz planes calculated by with 6' resolution, while the results from different methods with the real antenna are exactly coincident.

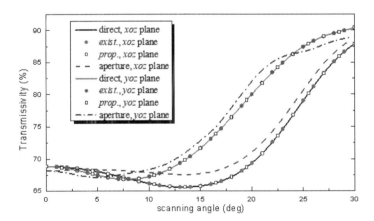

Fig. 3 Radome transmissivity at xoz and yoz planes from different methods, calculated by (14).

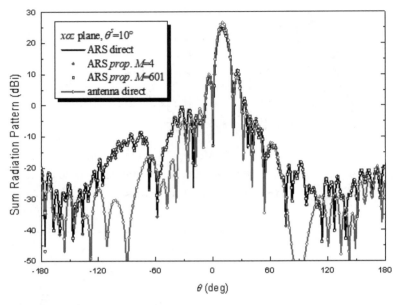

Fig. 4 Sum radiation patterns from antenna and ARS at xoz plane when $\theta^s=10°$, while the proposed ID method with M = 4 or 601 is used.

Table I

Time and memory costs for different methods when calculating the sum patterns at xoz plane

Model	method	Mem_{peak}	T_{tot}	M_s	Mem_{ID}	T_{ID}
Antenna	direct	101 MB	174 sec	—	—	—
	exist.	109 MB	30.7 sec	9	2.0 MB	5.1 sec
	prop.	109 MB	27.1 sec	8	0.98 MB	0.2 sec
ARS	direct	24.8 GB	194 min	—	—	—
	exist.	25.1 GB	40.6 min	9	105 MB	46 sec
	prop.	25.1 GB	36.5 min	8	60.3 MB	0.2 sec

Note: When using the existing ID method, the numbers of skeleton basis functions for antenna and for ARS are 496 and 2,535, respectively.

Table II

Time and memory costs by different ID implementations with various intervals

Interval of scan angles		10°	2°	1°	0.25°	0.1°	0.05°
M for direct method		4	16	31	121	301	601
exist.	M_s	4	9	9	9	9	10
	T_{ID} (sec)	25.6	25.6	25.7	25.9	26.2	26.5
	Mem_{ID} (MB)	67	99	101	104	106	131
prop.	M_s	4	8	8	8	9	9
	T_{ID} (sec)	0.01	0.04	0.13	0.31	0.40	0.52
	Mem_{ID} (MB)	60.2	60.2	60.3	60.6	61.7	71.5

Note: M denotes the number of total RHS vectors during the direct method. Ms denotes the number of skeleton RHS vectors during the ID method.

Next, it is investigated when the interval of scan angles narrows or widens, how the ID performance changes. During the calculation of the ARS sum pattern at xoz plane, we keep the start and end scan angles 0° and 30° unchanged, whereas modifying the interval to 10°, 2°, 1°, 0.25°, 0.1°, and 0.05°, respectively. Table II lists the number of RHS vectors M, the number of skeleton vectors Ms for different ID methods and their consumption. Clearly, the Ms value is stable and almost independent of the interval, i.e, the value of M. In other words, when the start and end scan angles are fixed, the value of Ms to a specific problem is almost determined. Therefore, if the ID technique is used, the concerned results can be calculated in dense scan angles without worrying about extra consumption. When the scan angle θ^s is 10°, the sum radiation patterns from the proposed ID method with two extreme M values 4 and 601 are calculated and shown in Fig. 4. Excellent agreements are observed among the results from the proposed method and those from the direct method, illustrating that the ID accuracy is not affected by the value of interval. Therefore, the ID technique can always be safely used.

4. Conclusion

The proposed ID method can successfully accelerate the MoM solution of radiation problems with multiple RHS vectors. The denser the scan angles are, the larger the speed-up ratio is. The radiation problems have a unique feature, that is, its RHS matrix contains lots of null rows. Utilizing this feature, the RHS matrix can be first reduced into a compressed form, then handled by the ID technique to find its skeleton vectors. In this way, the ID procedure is greatly simplified and more accurate compared with the existing ID method. When calculating the radome BSE and transmissivity, numerical results show that the proposed ID method can yield reliable results at all scan angles with high efficiency.

References

[1] KOZAKOFF D J. Analysis of radome-enclosed antennas[M]. Norwood: Artech House, 2010.

[2] VOLAKIS J L, VOLAKIS J L. Antenna engineering handbook[M]. New York: McGraw-hill, 2007.

[3] XU W, DUAN B Y, LI P, et al. Multiobjective particle swarm optimization of boresight error and transmission loss for airborne radomes[J]. IEEE Transactions on Antennas and Propagation, 2014, 62(11): 5880-5885.

[4] SHIFFLETT J A. CADDRAD: A physical optics radar/radome analysis code for arbitrary 3D geometries[J]. IEEE Antennas and Propagation Magazine, 1997, 39(6): 73-79.

[5] POULSEN S. Shooting and bouncing rays in radomes[C]. Granada, Spain: 2019 International Conference on Electromagnetics in Advanced Applications (ICEAA). IEEE, 2019: 0449-0452.

[6] CHEW W C, MICHIELSSEN E, SONG J M, et al. Fast and efficient algorithms in computational electromagnetics[M]. Norwood: Artech House, 2001.

[7] HARRINGTON R F, HARRINGTON J L. Field computation by moment methods[M]. Oxford: Oxford University Press, 1996.

[8] ZHOU P, ZHANG Z, HE M. Radiation pattern recovery of the impaired-radome-enclosed antenna array[J]. IEEE Antennas and Wireless Propagation Letters, 2020, 19(9): 1639-1643.

[9] XU W, DUAN B Y, Li P, et al. Study on the electromagnetic performance of inhomogeneous radomes for airborne applications—Part I: Characteristics of phase distortion and boresight error[J]. IEEE Transactions on Antennas and Propagation, 2017, 65(6): 3162-3174.

[10] LIU L, NIE Z. Performance improvement of antenna array-radome system based on efficient compensation and optimization scheme[J]. IEEE Antennas and Wireless Propagation Letters, 2019, 18(5): 866-870.

[11] WANG B, HE M, LIU J, et al. An efficient integral equation/modified surface integration method for analysis of antenna-radome structures in receiving mode[J]. IEEE Transactions on Antennas and Propagation, 2014, 62(9): 4884-4889.

[12] WANG B, HE M, LIU J, et al. Fast and efficient analysis of radome-enclosed antennas in receiving mode by an iterative-based hybrid integral equation/modified surface integration method[J]. IEEE Transactions on Antennas and Propagation, 2017, 65(5): 2436-2445.

[13] MANICA L, ROCCA P, MASSA A. On the synthesis of sub-arrayed planar array antennas for tracking radar applications[J]. IEEE Antennas and Wireless Propagation Letters, 2008, 7: 599-602.

[14] MA M T. Theory and Application of Antenna Arrays[M]. Hoboken, NJ, USA: Wiley, 1974.

[15] PAN X M, SHENG X Q. Accurate and efficient evaluation of spatial electromagnetic responses of large scale targets[J]. IEEE Transactions on Antennas and Propagation, 2014, 62(9): 4746-4753.

[16] PAN X M, GOU M J, SHENG X Q. Prediction of radiation pressure force exerted on moving particles by the two-level skeletonization[J]. Optics Express, 2014, 22(8): 10032-10045.

[17] PAN X M, SHENG X Q. Fast solution of linear systems with many right-hand sides based on skeletonization[J]. IEEE Antennas and Wireless Propagation Letters, 2015, 15: 301-304.

[18] PAN X M, HUANG S L, SHENG X Q. Wide angular sweeping of dynamic electromagnetic responses from large targets by MPI parallel skeletonization[J]. IEEE Transactions on Antennas and Propagation, 2018, 66(3): 1619-1623.

[19] LIU W, HE M. Accelerating solution of volume-surface integral equations with multiple right-hand sides by improved skeletonization techniques[J]. IEEE Antennas and Wireless Propagation Letters, 2019, 18(10): 2006-2010.

[20] LIBERTY E, WOOLFE F, MARTINSSON P G, et al. Randomized algorithms for the low-rank approximation of matrices[J]. Proceedings of the National Academy of Sciences, 2007, 104(51): 20167-20172.

[21] HALKO N, MARTINSSON P G, TROPP J A. Finding structure with randomness: Probabilistic algorithms for constructing approximate matrix decompositions[J]. SIAM review, 2011, 53(2): 217-288.

[22] RAO S, WILTON D, GLISSON A. Electromagnetic scattering by surfaces of arbitrary shape[J]. IEEE Transactions on antennas and propagation, 1982, 30(3): 409-418.

[23] SCHAUBERT D, WILTON D, GLISSON A. A tetrahedral modeling method for electromagnetic scattering by arbitrarily shaped inhomogeneous dielectric bodies[J]. IEEE Transactions on Antennas and Propagation, 1984, 32(1): 77-85.

[24] MAKAROV S N. Dipole and monopole antennas: the radiation algorithm[M]. Hoboken: Wiley, 2002.

[25] MAKAROV S N, KULKARNI S D, MARUT A G, et al. Method of moments solution for a printed patch/slot antenna on a thin finite dielectric substrate using the volume integral equation[J]. IEEE Transactions on Antennas and Propagation, 2006, 54(4): 1174-1184.

[26] SAAD Y, SCHULTZ M H. GMRES: A generalized minimal residual algorithm for solving nonsymmetric linear systems[J]. SIAM Journal on scientific and statistical computing, 1986, 7(3): 856-869.

[27] CARPENTIERI B, DUFF I S, GIRAUD L. Sparse pattern selection strategies for robust Frobenius-norm minimization preconditioners in electromagnetism[J]. Numerical linear algebra with applications, 2000, 7(7-8): 667-685.

[28] OVERFELT P L. Superspheroids: A new family of radome shapes[J]. IEEE transactions on antennas and propagation, 1995, 43(2): 215-220.

Extremely Wideband and Omnidirectional RCS Reduction for Wide-Angle Oblique Incidence[*]

1. Introduction

In recent years, different methods have been proposed to reduce the RCS of metal objects for stealth application, such as radar absorbing materials, shaping methods, and passive or active cancellation. Currently, metasurfaces put forward the RCS reduction and stealth to a new stage due to their ability to manipulate wave front flexibility, such as checkerboard metasurfaces, coding metasurfaces, phase gradient metasurfaces and polarization conversion metasurfaces. Most research based on metasurfaces focuses on reducing the RCS of metal objects under normal incidence. Namely, their performance of RCS reduction is poor for oblique incidence. In [11], a shared aperture metasurface divided into two dimensions (horizontal and vertical) was designed with enhanced scattering performance. However, the 10 dB RCS reduction can only be achieved from 4.8 to 16.4 GHz (109.4%) when the incident angle was less than 30°. A cross frame and a cross dipole were chosen as the patterns of two different artificial magnetic conductors (AMCs) in [5]. The bandwidth of the 10 dB RCS reduction was approximately 90% only when the incident angle was less than 10°. When the incident angle was 20°, the 10 dB RCS reduction bandwidth reduced dramatically for TM polarization. Surface wave suppression technology was applied in [12] to extend the 10 dB RCS reduction bandwidth. The high frequency of the 10 dB RCS reduction increased from 14.9 to 16.32 GHz with the fractional bandwidth of 68.6%. However, in the case of TM polarization, the RCS reduction characteristic at the high frequency worsened when the incident angle was 15°. In [13], diffraction in periodic structures and symmetrical pixelated geometry of unit cells were employed to achieve ultra-wideband and omnidirectional RCS reduction. However, as the incident angle increased, the RCS reduction performance in the low frequency range rapidly deteriorated and only obtained ~87% bandwidth (6.5-16.5 GHz) at a 30° oblique incidence. Therefore, the angle stability of the aforementioned designs needs to be improved.

A metasurface constructed by coding eight different linear-phase gradients in a spiral

---

[*] The paper was originally published in *IEEE Transactions on Antennas and Propagation*, 2022, 70 (8), and has since been revised with new information. It was co-authored by Meijun Qu, Chenyang Zhang, Jianxun Su, Jinbo Liu, and Zengrui Li.

layout was proposed in [14]. The bandwidth of 8.8 dB RCS reduction was 12-23 GHz (62.9%) at a large incidence angle of 60°. In [15], the square ring patch was chosen as the basic AMC element based on OMEPC. The 10 dB RCS reduction was from 11.91-21.17 GHz (56%) for TE polarization at a 45° oblique incidence angle. In [16], RCS reduction was achieved based on a concave/convex-chessboard random parabolic-phased metasurface. The operation bandwidth was 68% measured by 8.4 dB RCS reduction when the incidence approached 45°. An inhomogeneous perforated dielectric superstrate was introduced on conventional periodic metallic patches to improve the performance of reflection-type AMCs, which could contribute to oblique incidence stability up to 50° with a 95.7% bandwidth (15.5-44 GHz). Nevertheless, the superstrate is not conducive to miniaturization and cost reduction. From [14-17], it is concluded that the RCS reduction bandwidth still needs to be improved while maintaining a large incident angle.

In summary, although significant progress has been achieved in RCS reduction by using metasurfaces, some issues are still not adequately addressed, such as the difficulty of bandwidth expansion, narrow incident angles, polarization sensitivity and complex design. From the aforementioned literature, under oblique incidence, the 10 dB RCS reduction bandwidth would decrease dramatically for TM polarization because the RCS reduction performance turns down at high frequencies. For TE polarization, the RCS reduction level cannot reach 10 dB in the middle operating frequency bands. In reality, for metal objects, the frequency, polarization and direction of incoming waves are unpredictable. Therefore, there is an urgent need to reduce the RCS in the extremely wideband range for both TE and TM polarizations under wide-angle oblique incidence.

These above designs are based only on the case of normal incidence. Therefore, the performance of oblique incidence is much worse than that of normal incidence. In this communication, the cases of normal incidence and 40° oblique incidence are co-designed for extremely wideband and omnidirectional RCS reduction under wide-angle oblique incidence. A new metasurface with 25 AMCs, which is designed by OMEPC, is proposed. The equivalent transmission line model of the unit cell is analyzed under both normal and oblique incidences. A single thickness AMC structure could be seen as a short-circuited structure that causes in-phase reflections at certain frequencies. Therefore, a metasurface with multiple unit cells of different thicknesses is adopted to produce the phase difference to solve this problem. The OMEPC, which adopts varied basic unit cells to expand the bandwidth of RCS reduction, is employed to optimize and select the appropriate parameters for unit cells for extremely wideband and wide-angle RCS reduction. When the incident angle ranges from 0° to 40°, the proposed multiple element metasurface achieves the 10 dB RCS reduction from 7.34 to 64.85 GHz with a ratio bandwidth of 8.84:1 for both TE- and TM-polarization, and the 8 dB RCS reduction bandwidth is almost from 6.76 to 83.70 GHz. The theoretical analysis, simulation and measurement results verify the excellent performance of the proposed multiple element metasurface, which realizes extremely wideband and omnidirectional RCS reduction under wide-angle oblique incidence.

This communication is organized as follows: Section II describes the RCS reduction analysis for oblique incidence. In Section III, the equivalent circuit model of an AMC is analyzed under both normal and oblique incidences based on transmission line theory. In addition, the unit cell design and optimal process of the proposed multiple element metasurface are also described. Then, the simulation and the measurement results of the proposed multiple element metasurface are shown in Section IV. Finally, the conclusion is drawn in Section V.

2. RCS Reduction Analysis for Oblique Incidence

This communication focuses on the suppression of specular scattering for wide-angle oblique incidence. It is necessary to control and suppress the scattered field simultaneously under different incident angles in a wide frequency band. A metasurface consists of M×N lattices evenly distributed with dx in the x direction and dy in the y direction, as shown in Fig. 1.

Under oblique incidence, based on array pattern synthesis, the array factor (AF) of scattered field can be accurately synthesized by [20]:

$$AF(\theta,\varphi) = \sum_{m=1}^{M}\sum_{n=1}^{M} A_{m,n} \exp\{j[2\pi\sin\theta(md_x\cos\varphi + nd_y\sin\varphi)/\lambda + 2\pi\sin\theta^i(md_x\cos\varphi^i + nd_y\sin\varphi^i)/\lambda + \varphi_{m,n}]\} \quad (1)$$

where θ and φ are elevation angle and azimuth angle of the scattered fields, respectively. θ^i and φ^i are the elevation angle and azimuth angle of the arbitrary incident fields, respectively. Am,n and φm,n are the amplitude and phase of the reflection coefficient Γm,n(θ^i, φ^i) = Am,nejφm,n under oblique incidence, respectively. The first item in the middle bracket is the phase difference φ^2 of each lattice relative to the reflected waves, which is caused by the wave-path difference Δr2 shown in Fig. 2. The second item in the middle bracket is the phase difference φ^1 of each lattice relative to the incoming wave and is caused by the wave-path difference Δr1 shown in Fig. 2. It can be seen that Δr1 = Δr2, therefore $\varphi^1 = \varphi^2$, and the specular scattered field is independent of the arrangement of lattices, which only relates to the reflection coefficient of lattices. Thus, the AF in the specular direction can be simplified as:

$$AF(\theta^i, \varphi^i + 180^0) = \sum_{m=1}^{M} \sum_{n=1}^{N} \Gamma_{m,n}(\theta^i, \varphi^i) \qquad (2)$$

The RCS reduction of oblique incidence can be obtained by Eq. (3) in [21], which can be approximated by:

$$\sigma_R = 10\log\left|\frac{AF(\theta^i, \varphi^i + 180^0)}{M \times N}\right|^2 = 10\log\left|\frac{\sum_{k=1}^{M \times N} A_k e^{j\varphi_k}}{M \times N}\right|^2 \qquad (3)$$

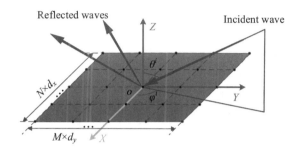

Fig. 1 Specular reflection of electromagnetic waves under oblique incidence on a checkerboard metasurface.

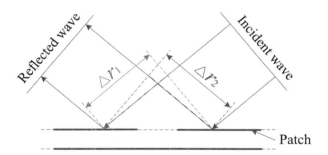

Fig. 2 Wave-path differences on the metasurface under oblique incidence.

For a lossless ground plane, the amplitude of the reflection coefficient can be approximated to unity (Am, n ≈ 1). The OMEPC combined with Eq. (3) is used to optimize and select the geometric parameters of basic AMC elements to find the maximum bandwidth for 10 dB RCS reduction at multiple incident angles in this work.

3. AMC Surface Design

3.1 Unit Cell Design

The reflection phase in a specific frequency range is determined by the shape and size of the basic unit cell. When the shape of the basic unit cell remains the same, the reflection phases are different due to the varying sizes of the unit cells. Meanwhile, the far scattered fields are also different in this frequency range. In this design, a square patch that can be simply designed and easily manufactured is selected as the basic AMC element but with variable geometric parameters, which results in different reflection coefficients. The geometry of the unit cell is shown in Fig. 3(a). The substrate used is Polytetrafluoroethylene Woven Glass (F4BM-2) with an effective dielectric constant of 2.65 and a loss tangent of 0.001.

According to the equivalent circuit model of the unit cell, the metal ground is equal to the transmission line with terminal shorted, as shown in Fig. 3(b). Transmission line theory is applicable for arbitrary incident angles and polarization. Figs. 3(c-d) are the side views of the unit cell under normal and oblique incidences, respectively. The thickness of the dielectric layer is expressed by h, and the relative thickness of the dielectric layer when the plane wave passes through is expressed by h'. Under normal incidence, h'= h. Under TE- or TM-polarized oblique incidence, h' can be obtained from the law of refraction as shown in Eq. (4):

$$h' = h \cdot \cos\gamma = h \cdot \sqrt{1-\sin^2\gamma} = h \cdot \sqrt{1-\frac{\sin^2\theta^i}{\varepsilon_r}} \quad (4)$$

Transmission line theory is utilized to explain why ultra-wideband RCS reduction can be realized by using the proposed unit cells of different thicknesses. From Fig. 4(a), it can be found that when the length of the short-circuit transmission line is between $i\times\lambda/2$ and $i\times\lambda/2+\lambda/4$, the short-circuit transmission line is equivalent to an inductance L_i. Symbol i is a natural number. The equivalent circuit can evolve to Fig. 4(b) at this time. When the length of the short-circuit transmission line is between $i\times\lambda/2+\lambda/4$ and $i\times\lambda/2+\lambda/2$, the short-circuit transmission line is equivalent to a capacitance C_i. Here, the equivalent circuit can evolve to Fig. 4(c). The parallel resonant frequencies (fp) of the equivalent circuit could be calculated as:

$$f_p = 1/(2\pi\sqrt{(L+L_i)C}), i\times\lambda/2 < i\times\lambda/2+\lambda/4$$
$$f_p = 1/\left(2\pi\sqrt{\frac{LCC_i}{C+C_i}}\right), \lambda/4+i\times\lambda/2 < h' < \lambda/2+i\times\lambda/2 \quad i=0,1,2,3\ldots \quad (5)$$

The impedance is equivalent to ∞ when parallel resonance occurs in the equivalent circuit. The unit cell can be seen as an AMC structure.

The series resonant frequencies (fs) of the equivalent circuit could be calculated as:

$$f_s = 1/(2\pi\sqrt{LC}) \quad (6)$$

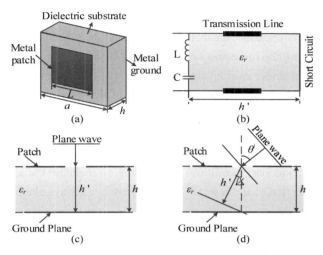

Fig. 3 The unit cell of the square patch: (a) the geometric structure, (b) the equivalent circuit model of the unit cell. The side view of the unit cell: (c) under normal incidence, (d) under oblique incidence.

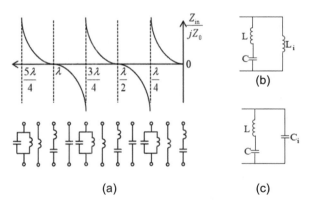

Fig. 4 (a) Impedance at any point along the short-circuit transmission line at the terminal. The equivalent circuit model of the unit cell when the length of the short-circuit transmission line is between (b) i×λ/2 and i×λ/2+λ/4, (c) i×λ/2+λ/4 and i×λ/2+λ/2.

The impedance is equivalent to zero when series resonance occurs in the equivalent circuit. The unit cell can be seen as a PEC structure.

The AMC and PEC structures have approximately 180° phase difference, thus opposite cancellation can be achieved. Note that substrates of the same thickness have different equivalent electrical lengths relative to electromagnetic waves of different frequencies and different directions. Meanwhile, we adopt substrates with various thicknesses in this work. In this way, numerous parallel and series resonant frequencies can be achieved, which helps to achieve wideband RCS reduction. According to Eqs. (4-6), the resonant frequencies can be determined by the layer thickness, the dielectric constant of the substrate and the incident angle.

In the simulation, the dielectric layer thickness h varies from 0.93 to 5.93 mm with an interval of 1 mm under normal incidence and 40° oblique incidence (TE and TM polarization). For each thickness, the side length L of the square patch changes from 0.8 to 7.8 mm with an interval of 0.1 mm. The periodicity a of the unit cell is 8 mm, and the thickness of the copper layer is 0.035 mm on both sides. The unit cell is simulated by the frequency domain solver (finite element method) of CST Microwave Studio® with appropriate periodic boundary conditions. Meanwhile, to maintain the same reference lane for the obtained reflection phase, a small "vacuum" patch is always placed on the plane (z = 7 mm) in the CST model.

3.2 Optimal Design of Multiple Element Metasurface

In this study, the multiple element metasurface aims to achieve 10 dB RCS reduction in an extremely wide frequency band under different incidence polarizations and angles. The proposed design is composed of 5 × 5 lattices. Each lattice is a sub-array of 25 basic unit cells. The OMEPC and Eq. (3) are used to optimize and select the optimal side lengths L and layer thicknesses h of the basic unit cells.

The flow chart of the RCS reduction optimization process is shown in Fig. 5. First, 25 combinations of L and h are selected randomly. For each combination, there are P optimization sample frequencies (f1, f2, …, fp). In this design, the reflection phases simulated in part A of Section III were pre-stored in three tables. Three groups of P reflection phases at the P optimization frequencies can be obtained by reading the three reflection phase tables. Then, three groups of P RCS reduction values can be calculated by Eq. (3). Finally, fitness can be obtained by calculating the P RCS reduction values of each group. Note that the fitness0°, fitness40°, TE and fitness40°, TM are fitness values under normal incidence, TE- polarized 40° oblique incidence and TM- polarized 40° oblique incidence, respectively. Hence, the fitness function fitnesssum in this work can be defined as:

$$fitness_{sum} = fitness_{0°} + fitness_{40°,TE} + fitness_{40°,TE} \quad (7)$$

$$fitness_n = \sum_{i=1}^{P} S_n(i), \quad S_n(i) = \begin{cases} 1, & \text{if } \sigma_R^i > -10\text{dB} \\ 0, & \text{if } \sigma_R^i \leq -10\text{dB} \end{cases} \quad (8)$$

Note that the subscript n represents 0°, "40°, TE", and "40°, TM". The sum of three fitness values (one for normal incidence and the other two for 40° TE- and TM-polarized incidence) is calculated as a score for evaluating the RCS reduction performance of each combination of L and h. The smaller the score is, the better the performance of RCS reduction.

In this study, the 25 best basic unit cell geometric parameters with the lowest specular

RCS can be obtained in the desired frequency band when the iterations of the three incident cases are completed at the same time. The optimized results of the side lengths L of the square patch and dielectric layer thicknesses h are shown in Table I. Since the arrangement of the 25 lattices has no effect on specular RCS reduction, lattices with the same thickness are randomly arranged together for convenient processing. To approximately satisfy the periodic boundary condition used in the simulation, each lattice is occupied by a subarray of 7 × 7 unit cells. The optimized RCS reducer has an overall dimension of 280×280 mm². The optimal distribution of 25 lattices and the proposed multiple element metasurface are depicted in Fig. 6. Since the geometric parameters and arrangement of each AMC element are known, the corresponding reflection phase can be obtained. Using Eq. (3), the specular and bistatic RCS reductions can be calculated theoretically, as shown in Figs. 7(a-c). The bandwidths of 10 dB RCS reduction are 7.16-62.55 GHz, 6.8-73.77 GHz, 7.16-70.80 GHz for normal incidence and 40° TE- and TM- polarized oblique incidence, respectively.

Compared with our previous work, the proposed design can achieve an extremely wideband RCS reduction at wide-angle oblique incidence with both TE and TM polarizations. This is because the scattered fields of the proposed multiple element metasurface under different frequencies, polarizations, and incident angles are jointly optimized and simultaneously suppressed by OMEPC. The optimization process in [19] only considers the different frequencies.

Table I

The optimized results of side lengths L (mm) of square patch and dielectric layer thickness h (mm)

unit cell	1	2	3	4	5
L/h	0.8/0.93	7.3/3.93	6.9/4.93	2.4/4.93	6.3/5.93
unit cell	6	7	8	9	10
L/h	7.0/0.93	7.8/5.93	1.4/2.93	5.7/4.93	6.2/2.93
unit cell	11	12	13	14	15
L/h	6.7/5.93	2.5/2.93	7.5/0.93	6.3/5.93	2.4/3.93
unit cell	16	17	18	19	20
L/h	3.9/0.93	4.8/5.93	3.4/3.93	3.1/0.93	0.8/0.93
unit cell	21	22	23	24	25
L/h	7.5/1.93	7.4/1.93	1.0/1.93	6.1/0.93	6.8/3.93

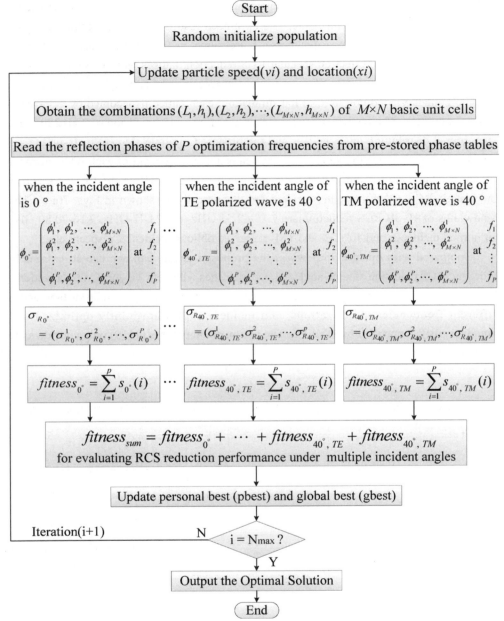

Fig. 5　Flow chart of the RCS reduction optimization process under multi-angle incidence.

Fig. 6 The model of the proposed multiple element metasurface. The inserted index is the unit cell number shown in Table I.

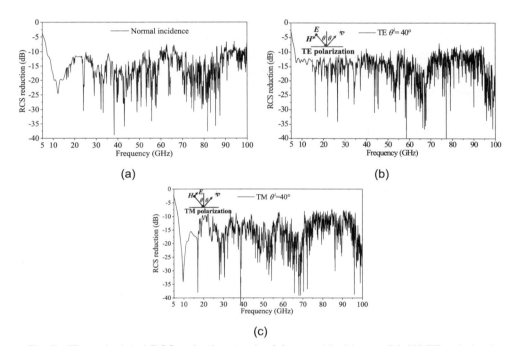

Fig. 7 The calculated RCS reductions under (a) normal incidence, (b) 40° TE-polarized oblique incidence, and (c) 40° TM-polarized oblique incidence, using Eq. (3) after the geometric parameters and arrangement of each AMC element are known. TE/TM: The direction of the electric/magnetic field is perpendicular to the plane of incidence.

4. Simulation and Measurement

4.1. Full-Wave Simulated Results

RCS reduction is simulated at oblique incidences of 0°, 20°, 40° and 60°. The RCS reduction of the proposed multiple element metasurface under TE- and TM-polarized incidence is shown in Fig. 8. When the incident angle is less than 40°, the 10 dB RCS reduction bandwidth is approximately 7.34 to 64.85 GHz, and the 8 dB RCS reduction bandwidth is approximately 6.76 to 83.70 GHz for both TE- and TM-polarization. The detailed bandwidths of 10 dB RCS reduction at four incident angles are summarized in Table II. The simulation results are in good agreement with the predicted results.

Furthermore, when the incident angle is 40° under TE-polarization, the specular RCS reductions at eight azimuth angles ($\varphi i = 0°$, 45°, 90°, 135°, 180°, 225°, 270° and 315°) are also simulated and compared, as shown in Fig. 9. The similarity between these eight curves is very high. This is because the specular scattered field is independent of the arrangement of lattices, which is proven in Section II. Hence, it can be inferred that when the incident angle is below 40°, the designed metasurface has excellent omnidirectional performance.

For TE- and TM-polarization, when the incident angles are 0°, 20°, and 40°, the three-dimensional (3D) specular RCS patterns of the proposed multiple element metasurface and equal-sized PEC surface at 8 GHz are presented in Figs. 10(a-c), respectively. It is obvious that the reflected fields are scattered more uniformly in many directions, and the scattered fields are suppressed at a low level by the proposed multiple element metasurface. The proposed multiple element metasurface also exhibits polarization-insensitive feature. Overall, it is demonstrated that the proposed multiple element metasurface has excellent monostatic and bistatic RCS reduction performance.

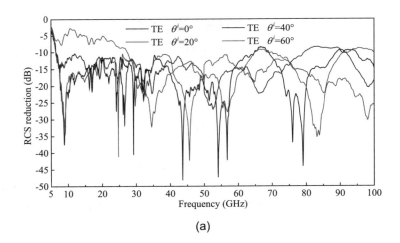

(a)

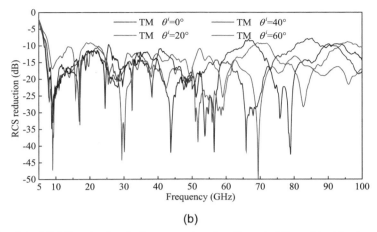

(b)

Fig. 8 Simulated RCS reduction in the specular direction at different incident angles for (a) TE-polarization and (b) TM-polarization.

Table II

The 10 dB RCS reduction bandwidths at four incident angles

Incident angle	TE		TM	
	OFB(GHz)	FBW (%)	OFB(GHz)	FBW (%)
0°	6.44-64.85	163.87	6.58-65.10	163.28
	69.05-89.05	25.30	68.80-89.10	25.71
20°	6.44-65.05	164.00	6.94-67.85	162.88
	68.35-92.65	30.19	72.70-93.60	25.14
40°	6.49-79.50	169.81	7.34-80.35	166.52
	91.35-100	9.04	87.55-100	13.28
60°	27.4-100	113.97	7.85-18.35	80.15
			21.5-100	129.22

OFB: Operating frequency band

FBW: The fractional bandwidth (FBW = (fH - fL)/fc, fc = (fH + fL)/2)

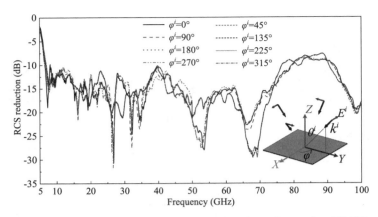

Fig. 9 Simulated RCS reductions at different azimuth angles under 40° TE-polarized incidence. ki and Ei are the incident direction and the electric field direction of the incident plane wave, respectively.

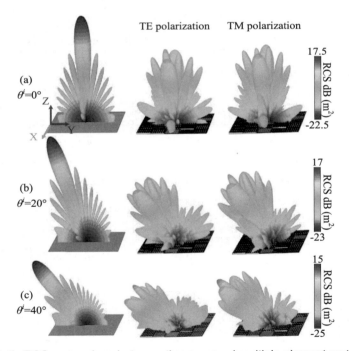

Fig. 10 Bistatic RCS comparison between the proposed multiple element metasurface and equal-sized PEC surface at 8 GHz.

4.2. Measured Results

To validate the predicted and simulated performance of the proposed multiple element metasurface, a sample is fabricated and measured. The sample is manufactured by printed circuit board (PCB) technology. Six pieces of the AMC surface boards with different thicknesses are processed separately and then stitched together. The fabricated multiple element metasurface, as is shown in Fig. 11, which is measured in the Compact Antenna Test Range (CATR) system.

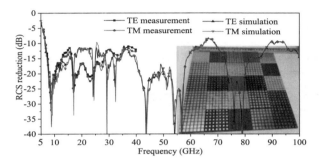

Fig. 11 Measurement and simulation results of RCS reduction under normal incidence for TE- and TM-polarization.

The measurement and simulation results of RCS reduction under normal incidence for TE- and TM-polarization are shown in Fig. 11. The proposed multiple element metasurface can realize an RCS reduction larger than 10 dB from 6.72 to 40 GHz for both TE- and TM-polarization. This is because the highest test frequency of our vector network analyzer only reaches 40 GHz. Nevertheless, it is clear that the reflection field of the proposed multiple element metasurface can be effectively suppressed in an extremely wide band. The deviation of the measurement result is due to fabrication and experimental error.

In Table III, a performance comparison of the proposed and state-of-the-art designs is presented. It is apparent that the proposed multiple element metasurface has an overwhelming advantage in the bandwidth of RCS reduction and stability of the incident angle.

Table III

Performance comparison of the proposed and state-of-the-art designs

Article (year)	Incident angle	σR (dB)	TE			TM		
			OFB (GHz)	FBW (%)	RBW f_H / f_L	OFB (GHz)	FBW (%)	RBW f_H / f_L
[5] (2019)	0°-30°	7	3.5-10	96.3	2.86:1	5-8	46.15	1.6:1
[7] (2019)	0°-20°	10	N/A	110	3.444:1	N/A	119	3.938:1

Table III Continued

Article (year)	Incident angle	σR (dB)	TE			TM		
			OFB (GHz)	FBW (%)	RBW f_H/f_L	OFB (GHz)	FBW (%)	RBW f_H/f_L
[10] (2019)	0°-20°	10	N/A	125.35	4.358:1	N/A	107.7	3.334:1
[12] (2019)	0°-15°	10	8-15.8	65.5	1.975:1	8-12.7	45.4	1.59:1
[19] (2019)	0°-20°	10	>(5.476-40)	>151.83	>7.305:1	10.52-19.66	60.57	1.869:1
[23] (2019)	0°-20°	10	N/A	109	3.396:1	N/A	107	3.301:1
[13] (2020)	0°-30°	10	8-14.5	57.78	1.81:1	N/A	N/A	N/A
[16] (2020)	0°-45°	8.4	9.9-20.2	68	2.04:1	N/A	N/A	N/A
[17] (2021)	0°-40°	10	16.5-49.5	100.0	3:1	N/A	N/A	N/A
This work	0°-40°	10	6.49-64.85	163.61	9.99:1	7.34-65.05	159.44	8.86:1
		8	6.31-100.00	176.26	15.85:1	6.76-83.70	170.11	12.38:1

5. Conclusion

A novel multiple element metasurface based on OMEPC is designed, fabricated, and tested for extremely wideband RCS reduction at wide-angle oblique incidence with both TE and TM polarizations. The transmission line model of the proposed unit cell is analyzed under both normal and oblique incidences. To eliminate the influence of blind zones and expand the bandwidth of RCS reduction, unit cells of multiple thicknesses are introduced in the design of the multiple element metasurface. The scattered fields of the proposed multiple element metasurface under different frequencies, polarizations, and incident angles are jointly optimized and simultaneously suppressed by OMEPC. Finally, when the incident angle ranges from 0° to 40°, the proposed metasurface achieves the 10 dB RCS reduction from 7.34 to 64.85 GHz with a ratio bandwidth of 8.84:1 for both TE- and TM-polarization, and the 8 dB RCS reduction bandwidth is approximately 6.76 to 83.70 GHz. The measurement results are in good agreement with the theoretical analysis and simulation results. This work effectively solves some unsettled issues in RCS reduction, such as difficulty of bandwidth expansion, incident angle sensitivity, polarization sensitivity, etc.

Acknowledgments

This work was supported by the National Natural Science Foundation of China (U2141233, 62171416 and 62101515) and the Fundamental Research Funds for the Central Universities (CUC210A001 and CUC210B012).

Author Information

Meijun Qu, Chenyang Zhang, Jianxun Su, Jinbo Liu and Zengrui Li are with the State Key Laboratory of Media Convergence and Communication, School of Information and Communication Engineering, Communication University of China, Beijing 100024, China. (e-mail: qumeijun@cuc.edu.cn, 13001942303@163.com, sujianxun_jlgx@163.com, liuj@cuc.edu.cn, zrli@cuc.edu.cn).

References

[1] SCHURIG D, MOCK J J, JUSTICE B J, et al. Metamaterial electromagnetic cloak at microwave frequencies[J]. Science, 2006, 314 (5801): 977-980.

[2] LIU R, JI C, MOCK J J, et al. Broadband ground-plane cloak[J]. Science, 2009, 323 (5912): 366-369.

[3] WEAR K A. The effect of phase cancellation on estimates of calcaneal broadband ultrasound attenuation in Vivo[J]. IEEE Trans. Ultrason. Ferroelect. Freq. Contr., 2007, 54 (7): 1352-1359.

[4] XUE J, JIANG W, GONG S. Chessboard AMC Surface Based on Quasi-Fractal Structure for Wideband RCS Reduction[J]. IEEE Antennas Wireless Propag. Lett, 2018, 17 (2): 201-204.

[5] SANG D, CHEN Q, DING L, et al. Design of Checkerboard AMC Structure for Wideband RCS Reduction[J]. IEEE Trans. Antennas Propag., 2019, 67 (4): 2604-2612.

[6] CUI T, QI M, WAN X, et al. Coding metamaterials, digital metamaterials and programmable metamaterials[J]. Light-Sci. Appl., 2014, 3 (10): e218.

[7] AKBARI M, SAMADI F, SEBAK A, et al. Superbroadband Diffuse Wave Scattering Based on Coding Metasurfaces: Polarization Conversion Metasurfaces[J]. IEEE Antennas Propag. Mag., 2019, 61 (2): 40-52.

[8] LI Y, et al. Wideband radar cross section reduction using two-dimensional phase gradient metasurfaces[J]. Appl. Phys. Lett., 2014, 104: 221110.

[9] ZAKER R, SADEGHZADEH A. A Low-Profile Design of Polarization Rotation Reflective Surface for Wideband RCS Reduction[J]. IEEE Antennas Wireless Propag. Lett., 2019, 18 (9): 1794-1798.

[10] AMERI E, ESMAELI S, SEDIGHY S. Ultra wideband radar cross section reduction by using polarization conversion metasurfaces[J]. Sci. Rep., 2019, 9 (478): 1-8.

[11] ZHENG Y, CAO X, GAO J, et al. Shared aperture metasurface with ultra-wideband and wide-angle low-scattering performance[J]. Optical Materials Express, 2017, 7 (8): 2706-2714.

[12] KIM S H, YOON Y J. Wideband radar cross-section reduction on checkerboard metasurfaces with surface wave suppression[J]. IEEE Antennas Wireless Propag. Lett., 2019, 18 (5): 896-900.

[13] ORUJI A, PESARAKLOO A, KHALAJ-AMIRHOSSEINI M. Ultrawideband and omnidirectional RCS reduction by using symmetrical coded structures[J]. IEEE Antennas Wireless Propag. Lett., 2020, 19 (7): 1236-1240.

[14] YUAN F, WANG G, XU H, et al. Broadband RCS reduction based on spiral-coded metasurface[J]. IEEE Antennas Wireless Propag. Lett., 2017, 16: 3188-3191.

[15] ZHANG C, SU J, LI Z. Wideband RCS reduction matesurface design[J]. ISEMC, 2019:1-4.

[16] YUAN F, XU H, JIA X, et al. RCS reduction based on concave/convex-chessboard random parabolic-phased metasurface[J]. IEEE Trans. Antennas Propag., 2020, 68 (3): 2463-2468.

[17] SAMADI F, MEMMBER S, SEBAK A. Wideband, very low RCS engineered surface with a wide incident angle stability[J]. IEEE Trans. Antennas Propag., 2021, 69 (3): 1809-1814.

[18] SINGH H, JHA R M. Active Radar Cross Section Reduction: Theory and Applications[M]. Delhi: Cambridge University Press, 2015.

[19] SU J, LU Y, LIU J, et al. A novel checkerboard metasurface based on optimized multielement phase cancellation for superwideband RCS reduction[J]. IEEE Trans. Antennas Propag., 2018, 66 (12):7091-7099.

[20] BALANIS C A. Antenna Theory: Analysis and Design[M]. 3rd ed. New York: Wiley, 2005.

[21] CHEN W, BALANIS C A, BIRTCHER C R. Checkerboard EBG surfaces for wideband radar cross section reduction[J]. IEEE Trans. Antennas Propag., 2015, 63 (6): 2636-2645.

[22] YANG F, RAHMAT-SAMII Y. Electromagnetic band gap structures in antenna engineering[M]. Cambridge, U.K.: Cambridge Univ. Press, 2009.

[23] SAMADI F, AKBARI M, CHAHARMIR M R, et al. Scatterer surface design for wave scattering application[J]. IEEE Trans. Antennas Propag., 2019, 67 (2): 1202-1211.

A Novel Wideband and High-Efficiency Electronically Scanning Transmitarray Using Transmission Metasurface Polarizer[*]

1. Introduction

With the development of 5th-generation (5G) cellular networks and beyond, beamforming (BF) is a crucial method that offers several advantages including low noise, high linearity, and high power handling capability. BF technology focuses electromagnetic waves toward a particular accepting device by the controllable device rather than spreading the signal altogether. The traditional BF realization of a phased antenna array requires many transmit/receive (T/R) modules that make the system more complicated and expensive. Additionally, the transmission line feed networks associated with phased arrays can become significantly lossy at high frequencies, especially in the mm-wave range.

In recent years, electronically reconfigurable reflectarray (RA) and transmitarray (TA) antennas, due to their low-cost electronic BF capabilities required by advanced communication systems, have attracted increasing interest. Compared to the TA, the RA design's most considerable challenge is feed blockage, while the feeding antenna and the radiated fields of the array are on the same side, leading to a beam scanning blind area.

Tunable devices such as micro-electromechanical systems (MEMS), varactor diodes, and positive-intrinsic-negative (PIN) diodes, are integrated on unit cells to achieve electronic reconfigurability. There is a trade-off between simplicity in the design of reconfigurable TA elements and TA gain when choosing the number of bits for phase quantization. A 1-bit TA antenna is considered as a fair tradeoff between performance and system cost.

Several reconfigurable TAs with 1-bit phase resolution have introduced in recent years. The receive-and-radiate method is widely used in the design of reconfigurable TA. In this solution, a typical array element is composed of receiving and radiating layers, which are connected through metallization via holes or coupling slots. Specific radiating structures, such as O-shaped slot patches, side shorted patches, asymmetric transmitting dipoles, and U-shaped microstrip lines, were proposed to realize current reversal leading to a 1-bit phase

---

[*] The paper was originally published in *IEEE Transactions on Antennas and Propagation*, 2022, 70 (4), and has since been revised with new information. It was co–authored by Hang Yu, Jianxun Su, Zengrui Li, and Fan Yang.

difference. In this case, the incoming space-wave is first coupled by an antenna to a guided-wave. The guided-wave is then phase shifted and finally reradiated, resulting in an antenna-phase-shifter-antenna topology. Intuitively, this results from the fact that in the receive-and-radiate approach all incoming power flows through the tunable devices (e.g., the PIN diodes), which causes a large ohmic loss. The intrinsic high loss characteristics of the receive-and-radiate method make it rather difficult to realize a low loss design over a wide frequency range. Actually, it is challenging to manipulate the TA element's phase response, as the phase response is usually related to the element's transmission performance. Our work aims to propose a novel strategy to improve the bandwidth and efficiency of the reconfigurable TA.

Metasurfaces, which are 2-D metamaterials, have attracted considerable attention due to their unique realization of many functionalities. It's worth noting that [24] proposed a method that can switch polarization conversion from high-efficiency and broadband reflection mode to transmission mode. The proposed reflection-type linear polarization converter design consists of a metal cut-wire array and a metal ground plane separated by a dielectric substrate. In order to switch polarization conversion in transmission mode, the metal ground plane is replaced by a metal grating, and another orthogonal metal grating is added in front of the cut-wire array. Besides, the polarization turning concept is also applied in reconfigurable RA design [26], [27] to obtain a broadband operating bandwidth and a flat phase-shift response. Based on the polarization conversion mode switching, we can migrate the reconfigurable RA design scheme to TA for realizing high-efficiency and wideband performance. Inspired by recent investigations of metasurfaces, we demonstrate a wideband electronically reconfigurable TA unit cell based on the polarization conversion mechanism, which can rotate a linear polarization state to its orthogonal one with two different phases. Unlike the receive-and-radiate method in [17]-[21], this unit cell design can consider a tunable resonator approach, where tunable devices are subjected to low induced currents and lower loss results. Through a surface current analysis, it is found that most of the induced current is concentrated in the vicinity of the diode in the OFF state. The presented element's transmission magnitude is a low insertion loss of 0.5 dB owing to little current flows through these diodes. An X-band 10×10-element TA fed by a pyramidal horn antenna is fabricated. Different beams can be switched by controlling the voltages of the diodes using a field programmable gate array (FPGA) and drive circuit board. Electronically reconfigurable design enables real-time switching between multiple beam modes in a digital manner. Our work provides a novel strategy for creating high-efficiency and broadband reconfigurable TA.

This work is organized as follows. Section II describes the theory of the TA, unit cell design, and array parameter optimization. Section III presents the implementation and measurements of the 10×10 element TA. Finally, Section IV concludes this work.

2. Transmitarray Design

2.1 Array Theory Analysis of TA Antenna

A planar square array of M × N transmitting elements forms the TA aperture. A feed antenna is placed on the z-axis at the point $z = -r_f$. To reconfigure the incident wave emitted by the source, tunable unit cells are used as phase shifters to form the required beams. Using array theory, the far-field radiation pattern of TA can be expressed as

$$E(\theta,\varphi) = \sum_{m=1}^{M}\sum_{n=1}^{N} \cos^{qe}(\theta) \frac{\cos^{qf}(\theta_f(m,n))}{\left|\vec{r}_{m,n} - \vec{r}_f\right|}$$

$$\cdot e^{-jk(|\vec{r}_{m,n}-\vec{r}_f|-\vec{r}_{m,n}\cdot\hat{u})}\left|T_{m,n}\right|e^{j\varphi_{m,n}} \quad (1)$$

$$\hat{u} = \hat{x}\sin\theta\cos\varphi + \hat{y}\sin\theta\sin\varphi + \hat{z}\cos\theta$$

where *qe* and *qf* are the unit cell and the feed pattern power factor, respectively, which are usually calculated by the full-wave simulation. $\theta_f(m,n)$ is the spherical angle in the feed's coordinate system, and $\vec{r}_f$ and $\vec{r}_{mn}$ are the position vector of feed antenna and the mn^{th} unit cell, respectively. k is the free space wavenumber. $|T_{m,n}|$ is the transmission magnitude of the mn^{th} element which is obtained from a unit-cell analysis, and $\varphi_{m,n}$ is the required phase shift of the mn^{th} element to set the required beam. To form a pencil-shaped beam in the direction ($\hat{u}_0$), the transmission phase of each unit cell should be set as

$$\varphi_{m,n} = k(\left|\vec{r}_{m,n} - \vec{r}_f\right| - \vec{r}_{m,n}\cdot\hat{u}_0) + \varphi_0 \quad (2)$$

where φ_0 is a constant phase, which is the optimized value for a 1-bit small-size TA antenna, indicating that a relative transmission phase rather than the absolute transmission phase is required for TA design. In this design, a 1-bit unit cell is adopted as the element, and the quantization rule is defined as

$$\varphi_{mn}^{q} = \begin{cases} 0°, & \text{if } -90° \leq \varphi_{mn} \leq 90° \\ 180°, & \text{other cases} \end{cases} \quad (3)$$

From (2) and (3), the acquired phase of each element may change as the constant phase φ_0 changes, resulting in different phase quantization errors. Therefore, it is necessary to select optimized values for the 1-bit TA radiation performance.

2.2 Reconfigurable Unit Cell Design

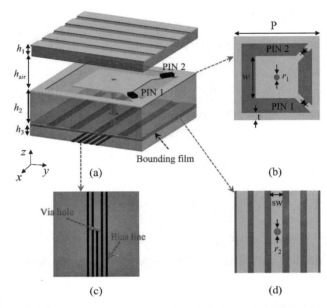

Fig. 1 Configuration of proposed structure. (a) Schematic. (b) Center patch layer. (c) DC bias line. (d) Upper grating layer. Dimensions are (in mm): p = 15, h_1 = 0.5, h_{air} = 2, h_2 = 1.5, h_3 = 0.2, t = 0.24, w = 8.7, sw = 2.4, r_1 = 0.5, and r_2 = 1.

In [24], a linear transmissive polarization conversion metasurface consisting of metallic resonators between a pair of orthogonally oriented subwavelength metal gratings was presented to realize wavefront shaping. By merely taking the mirror structure of the metallic resonator, one creates a new unit whose cross-polarized radiation has an additional phase shift. The PIN diode can be simply considered as a metal tab; switching between the ON and OFF, states is then just a matter of simulating the model with and without that piece of metal. Therefore, through the state switching of the two mirror-arranged diodes (ON/OFF or OFF/ON), an equivalent mirror unit can be produced, resulting in a phase difference.

Inspired by articles [23-25] on metasurfaces, the proposed 1-bit metasurface-based unit cell configuration is shown in Fig. 1, consisting of a reconfigurable anisotropic metallic resonator between a pair of orthogonally oriented subwavelength metal gratings. The unit cell periodicity is P = 15 mm ($\lambda_0/2$ at 10 GHz). The dielectric layers of substrates 1, 2, and 3 use the commonly utilized F4B substrate (ε_r=2.65 and tanδ=0.01). Substrate 2 and substrate 3 are bonded together by a 0.1 mm Rogers RO4450F (ε_r=3.52) bonding film. An intermediate air layer structure with a h_{air} thickness is used to load the PIN diodes in the middle patch. A rectangular patch and outer ring, which are linked by two mirror-arranged

A Novel Wideband and High-Efficiency Electronically Scanning Transmitarray Using Transmission Metasurface Polarizer

PIN diodes placed in reverse, act as the metallic resonator. The rectangular patch center is connected to a positive (State 0, PIN 1 is OFF and PIN 1 is ON) or negative (State 1, PIN 1 is ON and PIN 2 is OFF) voltage through a hole, while the outer ring is connected to the DC ground. The type of PIN diode is MACOM MADP-000907-14020 W, which is modeled as a series of lumped RLC elements: R = 7.8 Ω and L = 30 nH for the ON state (the bias voltage and current are 1.34 V and 10 mA, respectively); and C = 25 fF and L = 30 nH for the OFF state.

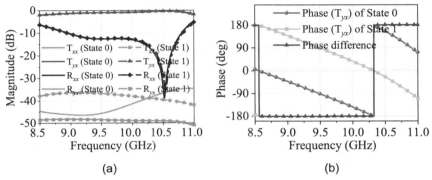

Fig. 2 Simulated performance of unit cell under normal incidence for two states: (a) magnitude and (b) phase.

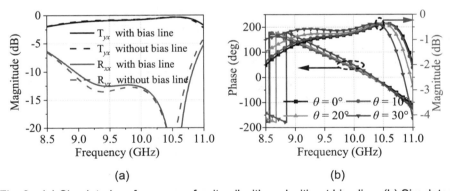

Fig. 3 (a) Simulated performance of unit cell with and without bias line. (b) Simulated transmission magnitude and phase of unit cell under oblique incidence on *xoz* plane.

The unit cell was analyzed by the full-wave simulation software CST Microware Studio® using the finite element method (FEM) with periodic boundary conditions. An *x*-polarized incident wave with -*z*-direction propagation is imping on the unit cell. The simulated performance under normal incident is shown in Fig. 2, which illustrates that the unit cell converts an *x*-polarized incident wave into a *y*-polarized transmitted wave with an insertion loss of less than -1 dB within 9.3-10.9 GHz for two states. The magnitude of the insertion loss at 10.0 GHz is only 0.5 dB regardless of the PIN states. The underlying reason for the enhanced polarization conversion is the interference between the multiple polarization couplings in the Fabry-Pérot-like cavity. The grating structures make the transmitted electromagnetic waves possess a good polarization isolation. The cross-polarization discrimination is > 30 dB for two states. By taking the mirror states of the PIN diodes to realize two unit cells with mirror structure properties, the cross-polarized radiation of the broadband converter has a 180° phase shift.

The DC bias line required by each array unit must exhibit minimal effects on its transmission properties. Therefore, the DC via hole is set at the center of the metallic resonator, which is the minimum electric field position, and the direction of the bias line is parallel to the bottom grating to reduce coupling. Fig. 3(a) illustrates the unit cell's simulated performance with and without 10 bias lines; the results show good isolation between RF and DC signals. This structure avoids designing complicated blocking components, such as DC capacitors, radial stubs, and meandering lines. Since not all unit cells in the TA receive a normal incident wave from the feed antenna, the unit cells arranged on the edge of the array are illuminated by a plane wave with oblique angles. Therefore, the performance of the unit cell under oblique incident should also be investigated. The simulated results are shown in Fig. 3(b), indicating that both the transmission magnitude and phase vary slightly when the incidence angle varies from 0 to 30°.

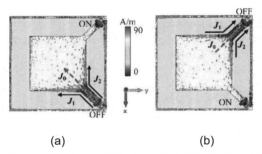

(a)   (b)

Fig. 4 Surface current on middle layer in (a) State 1 and (b) State 0.

To understand the physical mechanism of high-efficiency polarization conversion and phase difference in further detail, we investigated the surface current on the reconfigurable anisotropic metallic resonator as shown in Fig. 4. Strong surface currents (J_1, J_2) are mainly located on one side of the OFF state and flow to or away from the edge of the center patch. According to the vector synthesis principle, two black arrows can be combined into a red arrow. In this case, the synthetic current (J_0) can be considered equivalent to an LC resonant circuit, which is similar to the electric dipole resonance generated by a metal cut-wire structure. For the anisotropic metallic resonator, the incident x-polarized waves induce currents along the ±45° direction, providing both x- and y-polarized components.

The x-direction grating allows the newly generated y-polarized component to pass through while blocking the x-polarized component for forward scattering. For backscattering, the y-direction grating reflects the y-polarized component and allows the x-polarized component to pass. Through carefully optimizing the substrate thickness, a constructive interference enhances the polarization conversion and a destructive interference of the x-polarized reflections largely reduces the reflection loss. This process continues due to a multi-reflection process within this multilayer structure. The overall result is that the x-polarized incident waves can be efficiently converted to their orthogonal polarization in transmission mode. In addition, from Fig. 4, we can observe that the surface current distributions in the two cases in the y-direction show opposite directions, indicating that a 180° phase difference can be generated. Most of the induced current is concentrated in the vicinity of the diode in the OFF state. Little currents flow through these diodes, which results in a low ohmic loss of the unit cell.

2.3 Realization of Reconfigurable TA Antenna

To compose the reconfigurable TA antenna, in our case, a 10 × 10 element configuration array is adopted. The array has dimensions of 150 mm × 150 mm, and 200 PIN diodes are used to control the individual element by bias line and FPGA. Using the TA array theory described in part A, the F/D and phase constant (φ_0) are optimized.

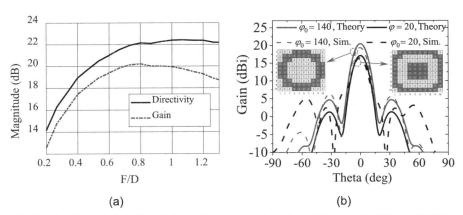

Fig. 5 (a) Calculated directivity and gain as a function of focal ratio F/D at 10 GHz. (b) Simulated and array theory calculated radiation patterns in E-plane with different phase offsets.

2.3.1 Feed Location Optimization

The F/D of the TA antenna can be optimized by choosing the feed location to balance the spillover efficiency and illumination efficiency. As the feed antenna is moved away from the aperture, the illumination efficiency increases because each element is equally illuminated. However, the spillover efficiency, which is the ratio between the feed energy and the array captured energy, decreases while F/D increases. In this work, a pyramidal horn antenna (Ainfoinc LB-90-10-A) with $qf = 5.5$, which is obtained by the full-wave simulation, is used as the feed antenna. Using the TA antenna array theory, the calculated directivity and gain as a function of F/D at 10 GHz are plotted in Fig. 5 (a). Therefore, the optimized focal length is selected at 100 mm, corresponding to an F/D of 0.67.

2.3.2 Phase Constant Optimization

The constant reference phase φ_0 in (2) provides additional design freedom to optimize the TA antenna. The optimal phase offset is determined by TA array theory. The simulated and calculated 2-D radiation patterns for the broadside beam in the E-plane with different phase offsets ($\varphi_0 = 140°$ and $20°$) are shown in Fig. 5 (b). The simulations are performed by the transient solver of the CST Microware Studio® based on the finite integration technique (FIT). There is an approximately 3 dB difference in the gain of the two different phase arrangements, and $\varphi_0 = 140°$ is the optimal value. At different beam scanning angles, there is also phase constant optimization.

3. Measured Results and Discussion

3.1 Fabrication of TA Antenna

For the purpose of realizing beam steering, a control circuit system is designed to control the states of each PIN diode independently. The TA antenna control circuit system, as shown in Fig. 6, includes personal computer (PC), stabilized voltage supply, FPGA board, and control circuit. The PC running the beam steering algorithm is used to generate a sequence for the required beams. The control data were first sent to an FPGA by a Registered Jack 45 (RJ45) connector. After obtaining valid information, the FPGA processes the data to regulate of the output port of the FPGA. In order to generate the DC current for the control of 100 bias lines, FPGA needs to provide 13 serial data outputs, 13 enable signal outputs, 2 reset control signal outputs, and a clock signal output. The front-end interface of the drive circuit is connected to the port of the FPGA, where the high and low levels can complete the logic control of the drive circuit, and the rear end interface can realize the voltage output control for the TA antenna. The control circuit is composed of 13 8-bit shift registers (SN74HC164), 13 8-bit latches (74HC573D), 26 monolithic industrial-CMOS analog switches (Analog Devices ADG1434, which comprise four independently selectable single-pole double-throw switches), and 100 protected resistance (the value is 330 Ω, which is selected to obtain ± 10 mA on each bias line with ± 5 V).

The shift registers convert data from serial to parallel formats and set the states of PIN diodes one by one. Furthermore, latches are utilized to retain the control data. Through 100 single pole double throw switches, +5 V or -5 V voltages provided by the stabilized power supply can be logically selected according to the control data. Connectivity between the control circuit system and the TA antenna is achieved using two 50-pin and a 3-m cables with adapters.

Based on the proposed design, a 10×10 TA prototype was fabricated using PCB technology, as shown in Fig. 7. The 2- mm middle air layer is supported by washers and nylon screws. At the same time, nylon support columns are used to support the array. The TA antenna includes 200 PIN di -field measurement system [see Fig. 7 (b)]. Since 100 PIN diodes work in the ON state with a biasing current of 10 mA, while it is negligible for the OFF state PIN diodes, the total power consumption is estimated to be 1.4 W.

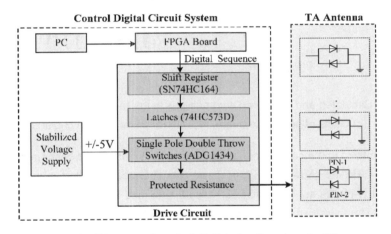

Fig. 6 Diagram of control digital circuit system for TA.

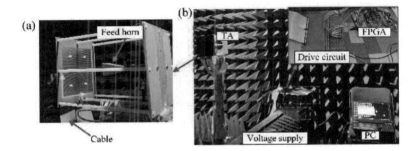

Fig. 7 Fabricated TA prototype. (a) System assembly.
(b) Measurement setup in anechoic chamber.

3.2 Broadside Radiation Performance and Difference-Beam

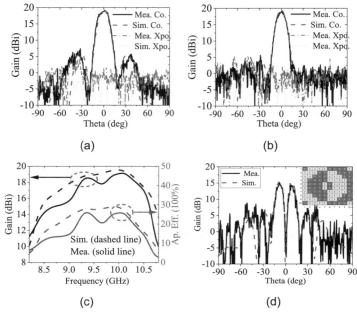

Fig. 8 Simulated and measured sum patterns in the (a) E-plane and (b) H-plane. (c) Gains of broadside beam. (d) Difference patterns in H-plane.

The simulated and measured radiation patterns of the broadside beam in both principal planes at 10 GHz are shown in Fig. 8(a) and (b). The measured half-power beamwidths in the E- and H-planes are 13.5° and 13.1°, with side-lobe levels of 12.4 dB and 13.5 dB, respectively. The measured relative cross-polarization level within the main lobe is below −18.1 dB. The measured maximum gain at 10.0 GHz is 19.1 dBi with an aperture efficiency of 25.8%, while the 3-dB gain bandwidth is from 9.0 to 10.55 GHz with a 15.9% fractional bandwidth as shown in Fig. 8(c). A detailed gain loss analysis is given in Table II. The ideal directivity is 25 dBi using following expression: $4\pi A/\lambda^2$, where A is the aperture area. The spillover is the part of the power from the horn antenna, which is not intercepted by the TA aperture, resulting in a 0.2 dB spillover loss. The amplitude and phase distribution on the TA aperture plane are nonuniform, resulting in a taper loss of 1.4 dB. The phase quantization loss of 2.1 dB is compared to the gain by assuming a continuous phase using TA array theory. Since the cross-polarization discriminations of the horn antenna are > 30 dB, the horn polarization loss is ignored. The measured reflection coefficient (Γ) is approximately -14 dB, and the horn mismatch loss can be calculated by $ML_{dB} = -10 \log_{10}(1-\Gamma 2) = 0.2$ dB. By comparing measured and simulated gains, the losses from the fabrication and measurement errors are 0.4 dB.

A Novel Wideband and High-Efficiency Electronically Scanning Transmitarray Using Transmission Metasurface Polarizer

Table I

Gain loss analysis of the broadside beam at 10.0 GHz

Aperture Size	150 mm × 150 mm
Ideal Directivity	25 dBi
Spillover Loss	0.2 dB
Taper Illumination Loss	1.4 dB
Phase Quantization Loss	2.1 dB
Horn Miss-Matching Loss	0.2 dB
Reflection Loss and Cross-Polarization Loss	0.8 dB
Element Insertion Loss	0.8 dB
Fabrication and Measurement Errors	0.4 dB
Measured Gain	19.1 dBi

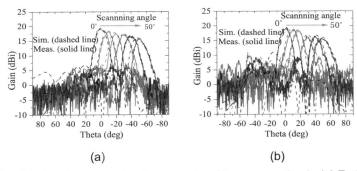

Fig. 9 Simulated and measured pattern results of beam scanning in (a) E-plane and (b) H-plane at 10 GHz.

Table II

Comparison between this work and previous research

Ref. No.	Freq. (GHz)	Array Size (λ×λ)	Max. Gain (dBi)/ Max. Ap. Eff.	3-dB Gain BW
[18]	13.5	8.6×8.6	21.4/15.4%	15.8%
[20]	12.5	5.3×5.3	17.0/14.0%	9.6%
[17]	9.8	9.8×9.8	22.7/15.4%	15.8%
[21]	25.0	4.8×4.8	17.5/20%	8.2%
This work	10.0	5×5	19.1/25.8%	15.9%

Because the unit cell was analyzed using periodic boundary conditions with a single state. In the TA array, the states of the surrounding elements may be different for switching multiple beam modes; periodic boundary conditions are broken, leading to an additional reflection loss. This loss term can be reduced by increase the array's size to approximate periodic boundary condition as much as possible. In addition, the performances of the edge unit cells also vary. They all cause TA's performance to deviate from its nominal values, resulting in higher reflection loss and higher cross-polarization levels. Thus, reflection loss and cross-polarization loss can be calculated by

$$\eta_r = \frac{P_{\text{total}} - P_{\text{cross}}^+ - P_{\text{co.}}^- - P_{\text{cross}}^-}{P_{\text{total}}} \quad (4)$$

where P_{total} is the total radiated power of the full space, $P_{\text{co.}}^+$ and P_{cross}^+ are the radiated power of co- and cross-polarization over the upper hemisphere, respectively, $P_{\text{co.}}^0$ and P_{cross}^0 are the radiated power of co- and cross-polarization over the lower hemisphere, respectively, which can be obtained by full-wave simulation. The reflection loss and cross-polarization loss are calculated to be 0.8 dB, and the element insertion loss is estimated to be 0.8 dB, which is the average value of multiple angle incidence.

In addition, the difference-beam patterns can be dynamically formed in either the E-plane or H-plane by controlling the bias voltages. For example, the dual-beam in the H-plane is obtained by setting the bias voltages of unit cells in the left half to be fully reversed. The simulated and measured dual-beam patterns at 10 GHz are plotted in Fig. 8(d).

3.3 Beaming Scanning

Generally, reconfigurable TA antennas receive incoming waves fed by a horn antenna to generate the required beams. To form the high-gain pencil beams of the desired direction, the PIN diode states should follow the arrangement calculated by (2) and (3). The phase constant of different scanning angles is also optimized by array theory to realize the maximum gain. The simulated and measured pattern results of beam scanning are illustrated in Fig. 9. The main lobe angles are in good agreement with the desired scanning angles of 0 to 50° in both the E- and H-planes. The measured maximum scan gain losses of the E- and H-planes are 2.8 and 3.4 dB, respectively, which demonstrates the good wide-angle beam scanning capabilities of the proposed TA antenna. Since the TA array is symmetrical in both the E- and H-planes, a steering range of ±50° can be obtained by the mirror phase arrangement. A comparison between this work and previous research of 1- bit TA antenna is shown in Table III. It can be concluded that this TA has the advantages of high efficiency and wide gain bandwidth compared to the published designs. The TA aperture efficiency reduction in [17-21] is due to the inherent drawback of the receive-and-radiate method, in which most current passing through the PIN diode causes a large ohmic loss.

4. Conclusion

A metasurface-inspired, electronically reconfigurable TA operating in the X-band with 2-D beam steering capability was proposed. The unit cell's transmission magnitude is enhanced with a low insertion loss of 0.5 dB by multiple interferences. A TA prototype consisting of 100 elements was experimentally validated. The beam reconfigurability is provided by switching the states of diodes employing a digital control circuit system. The experimental results have demonstrated its 2-D beam steering capability in both the E- and H-planes. Since its structure is simple, light-weight, low-cost, easy to process, and portable, this antenna with 2-D beam scanning capability is especially suitable for implementing high mobility applications, e.g., aircraft and unmanned aerial vehicles (UAV).

Acknowledgments

Manuscript received March 29, 2021. This work was supported by the National Natural Science Foundation of China (61701448 and 62071436) and the Fundamental Research Funds for the Central Universities (CUC200D053).

Author Information

H. Yu, J. Su and Z. Li are with the State Key Laboratory of Media Convergence and Communication, Communication University of China, Beijing 100024, China (e-mail: hangyu@cuc.edu.cn, sujianxun_jlgx@163.com, zrli@cuc.edu.cn).

F. Yang is with the Department of Electronic Engineering, Tsinghua University, Beijing 100084, China (e-mail: fanyang@tsinghua.edu.cn).

References

[1] SOLEIMANI M, ELLIOTT R C, KRZYMIEŃ W A, et al. Hybrid beamforming for mmWave massive MIMO systems employing DFT-assisted user clustering[J]. IEEE Trans. Veh. Technol., 2020, 69(10): 11646-11658.

[2] LI Y R, GAO B, ZHANG X, et al. Beam management millimeter wave communications for 5G and beyond[J]. IEEE Access, 2020, 8: 13282-13293.

[3] MAILLOUX R. Phased array antenna handbook[M]. Norwood: Artech House, 2005.

[4] KOPP B A, BORKOWSKI M, JERINIC G. Transmit/receive modules[J]. IEEE Trans. Microw. Theory Tech., 2002, 50(3): 827-834.

[5] MIURA A, et al. S-band active phased array antenna with analog phase shifters using double-

balanced mixers for mobile SATCOM vehicles[J]. IEEE Trans. Antennas Propag., 2005, 53(8): 2533-2541.

[6] HUM S V, PERRUISSEAU-CARRIER J. Reconfigurable reflectarrays and array lenses for dynamic antenna beam control: a review[J]. IEEE Trans. Antennas Propag., 2014, 62(1): 183–198.

[7] HAN J, LI L, LIU G, et al. A wideband 1 bit 12 × 12 reconfigurable beam-scanning reflectarray: design, fabrication, and measurement[J]. IEEE Antennas Wirel. Propag. Lett., 2019, 18(6): 1268-1272.

[8] KONG G, LI X, WANG Q, et al. A wideband reconfigurable dual-branch helical reflectarray antenna for high-power microwave applications[J]. IEEE Trans. Antennas Propag., 2021, 69(2): 825-833.

[9] XU H, XU S, YANG F etl al. Design and experiment of a dual-band 1 bit reconfigurable reflectarray antenna with independent large-angle beam scanning capability[J]. IEEE Antennas Wirel. Propag. Lett., 2020, 19(11): 1896-1900.

[10] ZHANG M, et al. Design of novel reconfigurable reflectarrays with single-bit phase resolution for Ku-band satellite antenna applications[J]. IEEE Trans. Antennas Propag., 2016, 64(5): 1634-1641.

[11] REIS J R, VALA M, CALDEIRINHA R F S. Review paper on transmitarray antennas[J]. IEEE Access, 2019, 7: 94171-94188.

[12] PERRUISSEAU-CARRIER J, SKRIVERVIK A K. Monolithic MEMS-based reflectarray cell digitally reconfigurable over a 360 phase range[J]. IEEE Antennas Wireless Propag. Lett., 2008, 7: 138–141.

[13] MOGHADAS H, DANESHMAND M, MOUSAVI P. MEMS-tunable half phase gradient partially reflective surface for beam-shaping[J]. IEEE Trans. Antennas Propag., 2015, 63(1): 369–373.

[14] LAU J Y, HUM S V. Reconfigurable transmitarray design approaches for beamforming applications[J]. IEEE Trans. Antennas Propag., 2012, 60(12): 5679–5689.

[15] PAN W, et al. A beam steering horn antenna using active frequency selective surface[J]. IEEE Trans. Antennas Propag., 2013, 61(12): 6218–6223.

[16] REIS J R, CALDEIRINHA R F S, HAMMOUDEH A, et al. Electronically reconfigurable FSS-inspired transmitarray for 2-D beamsteering[J]. IEEE Trans. Antennas Propag., 2017, 65(9): 4880-4885.

[17] CLEMENTE A, DUSSOPT L, SAULEAU R, et al. Wideband 400-element electronically reconfigurable transmitarray in X-band[J]. IEEE Trans. Antennas Propag., 2013, 61(10): 5017–5027.

[18] WANG Y, XU S, YANG F, et al. A novel 1 bit wide-angle beam scanning reconfigurable transmitarray antenna using an equivalent magnetic dipole element[J]. IEEE Trans. Antennas Propag., 2020, 68(7): 5691-5695.

[19] WANG Y, XU S, YANG F, et al. 1 bit dual-linear polarized reconfigurable transmitarray antenna using asymmetric dipole elements with parasitic bypass dipoles[J]. IEEE Trans. Antennas Propag., 2021, 69(2): 1188-1192.

[20] WANG M, XU S, YANG F, et al. Design and measurement of a 1-bit reconfigurable transmitarray with subwavelength H-shaped coupling slot elements[J]. IEEE Trans. Antennas Propag., 2019, 67(5): 3500-3504.

[21] VILENSKIY A R, MAKURIN M N, LEE C, et al. Reconfigurable transmitarray with near-field coupling to gap waveguide array antenna for efficient 2-D beam steering[J]. IEEE Trans. Antennas Propag., 2020, 68(12): 7854-7865.

[22] KAMODA H, IWASAKI T, TSUMOCHI J, et al. 60-GHz electronically reconfigurable large reflectarray using single-bit phase shifters[J]. IEEE Trans. Antennas Propag., 2011, 59(7): 2524-2531.

[23] AZAD A K, EFIMOV A V, GHOSH S, et al. Ultra-thin metasurface microwave flat lens for broadband applications[J]. Appl. Phys. Lett. 2017, 110(22): 4101.

[24] GRADY N K, HEYES J E, CHOWDHURY D R, et al. Terahertz metamaterials for linear polarization conversion and anomalous refraction[J]. Science, 2013, 340(6138): 1304–1307.

[25] GE Y, LIN C, LIU Y. Broadband folded transmitarray antenna based on an ultrathin transmission polarizer[J]. IEEE Trans. Antennas Propag., 2018, 66(11): 5974-5981.

[26] ZHANG M, et al. Design of novel reconfigurable reflectarrays with single-bit phase resolution for Ku-band satellite antenna applications[J]. IEEE Trans. Antennas Propag., 2016, 64(5): 1634-1641.

[27] MONTORI S, CACCIAMANI F, TOMASSONI C, et al. Novel 1-bit elementary cell for reconfigurable reflectarray antennas[C]. Manchester, the UK: Proc. Eur. Microw. Conf. (EuMC'11), 2011.

[28] ABDELRAHMAN A H, YANG F, ELSHERBENI A Z, et al. Analysis and design of transmitarray antennas[M]. San Rafael, USA: M&C Publishers, 2017.

[29] RUTSCHLIN M, SOKOL V. Reconfigurable antenna simulation: design of reconfigurable antennas with electromagnetic simulation[J]. IEEE Microw. Mag., 2013, 14(7): 92-101.

[30] CHEN H T, TAYLOR A J, YU N. A review of metasurfaces: physics and applications[J]. Rep. Prog. Phys., 2016, 79(7): 6401.

[31] https://www.ainfoinc.cn/amfilerating/file/download/file_id/1485/

[32] BALANIS C A. Antenna Theory: Analysis and Design[M]. Hoboken: John Wiley, 2016.

Design of a Dielectric Dartboard Surface for RCS Reduction*

1. Introduction

Radars cross section (RCS) is a critical characteristic of defense targets, therefore, low RCS is necessary for stealthy military platforms. Besides the conventional approaches for reducing RCS include circuit loading, shaping, and absorbing material, various types of artificial periodic structures and metasurfaces have been proposed for the application of RCS reduction.

In recent years, the chessboard surface has received widespread attention. Based on a combination of the artificial magnetic conductor (AMC) and perfect electric conductor (PEC) cells in a chessboard like configuration, RCS reduction in a narrow band is achieved in [1]. Several designs realized with the chessboard surfaces have been presented to expand the bandwidth of RCS reduction. The basic principle of chessboard surface is phase cancellation, namely, the reflection phase difference between two types of elements meets a necessary condition of $180 \pm 37°$ for achieving a 10 dB RCS reduction. Although the chessboard surfaces have good monostatic RCS reduction performance by using a simple and regular array of two elements, the bistatic scatterings in certain directions are increased because the total scattered energy almost remains the same for a low loss structure. Four peak scatterings are often found at the four corners of a chessboard structure.

In order to reduce the bistatic scattering in certain directions, the concept of coding metasurface is proposed and many designs have been presented. The surfaces can be designed with varied types of metal-substrate-ground structures, or realized by a single dielectric-ground structure. The FR-4 dielectric of different heights are used in [12], dielectric cubes of different sizes are utilized in [13], and drilled square holes of multiple sizes in dielectric elements are applied in [14]. In such type of design, multi-elements with different reflection phases are applied and irregularly arranged in chessboard to realize monostatic RCS reduction in wideband, meanwhile, the peak scatterings at the four corners are mitigated. Some designs assert that the used elements are randomly arranged, but actually, the position for elements with certain reflection phases must be placed in unique positions in the array, and sophisticated computer optimization algorithms are indispensably

---

* The paper was originally published in *IEEE Antennas and Wireless Propagation Letters*, 2022, 21 (2), and has since been revised with new information. It was co-authored by Shihao Qiu, Qingxin Guo, Jianxun Su, and Zengrui Li.

Design of a Dielectric Dartboard Surface for RCS Reduction

employed to obtain more evenly distributed scattering pattern in wideband. The design process is complicated, and the optimization time is lengthy. For the dielectric-ground structures realized with irregular arrangement, varied sizes or thicknesses of dielectric elements, the manufacturing process is also complicated.

In this letter, we propose a metasurface realized with two different dielectrics, which are arranged as a dartboard configuration. Monostatic RCS reduction in wideband and evenly distributed bistatic scattering pattern are achieved without complicated optimization algorithm. After the introduction, the design principle and simulation of the dartboard surface are shown in Section II, the prototype and measurements are shown in Section III. Finally, the conclusion is drawn in Section IV.

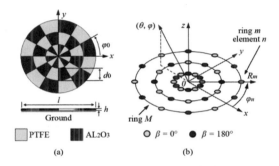

Fig. 1 The dartboard surface and concentric ring array. (a) Structure.
(b) Diagrammatic sketch of concentric ring array. (d_0 = 50 mm, φ_0 = 30°,
M = 3, N = 12, l = 300 mm and h = 6.8 mm).

2. Design Principle and Simulation

2.1 Design of Dielectric Elements

The structure of the dartboard surface is shown in Fig. 1(a), which consists of two types of elements realized by using PTFE and AL2O3, respectively. The whole structure contains M rings, and each ring has N elements. In this work, M = 3 and N =12. The spacing between two adjacent rings is d0, and the angular interval between elements in a same ring is φ0. The permittivity and the centertop loss tangent of PTFE are 2.2 and 0.001, while for AL2O3, they are 9 and 0.004. The loss of dielectrics does not affect the results of RCS reduction because the loss tangent value of the used dielectrics is small. A grounded metallic plane is attached on one side of the dielectrics to reflect the incident wave.

The configuration can be regarded as a concentric ring array (CRA) shown in Fig. 1(b). The array factor can be expressed by [17], [18].

$$AF(\theta,\varphi) = \sum_{m=1}^{M} w_m \sum_{n=1}^{N} e^{j[kR_m \sin(\theta)\cos(\varphi-\varphi_n)+\beta_{mn}]} \qquad (1)$$

where k is wave number, $R_m = (m-0.5) \times d_0$ is the radius of the mth loop, $\varphi_n = (n-0.5) \times \varphi_0$ is the angle of the nth element and βmn is the phase difference between two elements. Note that the w_m is the weight, which is related to the size of the elements and can be formulated with

$$w_m = \frac{1}{N} \cdot \frac{R_m^2 - R_{m-1}^2}{R_M^2} \qquad (2)$$

The RCS reduction is achieved by choosing β_{mn}, which is related to the permittivities and thicknesses. According to the previous studies, the phase difference of $180 \pm 37°$ between elements is a necessary condition to achieve 10 dB RCS reduction. The next is to choose a same thickness for two dielectrics to generate the necessary phase difference.

Fig. 2(a) shows the scenario of the plane wave transmitting to a grounded dielectric slab. In the figure, θ_0 and θ_1 are the incidence and refraction angle, which meet the Snell's law. The total reflection coefficient can be given by [19].

$$\Gamma = \frac{(Z_1 - Z_0) - (Z_1 + Z_0)e^{-j2k_1 \cos\theta_1 h}}{(Z_1 + Z_0) - (Z_1 - Z_0)e^{-j2k_1 \cos\theta_1 h}} \qquad (3)$$

where Z_0 and Z_1 are the impedances of the free space and the dielectric slab, k_1 is the wave number in the dielectric and h is the thickness of the dielectric. Suppose the dielectric is non-magnetic ($\mu_1 = 1$), we have

$$Z_1 = Z_0 / \sqrt{\varepsilon_1} \qquad (4)$$

Under the normal incidence, namely, $\theta_1 = \theta_i = 0°$, the phase of (3) can be simplified as

$$\phi = 2\arctan[\sqrt{\varepsilon_1} \cot(k_1 h)] \qquad (5)$$

Give a thickness and two permittivities, the phase difference at certain frequency can be obtained. Fig. 2(b) shows the reflection phase and the phase difference of two types of dielectrics with same thickness of $h = 6.8$ mm. At 7.5 GHz, the perfect phase cancellation condition of 180° is reached, namely, β_{mn} in (1) is equal to 180°. Within 4.1-10.8 GHz, the necessary condition ($180 \pm 37°$) is satisfied, which implies that a 10 dB RCS reduction can be achieved in this band.

Under the oblique incidences, the Z_0 and Z_1 in (3) are different for two polarizations. For the parallel polarization,

$$Z_0^{\|} = \eta_0 \cos\theta_i \qquad (6)$$

$$Z_1^{\|} = \eta_1 \cos\theta_1 \qquad (7)$$

where $\eta_0 = \sqrt{\mu_0/\varepsilon_0}$ and $\eta_1 = \sqrt{\mu_1/\varepsilon_1}$ are the intrinsic impedance of the free space and the dielectric. While for the perpendicular polarization,

$$Z_0^\perp = \frac{\eta_0}{\cos\theta_i} \tag{8}$$

$$Z_1^\perp = \frac{\eta_1}{\cos\theta_1} \tag{9}$$

The reflection phase of the reflection coefficient will be different for two polarizations, because the equivalent impedances of the dielectric and the free space are discrepant. Therefore, the performance of RCS reduction will be different for polarizations, too.

2.2 Performance of the Dartboard Surface

The simulated RCS pattern of the dartboard surface at 4.75 GHz, 7.5 GHz and 10.75 GHz under the normal incidence are shown in Fig. 4. It can be seen that the dartboard surface has approximate omnidirectional RCS reduction performance in these frequencies. The number of reflected beams is far more than 4 beams, which is a common pattern of the chessboard surfaces. The evenly distributed reflected beams in all directions are at least 10 dB lower than the main beam reflected by a same size of PEC, although the flatness gradually weakens as the frequency increases. At 4.75 GHz, RCS changes relatively smooth as θ increases from 0 to 90°; at 7.5 GHz, RCS decreases significantly after θ exceeds 50°; while at 10.75 GHz, RCS decreases rapidly after θ exceeds 30°. The reason for the performance degradation caused by the increase of frequency is that the sizes of the dielectric elements are different with regard to the wavelength.

Fig. 4 shows the simulated RCS patterns of the dartboard surface and PEC at 7.5 GHz under the oblique incidences of 30°, 45° and 60°. Similar to the results shown in Fig. 3, the reflected beams distributed in mirror directions of the incidence have at least 10 dB reduction, compare to the main beam reflected by the same size of PEC which shows on the bottom.

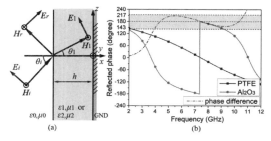

Fig. 2 Calculation of the reflected phase. (a) The schematic diagram. (parallel polarization showed only). (b) The reflection phase of PTFE and Al$_2$O$_3$ dielectric under normal incidence. (θ_i = 0°, h = 6.8mm).

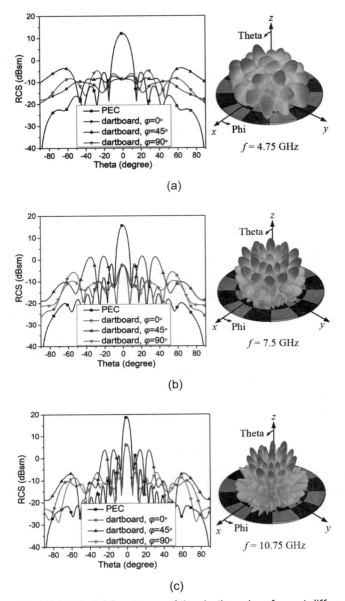

Fig. 3 The simulated bistatic RCS patterns of the dartboard surface at different frequencies. (a) 4.75 GHz. (b) 7.5 GHz. (c) 10.75 GHz.

The simulated frequency domain RCS reduction results under the normal and the oblique incidences are shown in Fig. 5. In the case of parallel polarization, when $\theta = 0°$, the 10 dB RCS reduction frequency bandwidth is 4.5-11 GHz (84%), which is in good agreement with the calculated reflected phase shown in Fig. 2. As the incident angle increases, the operation bandwidth narrows a little bit. When $\theta = 45°$, the 10 dB RCS reduction frequency bandwidth is 4.3-10.2 GHz (81%). For the perpendicular polarization, the performance of RCS reduction is obviously degraded if the incident angle is greater than 45°. According to equations (6)-(9), the influence of the incident angle on the phase difference is not the same, the calculation results are shown in Fig. 6. For parallel polarization, the phase differences mostly limit within $180 \pm 37°$ if $\theta \leq 45°$, although they increase in the high frequency and change slightly in the low frequency. While for the perpendicular polarization, the phase differences exceed $180 \pm 37°$ if $\theta \geq 30°$. Consequently, the performance under the parallel polarization maintain stability, while that of the perpendicular polarization is decreased.

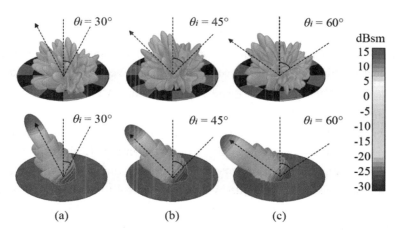

Fig. 4 The RCS patterns of the dartboard surface and PEC. (f = 7.5 GHz, parallel polarization). (a) θ_i = 30°. (b) θ_i = 45°. (c) θ_i = 60°.

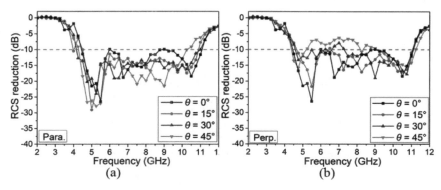

Fig. 5 The simulated RCS reduction results of the dartboard surface under normal and oblique incidence. (a) Parallel polarization. (b) Perpendicular polarization.

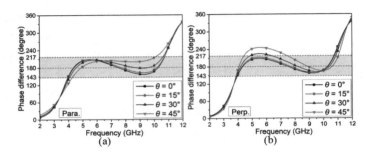

Fig. 6 Calculation of the phase difference under oblique incidence. (a) Parallel polarization. (b) Perpendicular polarization.

3. Measurement and Discussion

To verify this design, a prototype was fabricated with a diameter of 300 mm, as shown in Fig. 7. The dielectric elements are processed by computerized numerical control machine, the raw materials are 99% purity PTFE and AL_2O_3 ceramic. All the dielectric elements are stuck to a 2 mm thickness aluminum circular plate.

The schematic diagram of the measurement step is shown in Fig. 8. The measurement system includes a transmitting antenna (TX), a receiving antenna (RX) and a VNA (Agilent 5071C). For increasing the measurement accuracy, time-gating techniques are applied. In Fig. 8(a), two antennas are placed side-by-side and vertical to the targets to measure the monostatic RCS under the normal incidence. The monostatic RCS reduction is obtained by calculating the difference of RCS between the presented metasurface and a same size of metallic plate. Move the transmitting antenna and the receiving antenna to the mirror directions, as shown in Fig. 8(b), the RCS reduction under the oblique incidence can be measured. The bistatic RCS can be obtained by fixing the transmitting antenna and moving the receiving antenna, just as shown in Fig. 8(c). Fig. 8(d) is the photograph of the measurement environment.

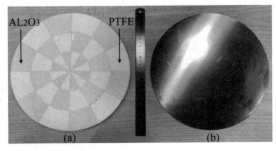

Fig. 7 Photograph of the test samples. (a) Dartboard. (b) Metallic plate.

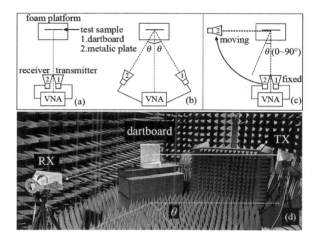

Fig. 8 Illustration of the measurement setup. (a) Monostatic RCS under normal incidence. (b) RCS reduction under oblique incidence. (c) Bistatic RCS. (d) Photograph of the measurement environment.

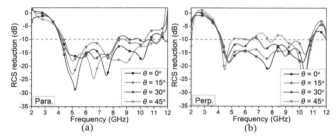

Fig. 9 The measured RCS reduction results of the dartboard surface prototype under normal and oblique incidence. (a) Parallel polarization. (b) Perpendicular polarization.

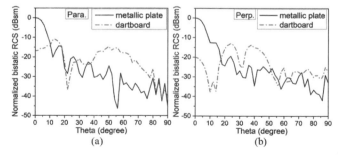

Fig. 10 The measured and normalized bistatic RCS of the prototype and metallic plate. ($\varphi = 0°$, $f = 7.5$ GHz). (a) Parallel polarization. (b) Perpendicular polarization.

The measured frequency domain RCS reduction results are shown in Fig. 9. When $\theta = 0°$, the 10 dB RCS reduction frequency range is 4.3-10.8 GHz (86%). The measured results are reasonably in agreement with the simulated results, meanwhile, deviation between them is observed. The reasons include the simulation and measurement errors, environmental noise, post processing to the extract measurement results. Another possible reason is that the space of 3 m between the antennas and the dartboard surface does not strictly meet the far-field condition, especially at high frequency. The placement accuracy of the prototype and metallic plate for comparison also result in the deviation.

Table I

Performance comparison

Ref.	Array	Elements	FBW	Bistatic Peaks	Optimization Method
[2]	Chessboard	AMC+PEC	42%	Y	Moderate
[3]	Chessboard	AMC+PEC	60%	Y	Moderate
[4]	Chessboard	AMC+PEC	83%	Y	Moderate
[5]	Irregular	Multi-elements	54%	N	Complicate
[6]	Irregular	Multi-elements	53%	N	Complicate
[11]	Irregular	Dielectric+PEC	83%	N	Complicate
[12]	Irregular	Dielectric	87%	N	Complicate
This work	Dartboard	Dielectric	84%	N	Moderate

The normalized bistatic scatterings for both polarizations are shown in Fig. 10, which exhibits the bistatic RCS characteristics of the dartboard surface. Compared to the metallic plate, the dartboard surface makes the reflected energy more evenly distributed along the θ angle, which is at least 10 dB lower than the main beam reflected by the metallic plate.

Table I compares the performances between some designs and this work, all of which realize monostatic RCS reduction. Wide fractional bandwidth (FBW) of 10 dB monostatic RCS reduction can be achieved by using three types of metasurfaces after meticulously designing and optimizing. Compared with the chessboard surfaces which have four scattering peaks appearing at the four corners of chessboard, the improvement of this design is that the bistatic scattering evenly distributed along φ direction. Compared with the irregular array realized with multi-elements, this design achieves wideband characteristic with two types of dielectrics without using complicate optimization method.

4. Conclusion

A novel and simple dartboard surface for RCS reduction is proposed in this letter. The condition of phase cancellation for RCS reduction is realized with a dart-shaped arrangement of two dielectrics elements on a metallic plate. Simulated and measured results show that the dartboard surface realizes an 84% fractional bandwidth of 10 dB monostatic RCS reduction and evenly distributed bistatic scattering pattern. Results also show that the performance of monostatic RCS reduction is stable under the oblique incidence for the parallel polarization.

Acknowledgments

This work was supported in part by the National Natural Science Foundation of China (NSFC) under Grant 62071436, and Grant 62071435; in part by the Fundamental Research Fund for the Central Universities under Grant CUC19ZD001; and in part by the Key Laboratory of All Optical Network and Advanced Telecommunication Network, Ministry of Education, Beijing Jiaotong University under Grant ZG19002.

Author Information

The authors are with the State Key Laboratory of Media Convergence and Communication, Communication University of China, Beijing 100024, China (e-mail: qxguo@cuc.edu.cn; qiush79@126.com; sujianxun_jlgx@163.com; zrli@cuc.edu.cn;).

References

[1] PAQUAY M, IRIARTE J, EDERRA I, et al. Thin AMC structure for radar cross-section reduction [J]. IEEE Trans. Antennas Propag., 2007, 55(12): 3630-3638.

[2] IRIARTE GALARREGUI J C, TELLECHEA PEREDA A, DE FALCÓN J L M, et al. Broadband radar cross-section reduction using AMC technology[J]. IEEE Trans. Antennas Propag., 2013, 61(12): 6136-6143.

[3] CHEN W, BALANIS C A, BIRTCHER C R. Checkerboard EBG surfaces for wideband radar cross section reduction[J]. IEEE Trans. Antennas Propag, 2015, 63(6): 2636-2645.

[4] MODI A Y, BALANIS C A, BIRTCHER C R, et al. Novel design of ultrabroadband radar cross section reduction surfaces using artificial magnetic conductors[J]. IEEE Trans. Antennas Propag., 2017, 65(10): 5406-5417.

[5] CUI T J, et al. Coding metamaterials, digital metamaterials and programmable metamaterials[J]. Light: Sci. Appl., 2014, 3(10): 218.

[6] SONG Y, DING J, GUO C, et al. Ultra-broadband backscatter radar cross section reduction based on polarization-Insensitive metasurface[J]. IEEE Antennas Wireless Propag. Lett., 2016, 15: 329-331.

[7] YANG P, YAN F, YANG F, et al. Microstrip phased-array in-band RCS reduction with a random element rotation technique[J]. IEEE Trans. Antennas Propag., 2016, 64(6): 2513-2518.

[8] YUAN F, WANG G, XU H, et al. Broadband RCS reduction based on spiral-coded metasurface[J]. IEEE Antennas Wireless Propag. Lett., 2017, 16: 3188-3191.

[9] LU Y, et al. Ultrawideband monostatic and bistatic RCS reductions for both copolarization and cross polarization based on polarization conversion and destructive interference[J]. IEEE Trans. Antennas Propag., 2019, 67(7): 4936-4941.

[10] YU H, SU J, GUO Q, et al. Dual wideband, polarization, angle insensitive diffusion electromagnetic surfaces for radar cross section reduction[J]. J. Phys. D: Appl. Phys., 2021, 54: 205102.

[11] HOU Y, LIAO W, KE J, et al. Broadband and broad-angle dielectric-loaded RCS reduction structures[J]. IEEE Trans. Antennas Propag., 2019, 67(5): 3334-3345.

[12] LIAO W, HOU Y, CHEN S. Dielectric-loaded ultrawideband RCS reduction structures[J]. IEEE Trans. Antennas Propag, 2020, 68(3): 2277-2289.

[13] SHAO L, ZHU W, LEONOV M Y, et al. Dielectric 2-bit coding metasurface for electromagnetic wave manipulation[J]. Appl. Phys., 2019, 125(20): 3101.

[14] SAIFULLAH Y, WAQAS A B, YANG G M, et al. Multi-bit dielectric coding metasurface for EM wave manipulation and anomalous reflection[J]. Opt. Express, 2020, 28(2): 1139-1149.

[15] GHAYEKHLOO A, AKBARI M, AFSAHI M, et al. Multifunctional transparent electromagnetic surface based on solar cell for backscattering reduction[J]. IEEE Trans. Antennas Propag., 2019, 67(6): 4302-4306.

[16] SHAO L, PREMARATNE M, ZHU W. Dual-functional coding metasurfaces made of anisotropic all-dielectric resonators[J]. IEEE Access, 2019, 7: 45716-45722.

[17] BALANIS C A. Antenna Theory: Analysis and Design[M]. 3rd ed. New York: Wiley, 2005.

[18] ZHAO X, YANG Q, ZHANG Y. A hybrid method for the optimal synthesis of 3-D patterns of sparse concentric ring arrays[J]. IEEE Trans. Antennas Propag., 2016, 64(2): 515-524.

[19] BALANIS C A, WILEY J. Advanced Engineering Electromagnetics[M]. New York: Wiley, 2012.

[20] SOLTANE A, ANDRIEU G, REINEIX A. Monostatic radar cross-section estimation of canonical targets in reverberating room using time-gating technique[C]. Amsterdam, Netherlands: 2018 International Symposium on Electromagnetic Compatibility (EMC Europe), 2018: 355-359.

[21] JARVIS RACHEL E, MATTINGLY RYLEE G, MCDANIEL JAY W. UHF-Band radar cross section measurements with single-antenna reflection coefficient results[J]. IEEE Trans. Instrum. Meas., 2021, 70: 1-4.

[22] JARVIS R E, METCALF J G, RUYLE J E, et al. Measurement and signal processing techniques for extracting highly accurate and wideband RCS[C]. Glasgow, the UK: Proc. IEEE Int. Instrum. Meas. Technol. Conf. 2021: 1-6.

Solving the Surface Current Distribution for Open PEC-Dielectric Objects Using the Volume Surface Integral Equation[*]

1. Introduction

The accurate analysis of current distribution is fairly necessary in many fields of electromagnetics, such as the designs of antennas, waveguides, radar absorbers and electromagnetic (EM) stealth, and the solution of inverse radiation problems. All the concerned near and far field quantities can be derived from the current distribution. Among abundant methods of numerical analysis, the method of moments (MoM) solution of appropriate integral equations is an attractive one. In various integral equations, for arbitrarily shaped perfect electric conductor (PEC)-inhomogeneous dielectric objects, the volume surface integral equation (VSIE) is widely used because of its generality. The traditional VSIE, formed by combining the electric field integral equation (EFIE) with the volume integral equation (VIE), can yield the summation of the surface currents on both sides of open infinitesimally thin PEC surfaces, and result in right scattered fields. However, if the surface currents on either side of open PEC surfaces are needed, the traditional VSIE is incompetent. For example, He et al. stated that if the surface current on either side of open PEC surfaces cannot be distinguished, the continuity condition of electric flux will not be able to be enforced on the PEC-dielectric interfaces to reduce the volume unknowns. In areas of EM shielding and EM interference, the fields inside are needed to estimate the shielding or interference. The tangential components of magnetic fields near the metal surfaces are the same as the surface currents on either side of surfaces. In [14], by combining the electric and magnetic field integral equations, the currents on both sides of a planar PEC surface were solved, while non-planar surfaces were not discussed.

For open PEC-dielectric objects, to distinguish the surface current on two sides of arbitrarily shaped open PEC surfaces, a novel type of VSIE is presented in this letter. In the proposed method, both the electric and magnetic field integral equations are modeled, which can give the summation and difference of the surface currents on both sides,

---

[*] The paper was originally published in *IEEE Antennas and Wireless Propagation Letters*, 2022, 21 (1), and has since been revised with new information. It was co-authored by Jinbo Liu, Jiming Song, Hui Zhang, and Zengrui Li.

respectively. Further, by solving the summation and difference of the surface current asynchronously, compared with the traditional VSIE, the proposed method can give the surface current distributions on open PEC surfaces with a very limited added CPU time and the same memory usage.

2. Formulations

As shown in Fig. 1, an arbitrarily shaped PEC surface S with thickness h is partly covered by inhomogeneous dielectric region V, illuminated by an EM plane wave $(\vec{E}^i, \vec{H}^i)$, radiating the scattered fields $(\vec{E}^s, \vec{H}^s)$. Ignored the current on the broadsides, the scattered fields are generated by the equivalent volume current $\vec{J}_V$ in V and surface currents $\vec{J}_{S_p}$ and $\vec{J}_{S_m}$ on Sp and Sm sides of S as

$$\begin{cases} \vec{E}^s(\vec{r}) = -j\omega \left[\vec{A}_{S_p}(\vec{r}) + \vec{A}_{S_m}(\vec{r}) + \vec{A}_V(\vec{r}) \right] \\ \qquad\qquad - \left[\nabla \varphi_{S_p}(\vec{r}) + \nabla \varphi_{S_m}(\vec{r}) + \nabla \varphi_V(\vec{r}) \right] \\ \vec{H}^s(\vec{r}) = \dfrac{1}{\mu_0} \nabla \times \left[\vec{A}_{S_p}(\vec{r}) + \vec{A}_{S_m}(\vec{r}) + \vec{A}_V(\vec{r}) \right] \end{cases} \quad (1)$$

The potentials are expressed as the convolutions of equivalent current or their divergences and the Green's function G as [8].

$$\begin{cases} \vec{A}_T(\vec{r}) = \mu_0 \int_T \vec{J}_T(\vec{r}')G dT' \\ \varphi_T(\vec{r}) = \dfrac{j}{\omega\varepsilon_0} \int_T \nabla' \cdot \vec{J}_T(\vec{r}')G dT' \end{cases} \quad T = S_p, S_m \text{ or } V \quad (2)$$

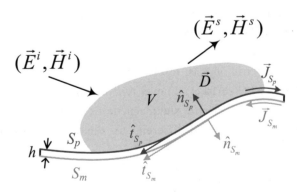

Fig. 1 Side view of an open PEC-dielectric object.

Solving the Surface Current Distribution for Open PEC-Dielectric Objects Using the Volume Surface Integral Equation

In the region V, according to the principle of volume equivalence, the VIE is written as

$$\frac{\vec{D}(\vec{r})}{\varepsilon(\vec{r})} - \vec{E}^s(\vec{r}) = \vec{E}^i(\vec{r}) \quad \vec{r} \in V \tag{3}$$

where $\vec{D}$ is the electric flux density in V, and ε is the $\vec{r}$-dependent permittivity. On the opposite sides Sp and Sm of the PEC surface S, by vanishing the tangential total electric field, the EFIE is formed as

$$\begin{cases} \hat{t}_{S_p}(\vec{r}) \cdot \vec{E}^i(\vec{r}) = -\hat{t}_{S_p}(\vec{r}) \cdot \vec{E}^s(\vec{r}) & \vec{r} \in S_p \\ \hat{t}_{S_m}(\vec{r}) \cdot \vec{E}^i(\vec{r}) = -\hat{t}_{S_m}(\vec{r}) \cdot \vec{E}^s(\vec{r}) & \vec{r} \in S_m \end{cases} \tag{4}$$

where $\hat{t}_{S_p}$ and $\hat{t}_{S_m}$ are any unit tangents of Sp and Sm, respectively. The EFIE can be combined with the VIE to form the EFIE-VIE, the commonly used VSIE type. Using Galerkin's MoM, $\vec{J}_S$ and $\vec{J}_V$ are expanded by the common RWG and SWG basis functions respectively, and the VSIE is converted into a generalized impedance matrix equation which is non-singular.

When h decreases to zero, Sp and Sm become coincident as well as $\hat{t}_{S_p} = \hat{t}_{S_m}$. Consequently, the two formulations in (4) are identical, in which situation the EFIE-VIE solution can only yield the summation of surface currents on the sides Sp and Sm as $\vec{J}_S = \vec{J}_{S_p} + \vec{J}_{S_m}$. To independently solve $\vec{J}_{S_p}$ and $\vec{J}_{S_m}$, the MFIE needs to be introduced. On both sides, according to the magnetic field boundary condition, the MFIE

$$\begin{cases} \vec{J}_{S_p}(\vec{r}) - \hat{n}_{S_p}(\vec{r}) \times \vec{H}^s(\vec{r}) = \hat{n}_{S_p}(\vec{r}) \times \vec{H}^i(\vec{r}) & \vec{r} \in S_p \\ \vec{J}_{S_m}(\vec{r}) - \hat{n}_{S_m}(\vec{r}) \times \vec{H}^s(\vec{r}) = \hat{n}_{S_m}(\vec{r}) \times \vec{H}^i(\vec{r}) & \vec{r} \in S_m \end{cases} \tag{5}$$

is established, where $\hat{n}_{S_p}$ and $\hat{n}_{S_m}$ are the outward unit normal of Sp and Sm, respectively. When h→0, Sp and Sm are very close, while the surface integrals about $\vec{J}_{S_p}$ over Sm and $\vec{J}_{S_m}$ over Sp will encounter the singularity problem. Substituting (1) and (2) into (5) and extracting the singularity result in

$$\begin{cases} \dfrac{1}{2} \vec{J}_D - \hat{n}_{S_p} \times \nabla \times \begin{bmatrix} PV \int_{S_p} \vec{J}_S G dS' \\ + \int_V \vec{J}_V G dV' \end{bmatrix} = \hat{n}_{S_p} \times \vec{H}^i & \vec{r} \in S_p \\ -\dfrac{1}{2} \vec{J}_D - \hat{n}_{S_m} \times \nabla \times \begin{bmatrix} PV \int_{S_m} \vec{J}_S G dS' \\ + \int_V \vec{J}_V G dV' \end{bmatrix} = \hat{n}_{S_m} \times \vec{H}^i & \vec{r} \in S_m \end{cases} \tag{6}$$

where $\vec{J}_D = \vec{J}_{S_p} - \vec{J}_{S_m}$, and PV denotes the principle value integral. The variable $\vec{r}$ is omitted for brevity. For the planar PEC surfaces, the second term including $\vec{J}_S$ of left-hand side will be eliminated. In (6), both the summation $\vec{J}_S = \vec{J}_{S_p} + \vec{J}_{S_m}$ and difference $\vec{J}_D = \vec{J}_{S_p} - \vec{J}_{S_m}$ appear. Because $\hat{n}_{S_p} = -\hat{n}_{S_m}$ in this situation, the difference between the two formulations in (6) is just a sign.

Even though h→0, we can still deem that Sp and Sm form a "closed" PEC surface. Then the CFIE

$$\text{CFIE} = \alpha\, \text{EFIE} + (1-\alpha)\eta_0\, \text{MFIE} \tag{7}$$

the linear combination of EFIE and MFIE, can be applied on the "closed" surface, where α is a real constant and 0<α<1, η0 is the intrinsic impedance of free space. Substituting (4) and (6) into (7), it is found that the CFIE is reduced to a summation and difference of the EFIE and MFIE on opposite sides Sp and Sm. Since the summation and difference of the two equations are linear independent, the CFIE can be combined with VIE to form the CFIE-VIE, another VSIE type, to determine the current on both sides of open PEC surfaces.

However, compared with the EFIE-VIE implementation which only yields the summation of surface current $\vec{J}_S$, because $\vec{J}_{S_p}$ and $\vec{J}_{S_m}$ are solved simultaneously, the number of surface unknowns for the CFIE-VIE is doubled, whereas the computational cost will increase exponentially. To overcome this drawback, we propose another solving scheme. Since the EFIE and MFIE are linearly independent, we can use them on one side but not both sides of PEC surfaces to solve the currents on both sides. We call this formulation as electric-magnetic integral equation (EMFIE)-VIE, which are described as follows:

1) Combining the VIE (3) with any one of (4) to form EFIE-VIE to solve $\vec{J}_V$ and $\vec{J}_S$. This step is actually the same as the traditional VSIE to calculate the open PEC-dielectric objects.

2) Substitute $\vec{J}_V$ and $\vec{J}_S$ into any one of (6) to calculate the difference of surface current $\vec{J}_D$ directly as

$$\vec{J}_D = 2\hat{n}_{S_x} \times \vec{H}^i + 2\hat{n}_{S_x} \times \nabla \times \begin{bmatrix} PV \int_{S_x} \vec{J}_S G dS' \\ + \int_V \vec{J}_V G dV' \end{bmatrix} \quad x = p \text{ or } m \tag{8}$$

It is worth pointing out that $\vec{J}_D$ can be solved by projecting it over a number of basis functions, which will lead to an extremely sparse matrix equation. Because each row or column of this matrix has a few nonzero entries, the computational complexity of solving the matrix equation is O(NS) order, where NS is the number of surface unknowns. Thus, the computational time of this step is very slight.

3) Calculate $\vec{J}_{S_p}$ and $\vec{J}_{S_m}$ on both sides of surface S by the summation $\vec{J}_S = \vec{J}_{S_p} + \vec{J}_{S_m}$ and difference $\vec{J}_D = \vec{J}_{S_p} - \vec{J}_{S_m}$ obtained from last two steps, respectively. The CPU time of this step is completely negligible.

It is seen that to reduce the computational cost, the proposed EMFIE-VIE solves the $\vec{J}_{S_p}$ and $\vec{J}_{S_m}$ asynchronously. For the PEC objects that only contain planar surfaces, $\vec{J}_V$ and the summation $\vec{J}_{S_p}$ does not appear in (8). Hence, the steps 1 and 2 are totally independent. On the contrary, if the volume regions or non-planar surfaces are included, the steps 1 and 2 must be executed in order. Moreover, since the second term of RHS in (8) contains high-order singularity, compared with the planar surface, it is more difficult to accurately compute the difference in the currents on the open non-planar surface. Because the steps 2 and 3 have very low computational complexity, the total cost on the EMFIE-VIE is at the same level as that of the traditional EFIE-VIE. Furthermore, after step1, the allocated memory for storing the impedance matrix generated from VIE-EFIE is released, part of which can be reallocated for the steps 2 and 3. Thus, the memory usage of the EMFIE-VIE is the same as that of the EFIE-VIE.

Particularly, at the border of S, there is current flowing over the edge. Different from the situation that the EFIE is applied on the surface S where only the tangential component of $\vec{J}_S$ exists at the edge, the EMFIE-VIE needs to consider the normal component of the current difference $\vec{J}_D$. As a consequence, the half-RWG basis function consisting of one single triangle is needed at the surface edge to solve the normal component of $\vec{J}_D$. Compared with the EFIE-VIE, the introduction of half-RWG basis functions will increase the surface knowns. Fortunately, at the edge, the normal component of the summation $\vec{J}_S$ is zero. Therefore, at the step 1 where the EFIE-VIE is formed to solve $\vec{J}_S$, the half-RWG functions can be eliminated. At the steps 2 and 3, the half-RWG functions need to be considered, whereas the evocable cost is distinctly slight.

3. Numerical Results

In this section, all computations serially run on a computer with 3.2 GHz CPU and 16 GB RAM in single precision. Restarted GMRES with the restart number 100 is used as the iterative solver to reach the target residual relative error of 0.001. For the EMFIE-VIE implementation, during the step 2, the GMRES is also used to iteratively solve the sparse matrix equation generated by the MFIE to yield the difference $\vec{J}_D$. For the comparison, the EM simulation software Altair FEKO based on the pure surface integral equation method is also shown as the baseline.

In the first case, a PEC semispherical shell of radius 1λ, coated with a 0.05λ thickness material whose relative permittivity is $\varepsilon r = 2.2-j0.00198$, is illuminated by an x-polarized EM plane wave from +z axis. After discretization, the numbers of triangles and tetrahedrons are 3,144 and 10,029, respectively. The bistatic radar cross section (RCS) at $\varphi = 0$ plane and surface current distribution are calculated using the EFIE-VIE, EMFIE-VIE and CFIE-VIE implementations, respectively, while for EMFIE-VIE, the number of half-RWG basis functions is 80. The RCS results from different implementations are shown in Fig. 2. Because the volume current $\vec{J}_V$ and the summation of surface current $\vec{J}_S$ solved by the EFIE-VIE and EMFIE-VIE are exactly the same, the numerical results from the two implementations are identical. Thus, for easy reading, the EFIE-VIE result is not drawn. The amplitude of surface current density at the axis line of the PEC semispherical shell which is in the $\varphi=0$ plane (the yellow line in the insert in Fig. 2) with respect to θ is shown in Fig. 3. It should be noted that since FEKO cannot distinguish the current density at two sides of an open PEC surface, the PEC semispherical shell simulated by FEKO has 0.01λ thickness, which is the main reason that cause the slight difference between two approaches as shown in Fig. 3. Solving by the EMFIE-VIE, the amplitude and direction at a certain moment of the surface current density on both sides of the PEC semispherical surface are shown in Fig. 4. Please note the color scales used by the two figures are different. It is observed that the surface currents on the two sides change periodically and show very different distribution. Meanwhile, there is a normal current flowing over the edge, expressed by the half-RWG basis functions. At the edge, the magnitude of the normal component on two sides are strictly equal. However, as required by current continuity, they are 180° out of phase. Thus, the normal component of the summation current is zero at the edge. In the EFIE-VIE implementation, since the half-RWG basis functions are ignored, this phenomenon cannot be shown. The computational details are given in Table I. It is found that although the number of unknowns from the EMFIE-VIE is the same as that from the CFIE-VIE, the CPU time and memory usage are much less, fully illustrating the efficiency of the proposed EMFIE-VIE implementation.

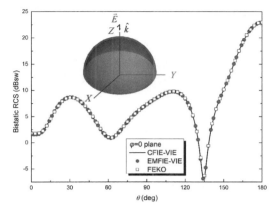

Fig. 2 Bistatic RCS of a PEC semispherical shell of radius 1λ coated with 0.05λ thick material at $\varphi = 0$ plane, whose relative permittivity is $\varepsilon_r = 2.2\text{-j}0.00198$.

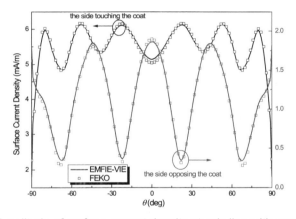

Fig. 3 Amplitude of surface current density at axis line with respect to θ.

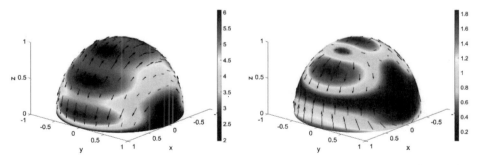

(a) the side touching the coat (top)   (b) the side opposing the coat (bottom)

Fig. 4 Amplitude and direction at a certain moment of the current density (unit: mA/λ) on both sides of the PEC semispherical shell, the length and direction of arrows denote the value and direction of the surface current.

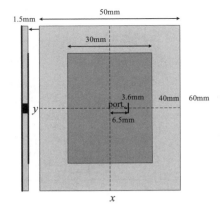

Fig. 5　Model of a linearly polarized rectangular microstrip patch antenna working at 3 GHz, relative permittivity of the dielectric substrate is εr = 2.55.

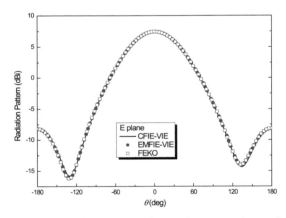

Fig. 6　Radiation pattern of the microstrip patch antenna in xz plane (E plane).

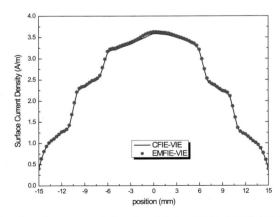

Fig. 7　Amplitude of surface current density at the midline parallel to x-axis on the side touching the substrate of the radiation patch.

Solving the Surface Current Distribution for Open PEC-Dielectric Objects Using the Volume Surface Integral Equation

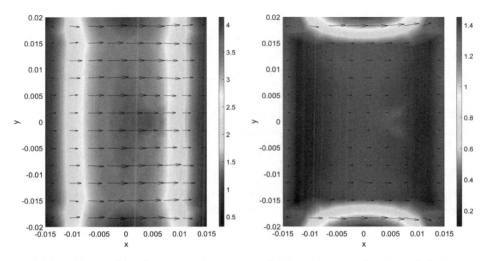

(a) the side touching the substrate (b) the side opposing the substrate

Fig. 8 Amplitude and direction at a certain moment of the surface current density (unit: A/m) on both sides of the radiation patch.

Table I

Computational details of different implementations

Object	Coated semispherical shell			Patch antenna		
Method	EFIE-VIE	EMFIE-VIE	CFIE-VIE	EFIE-VIE	EMFIE-VIE	CFIE-VIE
Triangles	3,144			1,440		
Tetrahedrons	10,029			3,100		
Unknowns	28,142	32,903		9,395	11,624	
CPU time (min)	10.2	11.3	17.8	1.8	2.1	3.7
Memory usage (MB)	6,063		8,286	681		1040

The second case is a linearly polarized rectangular microstrip patch antenna, whose structure is shown in Fig. 5. The relative permittivity of the dielectric substrate is $\varepsilon r = 2.55$, the radiation patch is fed by a narrow strip with a delta-function voltage source, and the working frequency is 3 GHz. A moderate mesh size is chosen to generate 1,440 triangles and 3,100 one-layer tetrahedrons, whereas the number of half-RWG basis functions using in the EMFIE-VIE is 138. Besides, when the MFIE is modeled on the PEC surfaces, the excitation at the port from the magnetic field is equal to zero as [24]. The computational details are also listed in Table I. The numerical radiation patterns at E plane are shown in Fig. 6, whereas the result from EFIE-VIE is exactly the same as the EMFIE-VIE and omitted. Computing by the CFIE-VIE and EMFIE-VIE, the amplitude of surface current density at the midline (the yellow line in Fig. 5) parallel to x-axis on the side touching the substrate of the radiation patch is shown in Fig. 7. Different results are overlapped everywhere, which illustrates that the solution of the EMFIE-VIE is almost the same as that of the CFIE-VIE. The amplitude of the surface current density and current direction at a certain moment on both sides of the radiation patch are shown in Fig. 8. Clearly, the current direction parallels the short side of the patch. Therefore, the E plane is parallel to the shorter edge, that is, the xz plane. The current on the side touching the dielectric substrate is much larger than that opposing the substrate. Except the regions nearby the two long edges, the currents on the two sides roughly have the same phase, whereas the edge currents are relatively small. Thus, the fields radiated by the two sides will be enhanced overlay but not cancelled out. Another observation is that the current nearby the feed port is somewhat larger than other regions and is not symmetrical left to right due to the location of the feeding port, which causes the slight asymmetry of the radiation pattern at E plane.

4. Conclusion

For open PEC-dielectric objects, how to yield the independent surface current distribution on both sides of open PEC surfaces has been investigated. Modeling both electric and magnetic field integral equations on PEC surfaces, a novel type of VSIE, called EMFIE-VIE, is proposed to achieve this goal. Owe to the asynchronous solution of surface current on either side, compared with the traditional EFIE-VIE, the proposed method has very limited extra CPU time and an equal memory usage. Due to the use of the EMFIE-VIE, the behavior of current distribution on the open PEC surfaces is observed for some numerical cases, which cannot be achieved by the traditional EFIE-VIE implementation.

References

[1] HE T, ZHANG X W, PAN W Y, et al. Current distribution and input impedance of a VLF linear antenna in an anisotropic plasma[J]. IEEE transactions on antennas and propagation, 2018, 67(3): 1519-1526.

[2] MENDIL S, AGUILI T. Analysis of electromagnetic scattering in waveguides by hybrid MOM-PO-GTD method[C]. Hammament, Tunisia: 2019 IEEE 19th Mediterranean Microwave Symposium (MMS). IEEE, 2019: 1-5.

[3] GUO Q, SU J, LI Z, et al. Miniaturized-element frequency-selective rasorber design using

characteristic modes analysis[J]. IEEE Transactions on Antennas and Propagation, 2020, 68(9): 6683-6694.

[4] GUO Q, LI Z, SU J, et al. Dual-polarization absorptive/transmissive frequency selective surface based on tripole elements[J]. IEEE Antennas and Wireless Propagation Letters, 2019, 18(5): 961-965.

[5] LI H, ZHOU J, ZHAO H, et al. Stealth treatment of turntable in ultra-wideband radio frequency simulation system[C]. Nanjing, China: 2010 IEEE International Conference on Ultra-Wideband. IEEE, 2010, 2: 1-4.

[6] ALVAREA Y, LAS-HERAS F, PINO M R. Reconstruction of equivalent currents distribution over arbitrary three-dimensional surfaces based on integral equation algorithms[J]. IEEE Transactions on Antennas and Propagation, 2007, 55(12): 3460-3468.

[7] HARRINGTON R F, HARRINGTON J L. Field computation by moment methods[M]. Oxford: Oxford University Press, 1996.

[8] CHEW W C, MICHIELSSEN E, SONG J M, et al. Fast and efficient algorithms in computational electromagnetics[M]. Norwood: Artech House, 2001.

[9] MORSEY J D, OKHMATOVSKI V I, CANGELLARIS A C. Finite-thickness conductor models for full-wave analysis of interconnects with a fast integral equation method[J]. IEEE transactions on advanced packaging, 2004, 27(1): 24-33.

[10] MARICARU M, CIRIC I R, HANTILA F I, et al. Fast and accurate analysis of thin shields with holes based on the current sheet integral equation[C]. Tel Aviv, Israel: 2011 IEEE International Conference on Microwaves, Communications, Antennas and Electronic Systems (COMCAS 2011). IEEE, 2011: 1-4.

[11] HE M, LIU J, WANG B, et al. On the use of continuity condition in the fast solution of volume-surface integral equation[J]. IEEE Antennas and Wireless Propagation Letters, 2016, 16: 625-628.

[12] ABELE M G. Generation and confinement of uniform magnetic fields with surface currents[J]. IEEE transactions on magnetics, 2005, 41(10): 4179-4181.

[13] DIAO Y L, SUN W N, LEUNG S W, et al. Prediction of magnetic field radiation using equivalent current distribution[C]. Dresden, Germany: 2015 IEEE International Symposium on Electromagnetic Compatibility (EMC). IEEE, 2015: 887-890.

[14] NEWMAN E, SCHROTE M. On the current distribution for open surfaces[J]. IEEE transactions on antennas and propagation, 1983, 31(3): 515-518.

[15] RAO S, WILTON D, GLISSON A. Electromagnetic scattering by surfaces of arbitrary shape[J]. IEEE transactions on antennas and propagation, 1982, 30(3): 409-418.

[16] SCHAUBERT D, WILTON D, GLISSON A. A tetrahedral modeling method for electromagnetic scattering by arbitrarily shaped inhomogeneous dielectric bodies[J]. IEEE transactions on antennas and propagation, 1984, 32(1): 77-85.

[17] HU F G, SONG J, YANG M. Errors in projection of plane waves using various basis functions[J]. IEEE antennas and propagation magazine, 2009, 51(2): 86-98.

[18] ERGUL O, GUREL L. Investigation of the inaccuracy of the MFIE discretized with the RWG basis functions[C]. Monterey, CA, USA: IEEE Antennas and Propagation Society Symposium, 2004. IEEE, 2004, 3: 3393-3396.

[19] GUREL L, ERGUL O. Singularity of the magnetic-field integral equation and its extraction[J]. IEEE antennas and wireless propagation letters, 2005, 4: 229-232.

[20] Hou Y, Xiao G, Tian X. A discontinuous Galerkin augmented electric field integral equation for multiscale electromagnetic scattering problems[J]. IEEE transactions on antennas and propagation, 2017, 65(7): 3615-3622.

[21] He Z, Li Y S, Zhao Y, et al. Uncertainty RCS computation for multiple and multilayer thin medium-coated conductors by an improved TDS approximation[J]. IEEE transactions on antennas and propagation, 2020, 68(12): 8053-8061.

[22] SAAD Y, SCHULTZ M H. GMRES: A generalized minimal residual algorithm for solving nonsymmetric linear systems[J]. SIAM journal on scientific and statistical computing, 1986, 7(3): 856-869.

[23] Altair, Troy. FEKO [EB/OL]. (2014-01-01) [2015-10-25]. https://www.altair.com/feko.

[24] LIU J, YUAN J, LUO W, et al. On the Use of Hybrid CFIE-EFIE for Objects Containing Closed-Open Surface Junctions[J]. IEEE antennas and wireless propagation letters, 2021, 20(7): 1249-1253.

Breaking the High-Frequency Limit and Bandwidth Expansion for Radar Cross Section Reduction*

1. Introduction

Wave interference is the superposition of two or more waves that result in a new wave pattern. Destructive interference occurs when the phase difference between the waves is an odd multiple of π. One of the potential applications in electromagnetics is to reduce the scattering field of a metal object. As a traditional destructive interference method, the opposite phase cancellation (OPC) produced by two unit cells with equal amplitude and opposite phase has been widely implemented to achieve radar cross section (RCS) reduction. The future development of low observable technology requires the target to have full-band, omnidirectional stealth performance to avoid detection by advanced radar systems. Bandwidth is the most important factor in stealth technology. In 2007, a thin chessboard structure combining perfect electric conductors (PECs) and artificial magnetic conductors (AMCs) was proposed to reduce RCS, but the bandwidth was found to be limited due to the narrow in-phase reflection bandwidth of the AMCs. Then, a 180° phase difference was obtained over a more than 40% fractional bandwidth (FBW) by combining two AMC cells in a following study. In addition, a hexagonal checkerboard surface combining two different electromagnetic bandgap (EBG) structures was displayed in [6], and the 10 dB RCS reduction bandwidth was 61%. In [7], a pixelated chessboard metasurface was designed over an ultrawide frequency band. The 95% bandwidth was obtained for 10 dB RCS reduction. In [8], a chessboard AMC surface based on the multilayer quasi-fractal structure was proposed for the wideband RCS reduction. The chessboard configuration was formed by two different cells with a 180°±37° phase difference over a more than 90% bandwidth. In [9] and [10], the equivalent circuit model was exploited to achieve an improved reflection phase difference between two unit cells. The checkerboard surfaces obtained bandwidths of 67% and 91.5%, respectively. In 2018, Ref. [11] proposed optimized multielement phase cancellation for achieving superwideband RCS reduction with a ratio bandwidth (f_H / f_L) of 5.87:1. In 2019, a new class of metasurfaces was proposed based on the phasor representation method for RCS

---

* The paper was originally published in *IEEE Antennas & Propagation Magazine*, 2021, 63 (6), and has since been revised with new information. It was co-authored by Jianxun Su, Hang Yu, Hongcheng Yin, Qingxin Guo, Zhihe Xiao, and Zengrui Li.

reduction. Furthermore, other EM surfaces have been presented for RCS reduction in previous studies. Another method based on polarization conversion is proposed for co-polarized RCS reduction by converting the co-polarized energy to the cross-polarized component. A unit cell of double arrow shape with 113% bandwidth and polarization conversion ratio (PCR) less than 60% was presented in [18]. However, the remaining co-polarized field component can lead to high RCS. The 10 dB RCS reduction requires the PCR to be higher than 90%. In [19], the proposed metasurface is composed of a square and L-shaped patches, and it can convert the polarization of the incident wave to its cross-polarized direction. A 98% fractional bandwidth was achieved for 10 dB RCS reduction. A linear-to-linear 2.5-dimensional polarization conversion metasurface (2.5D PCM) with a high polarization conversion ratio (PCR>96%) and a relative bandwidth of 99.5% was investigated in [20]. However, it is extremely challenging to design a polarization conversion element with a high PCR in an ultrawide frequency band (FBW >100%). In the above literature [4-10], [12-20], the working frequencies are within the so-called no grating lobe region such that only the dominant Floquet harmonic is propagating while the remaining harmonics are evanescent. In particular, Ref. [21] proposed a methodology to obtain wideband scattering diffusion based on the periodic artificial surface with a repetition period larger than one wavelength, which induces the excitation of multiple Floquet harmonics.

The main contribution of this paper was to increase the highest frequency of RCS reduction, which was extended to the band of the grating lobe appearance (the operating wavelength was less than the spatial periodicity) while ensuring the stability of the lowest frequency (the operating wavelength was greater than the spatial periodicity). The root cause of the bandwidth limitation for RCS reduction was discovered by analyzing the reflection phase of unit cells using the equivalent circuit model, and a novel strategy based on the AMC element with variable substrate thickness, multielement phase cancellation (MEPC) and the particle swarm optimization (PSO) algorithm was proposed for breaking through the bandwidth limits. The proposed 20-element AMC surface can achieve a 10 dB RCS reduction in an ultrawide frequency ranging from 5.7 to 40 GHz with a ratio bandwidth (f_H / f_L) of 7.02:1 under normal incidence for both polarizations.

The paper is structured as follows: The root cause of the bandwidth limit of RCS reduction was discovered and introduced in Section II. Section III describes a novel strategy based on AMC cells with a fixed or variable substrate thickness, using the MEPC and PSO algorithms for breaking the bandwidth limit and realizing multiband RCS reduction and ultrawideband RCS reduction. The simulations and measurements of the monostatic and bistatic scattering characteristics for both normal and oblique incidences are discussed in Section IV. Finally, this work is summarized in Section V.

2. The Bandwidth Limit and Its Solution

In this section, we analyzed the reflection phase characteristics based on the transmission line theory and discovered the bandwidth limit of the RCS reduction. To overcome the in-phase reflection problem at short-circuited frequencies, multiple substrate thicknesses are exploited to obtain the phase shift coverage.

2.1 Bandwidth Limit of RCS Reduction

As we know, the narrow phase changing range cannot satisfy the phase cancellation condition. The design of the AMC unit cell with a large range phase difference is a prerequisite for RCS reduction. The lowest frequency (f_L) and the highest frequency (f_H) of the RCS reduction limit are discussed. The bandwidth limit for RCS reduction is discussed by the analysis of unit cells based on the equivalent circuit model. Here, we proposed the configuration of the typical AMC to illustrate the phase limitation. The unit cell was placed on the *xy* plane. The incident wave was assumed to be propagating toward the -*z*-direction and was linearly polarized along the *x*-axis. The geometry of the unit cell and its corresponding equivalent circuit are shown in Fig. 1 and Fig. 2, respectively. The AMC unit cell with a periodicity of *P* was printed on a dielectric substrate backed by a conducting ground plane with a dielectric constant of ε_r and a thickness of *h*. The phase difference was obtained by changing the upper metallic structure. The equivalent circuit model for different periodic structures can be obtained from the analysis presented in [22]; additional transmission lines representing the dielectric layers need to be considered in retrieving the L and C parameters. Although the values of the lumped elements for those AMCs were different, the equivalent circuit models were the same. The circuit was composed of a series LC.

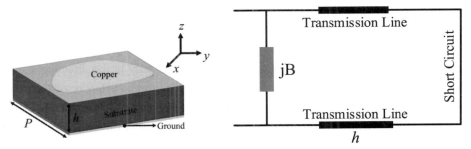

Fig. 1   Geometries of the AMC cell.   Fig. 2   Equivalent circuit model of the AMC cells.

The infinite metal ground was equivalent to the short-circuited terminal. Input admittance of the short-circuited transmission line is

$$Y_{in}(z) = -jY_1\cot(\beta_1 h) \qquad (1)$$

and

$$\beta_1 = \frac{2\pi}{\lambda_0/\sqrt{\varepsilon_r}} = \frac{2\pi f}{c/\sqrt{\varepsilon_r}} \qquad (2)$$

where $Y_1=\sqrt{\varepsilon_r}Y_0$ and $Y_0=\sqrt{\varepsilon_0/\mu_0}=260\text{mS}$ are the characteristic admittance in the dielectric substrate and free space, respectively, β_1 is the phase constant in the substrate, c is the speed of light and f is the operating frequency. We assumed the equivalent admittance of the patch was $Y_P = jB$, which depends on the shape and size of the metallic patch. Total admittance of the AMC cell can be expressed by

$$Y_t(h) = j[B - Y_1\cot(\beta_1 h)] \qquad (3)$$

The reflection coefficient of the AMC cell is

$$\Gamma = \frac{Y_0 - Y_t}{Y_0 + Y_t} = \frac{Y_0 - j[B - Y_1\cot^{f_0}(\beta_1 h)]}{Y_0 + j[B - Y_1\cot^{f_0}(\beta_1 h)]} \qquad (4)$$

The amplitude is unity ($|\Gamma|\approx 1$) due to an infinite ground surface at the no grating lobe region. The reflection phase is simplified as

$$\begin{aligned}\angle\Gamma &= 2\arg\{Y_0\text{-}j[B\text{-}Y_1\cot(\beta_1 h)]\} \\ &= 2\arg\left\{1\text{-}j\left[\tilde{B}-\sqrt{\varepsilon_r}\cot(\beta_1 h)\right]\right\} \\ &= -2\arctan\left[\tilde{B}-\sqrt{\varepsilon_r}\cot(\beta_1 h)\right]\end{aligned} \qquad (5)$$

where $j\tilde{B}(= jB/Y_0)$ is the normalized admittance of the metallic patch. The above analysis is valid up to the frequency at which grating lobes occur ($\lambda = P$ for normal incidence). This is because the above this limit, the energy is reflected not only in the direction stated by Snell's law but also in other directions, resulting in the nonlinear behavior of the periodic surface. The reflection coefficient of the fundamental harmonic appears less than 1.

The admittance of the metallic patch and short-circuited terminal jointly affect the reflection phase according to Eq. (5). The normalized admittance $j\tilde{B}$ depends on the structure of the patch and is a function of frequency. It can be divided into capacitive, inductive, parallel and series resonance states. The phase range can be accomplished by using the metallic patch with adjustable circuit parameters (inductance, capacitance). Consequently, increasing the spatial periodicity is a simple solution for increasing the lowest frequency bound. However, this effort will decrease the frequency of the grating lobes, which is the highest bound in opposite phase cancellation (OPC). OPC requires a phase difference of two unit cells to be stably distributed approximately 180° in a wide band, and the nonlinearity of the grating lobes results in a phase jump to break the stability. To solve this problem, multielement phase cancellation (MEPC) was proposed to deal with this nonlinearity.

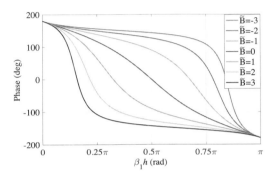

Fig. 3. The reflection phase versus $\beta_1 h$ as the different normalized admittance of the metallic patch in the case of $\varepsilon_r = 2.65$.

At the same time, near the low-frequency region, the reflection phase was mainly determined by the short-circuited terminal, rather than the admittance of the patch. For simplicity, we supposed the normalized admittance of the metallic patch was frequency-independent. Since the resonance states were impossible to implement in broadband, only the capacitive and inductive states was analyzed. The reflection phase versus $\beta_1 h$ of the different normalized admittance of the metallic patch is plotted in Fig. 3. From the above analysis, the lowest frequency minimization can be accomplished by using the metallic patch by changing broad circuit parameters (corresponding to increasing spatial periodicity) and increasing the thickness of the substance (in the case of a fixed dielectric constant). Furthermore, the admittance of the short-circuited terminal is a periodic function of frequency. When $\beta_1 h = n\pi (n = 1, 2, ...), \cot(\beta_1 h) = \pm\infty$. The reflection phase $\angle \Gamma$ in Eq. (5) is identically equal to π, and the AMC cell with any configuration is equivalent to the PEC surface ($\Gamma = -1$). Plugging Eq. (2) into $\beta_1 h = n\pi$, the short-circuited frequencies can be obtained as

$$f_s = \frac{nc}{2h\sqrt{\varepsilon_r}} \quad (n = 1, 2, ...) \quad (6)$$

It is worth mentioning that the short-circuited frequencies are valid in the full band. The short-circuited frequencies are another high-frequency limit. The in-phase reflection is produced at the short-circuited frequencies. The reflection phase is independent of the configuration of the patch. We have noticed that these structures, such as metasurface, frequency selective surface (FSS), artificial magnetic conductor (AMC), electromagnetic bandgap (EBG) and polarization conversion used for RCS reduction, also suffer from the short-circuited problem, which is insurmountable and makes bandwidth expansion extremely difficult.

In conclusion, the spatial periodicity and thickness of AMC determine the lowest frequency, and the highest frequency bound is limited by the grating lobes and short-circuited frequencies. In most works about RCS reduction, the operating frequency bands of RCS reduction are in the band from low frequency to the minimum value of frequency of the first grating lobe ($f_{H1} = \dfrac{c}{P}$) and short-circuited frequency ($f_{H2} = \dfrac{c}{2h\sqrt{\varepsilon_r}}$). Therefore, our work aimed to solve the highest frequency limit with the stability of the lowest frequency and expand the bandwidth. The multiple thickness elements and MEPC solve the high-frequency limit problems of the short-circuited frequencies and grating lobes.

2.2 Full-Wave Simulation Verification of AMC Unit Cell

The unit cell with a fixed substrate thickness will limit the bandwidth for RCS reduction because of the short-circuited problem at some frequencies, resulting in the bandwidth blind zone for RCS reduction. To overcome the in-phase reflection problem at short-circuited frequencies, multiple substrate thicknesses were introduced to impose additional wave path differences. The Minkowski Fractal patch illustrated in Fig. 4 was chosen as the AMC element.

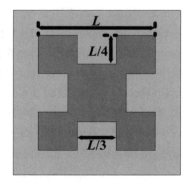

Fig. 4　The geometries of the Minkowski Fractal patch.

The AMC elements with a periodicity of 10.0×10.0 mm2 were printed on an F4B-2 dielectric substrate backed by a conducting ground plane with a dielectric constant $\varepsilon_r = 2.65$ (loss tangent $tan\delta = 0.001$) and a thickness of h. The substrate thickness includes the copper thickness of 0.035 mm×2 on both surfaces. The AMC cells were simulated by the finite element method (FEM) with the periodic boundary condition of CST Microwave

Studio 2018®. In this simulation, the side length L of the AMC cells varied from 0.1 to 10 mm with a step size of 0.1 mm, while there were six choices of layer thickness for the dielectric substrate: 0.93, 1.93, 2.93, 3.93, 4.93, and 5.93 mm. The reflection coefficients are illustrated in Fig. 5 (a)-(l), respectively. The reflection amplitude was approximately uniform ($|\Gamma_{m,n}| \approx 1$) below 30 GHz due to the low-loss substrate and the infinite PEC ground. Since the power scatters toward high order harmonics over 30 GHz, the reflection coefficient of the fundamental harmonic appeared to be less than 1. The reflection coefficients of AMC at the short-circuited frequencies were the same as the PEC ($\Gamma = -1$) regardless of whether the grating lobes appeared.

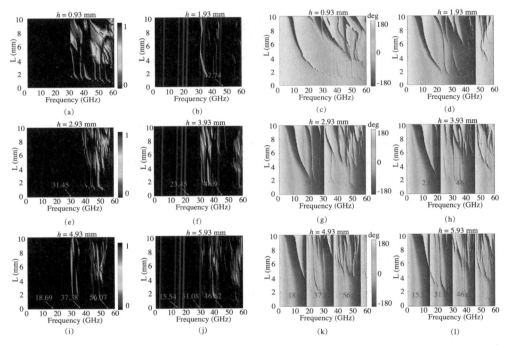

Fig. 5 The reflection coefficient for different substrate thickness h. The amplitudes (a, b, e, f, i, j) and the phases (c, d, g, h, k, l) with the change of the geometry parameter L for the layer thicknesses of 0.93, 1.93, 2.93, 3.93, 4.93, and 5.93 mm, respectively.

The short-circuited frequencies for different substrate thicknesses are listed in Table I. The phase term clearly indicates that the high-thickness AMC unit cell has a large change range at low frequencies, and the short-circuited frequency moved to a lower frequency accordingly. The phase changed discontinuously at high frequencies due to high-order resonances and grating lobes. The newly introduced substrate thicknesses themselves also had a short-circuited problem. However, the common short-circuited frequencies were moved to the high-frequency range.

Table I

The short-circuited frequencies (SCF) of the first four orders

h (mm)		0.93	1.93	2.93	3.93	4.93	5.93
SCF (GHz)	n=1	99.08	47.74	31.45	23.45	18.69	15.54
	n=2	198.16	95.48	62.9	46.9	37.38	31.08
	n=3	297.24	143.22	94.35	70.35	56.07	46.62
	n=4	396.32	190.96	125.8	93.8	74.76	62.16

In our design, the grounds of AMC were in the same plane, and the reflection phase of the reference plane with a height (h_{ref}) from the ground was calculated as

$$\varphi_{ref} = \varphi_{AMC} - 2\beta_0(h_{ref} - h_{AMC}) \tag{7}$$

where φ_{AMC} and h_{AMC} are the reflection phase and the thickness of AMC unit cell, respectively, β_0 is the phase constant in free space, and h_{AMC} is the height of AMC. There were two merits of utilizing various substrate thicknesses. First, the wave path difference between the AMC cells with different substrate thicknesses can extend the coverage range of the reflected phase over an ultrawide frequency band. Second, variable substrate thicknesses can effectively avoid the short-circuited frequencies, and the reflection phase can be manipulated by the configuration of the patch. The broader phase shift coverage would enhance the ability to manipulate EM waves. The highest frequency limitation caused by the short-circuited frequencies can be counteracted by mixing the Minkowski cells with different patch sizes and different substrate thicknesses to realize an ultrawideband destructive interference.

3. Bandwidth Expansion for RCS Reduction

This section introduces the physical mechanism and optimization process for superwideband RCS reduction. Compared with opposite phase cancellation, the advantages of multielement phase cancellation were investigated. RCS reduction performances with fixed substrate thickness and variable substrate thickness are discussed.

3.1 MEPC Synthesis Algorithm

The advantage of multielement phase cancellation (MEPC) was demonstrated by comparing the ability of RCS reduction with opposite phase cancellation (OPC). Figure 6 illustrates a multielement checkerboard AMC surface that consists of $P=M \times N$ lattices with same dimension. The AMC surface was placed on the xy plane. The incident wave propagated toward the -z-direction.

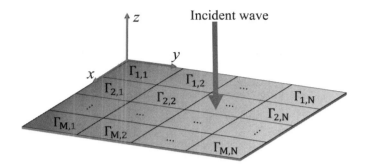

Fig. 6　The plane wave incident on an AMC surface consists of P=M × N lattices.

For the multielement AMC surface that is illuminated by the plane wave under normal incidence, P propagating waves produced from the AMC lattices traversed into the same space. According to the principle of the superposition of waves, the RCS of the entire AMC surface was the vector sum of the P scatterers. For a reference flat PEC plate, its RCS at normal incidence in the optical region is

$$\sigma_{REF} = \frac{4\pi A^2}{\lambda^2} \tag{8}$$

where A is the area of the reference PEC plate and λ is the operating wavelength. The monostatic RCS of the AMC surface can be approximated by

$$\sigma_{AMC} = \left| \sum_{p=1}^{P} \sqrt{\sigma_p} e^{j\phi_p} \right|^2 = \left| \sum_{p=1}^{P} \sqrt{\frac{4\pi A_p^2}{\lambda^2}} |\Gamma_p| e^{j\phi_p} \right|^2 \tag{9}$$

where A_p and Γ_p are area and the reflection coefficient of the p-th AMC lattice, respectively, while φ_p is the reflection phase of the AMC at reference plane. When the proposed surfaces have equal physical area ($A = A_p \times P$) with the reference PEC plate, the RCS reduction can be expressed as

$$\begin{aligned} RCSR(dB) &= 10\log_{10} \left| \frac{\sqrt{\sigma_{AMC}}}{\sqrt{\sigma_{REF}}} \right|^2 \\ &= 10\log_{10} \left| \frac{1}{P} \sum_{p=1}^{P} |\Gamma_p| e^{j\omega p} \right|^2 \end{aligned} \tag{10}$$

Equation (10) does not include the coupling between the AMC lattices and edge effects but provides an approximate calculation for RCS reduction of the AMC surface.

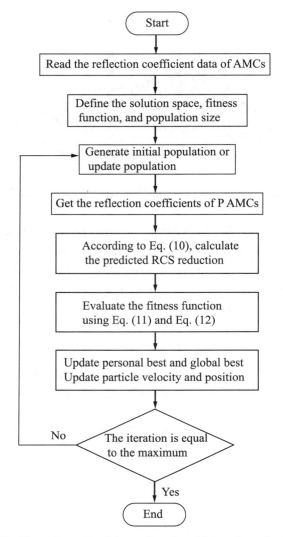

Fig. 7　The schematic diagram for ultrawideband manipulation and optimization of RCS reduction.

　　This work's goal was to achieve and compare the largest bandwidth for different AMC unit combinations. In this research, the RCS reduction surface consisted of P lattices. Finding optimal P sets of data from reflection phase curves to achieve maximum RCS reduction bandwidth is computationally complex. Thus, particle swarm optimization (PSO) was utilized to obtain the lattice sizes that need to be evaluated. The schematic diagram for ultrawideband control and optimization of RCS reduction by selecting the size of the AMC unit cell is depicted in Fig. 7. The fitness function used to assess the wideband RCS reduction performance is defined as

$$\text{fitness} = \sum_{k=0}^{K} \max\left\{[\sigma_R(f_k) + S_{dB}], 0\right\} \quad (11)$$

$$f_k = f_{min} + \frac{k(f_{max} - f_{min})}{K} \quad (12)$$

where f_{min} and f_{max} are the optimized minimum frequency and maximum frequency, respectively. K is the number of optimized frequency points, and S_{dB} is the optimized target value for RCS reduction. The value is set to $S_{dB} = 12$ dB. The redundancy of 2 dB was considered for the approximation error by Eq. (10) owing to the coupling between the lattices and their edge effects. The predicted RCS value of the optimized frequency band formed the basis of the fitness function. In this optimization, the smaller the value of the fitness function, the better effect of RCS reduction was. The iterative optimization process ended when the maximum number of iterations was met.

When the RCS reduction surface consisted of $P = 2$ lattices, this was the case of OPC. For the number of lattices greater than 2 ($P > 2$), this situation is called MEPC. For the traditional OPC with two unit cells, due to the problems of the grating lobes and the high-order resonance, it was impossible to obtain two reflection coefficients with an approximate 180° phase difference at multiple short-circuited frequencies. For the destructive interference capability, the MEPC is far more powerful to deal with these nonlinear problems than the OPC. Moreover, the reflection coefficient of the fundamental harmonic appeared to be less than 1 due to the appearance of high order harmonics. For OPC, which required two equal-amplitude reverse electric fields, it was impractical to achieve RCS reduction in the grating lobe region. However, for the MEPC method of multiwave cancellation, the above problems can be handled well. Therefore, MEPC is a very effective method to solve the highest frequency limitation caused by the grating lobes and high-order resonance. However, the actual destructive interference capabilities of OPC and MEPC are limited by the basic AMC cell and related to its reflection phase characteristic.

In this paper, MEPC was applied to achieve the multiband RCS reduction and ultrawideband RCS reduction. Multiple reflected waves were superimposed in space to achieve phase cancellation. A variable number of AMC cells and, in particular, the variable phase difference between them greatly increased the ability to achieve ultrawideband phase cancellation.

3.2 AMC Cells with a Fixed Substrate Thickness

In this part, the effect of substrate thickness on RCS reduction bandwidth is further discussed. During optimization, the substrate thickness was fixed, and the patch size was changed. The number of basic AMC elements for the MEPC was set to 20. In the optimization, the number of iterations was set to 1000. The optimization frequencies were, discretely, {4, 5, 6, …, 39, 40} GHz. After all iterations was completed, we obtained the largest bandwidth for 10 dB RCS reduction. For different substrate thicknesses, the maximum bandwidth of 10 dB RCS reduction was achieved by MEPC, as shown in Fig. 8. The AMC surface worked at multiple frequency bands. Multielement phase cancellation (MEPC) can effectively avoid the insurmountable blind zones and realize multiband RCS reduction of harmonics regions. For the substrate thickness of $h = 3.93$ mm, obvious RCS reductions were achieved in two frequency bands: 5.1-16.6 GHz and 28.6-40 GHz. A blind zone (16.6-28.6 GHz) was formed at the short-circuited frequency (23.45 GHz) and its vicinity, which is an atmospheric absorption band. Opposite phase cancellation usually operates at the first available frequency band. Due to the presence of short-circuited frequencies, a phase difference of approximately 180° between two AMC elements could not be guaranteed in multiple available frequency bands.

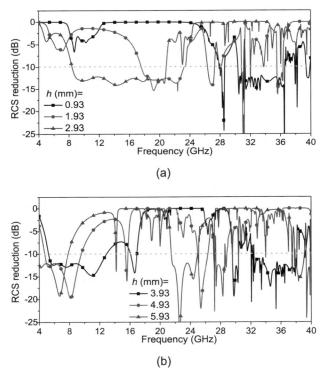

Fig. 8 Comparison of the simulated bandwidth of RCS reduction for different substrate thicknesses h.

The maximum bandwidths corresponding to different substrate thicknesses are listed in Table II. As Fig. 8 shows, the working frequency band moves toward the low frequency range as the substrate thickness h increases. When the substrate thickness h is thin, Eq. (5) shows that the phase change range is narrow as the configuration of the patch changes.

Thus, the AMC surface with a thin dielectric substrate operates at high frequency, while the AMC surface with thick dielectric substrate operates at low frequency. As shown in Fig. 8 and Table II, the AMC surface with substrate thicknesses of 2.93 and 3.93 mm can achieve the maximum RCS reduction bandwidth of approximately 84%. Since the short-circuited problem is insurmountable, the combination of the MEPC and AMC elements with a fixed substrate thickness does not significantly improve the bandwidth of the first operating band but can achieve RCS reduction in multiple available frequency bands outside the blind zones. Compared with traditional OPC, the MEPC demonstrates a powerful RCS reduction capability.

Table II

The maximum bandwidth for different substrate thicknesses

h (mm)	FSCF (GHz)	FOFB (GHz)	FBW (%)	RBW (f_H / f_L)
0.93	99.08	31.31-36.54	15.4	1.167
1.93	47.74	17.61-23.26	27.7	1.321
2.93	31.45	8.46-20.88	84.7	2.468
3.93	23.45	5.26-12.93	84.3	2.458
4.93	18.69	4.76-9.58	67.2	2.013
5.93	15.54	4.0-7.82	64.6	1.955

FSCF: First short-circuited frequency

FOFB: First operating frequency band, f_L f_H

FBW: The fractional bandwidth (FBW = $(f_H - f_L)/f_c$, $f_c = (f_H + f_L)/2$)

3.3 AMC Cells with a Variable Substrate Thickness

At the second optimization, the substrate thickness and the patch size are changeable. The number of basic AMC elements is variable. A total of 49 optimizations are performed for the different number of AMC elements from 2 to 50. The optimized results of the largest bandwidth for different numbers of AMC elements are shown in Fig. 9. As the number of AMC cells increases, MEPC can significantly expand the bandwidth. The side length L and substrate thickness h for the case of 20 AMC cells are listed in Table III. A comparison of bandwidth for 10 dB RCS reduction between traditional OPC (2 AMC cells) and new proposed MEPC (20 AMC cells) is made in Fig. 10. The highest frequency is greatly improved from 17.86 GHz to 40 GHz, while the lowest frequency is almost unchanged. The highest frequency extends from the nongrating lobe region to the frequency of the grating lobes. The ratio bandwidth is improved from 3.2:1 to 7.02:1. For the OPC with two electric fields, it is impossible to obtain two reflection coefficients with an approximate 180° phase difference at multiple short-circuited frequencies. Therefore, the combination of OPC and AMC elements with a variable substrate thickness just broadens the bandwidth a little. The AMC cells with different substrate thicknesses had different short-circuited frequencies. Mixing AMC cells with different substrate thicknesses and different patch sizes can produce phase differences at the short-circuited frequencies and effectively overcome the blind zone of RCS reduction. The RCS reduction bandwidth of three implementations is compared in Table IV. In conclusion, the perfect combination of AMC cells with variable substrate thickness, multielement phase cancellation and the PSO algorithm can effectively overcome the short-circuited problem and achieve the maximum bandwidth.

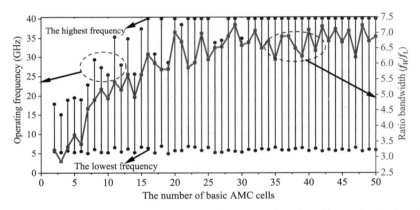

Fig. 9 The optimized results of the operating frequency band and largest ratio bandwidth for the number of AMC cells from 2 to 50.

Table III

The optimized parameters $(h\,|\,L, mm)$ of 20 basic AMC cells

(6.0 | 2.45)	(6.0 | 8.3)	(6.0 | 6.8)	(6.0 | 2.9)	(6.0 | 9.45)
(3.0 | 9.8)	(2.0 | 3.5)	(2.0 | 9.5)	(2.0 | 9.05)	(2.0 | 5.75)
(3.0 | 7.55)	(1.0 | 0.8)	(1.0 | 4.85)	(1.0 | 3.8)	(1.0 | 0.8)
(3.0 | 9.8)	(5.0 | 8.15)	(5.0 | 5.9)	(5.0 | 8.45)	(4.0 | 3.65)

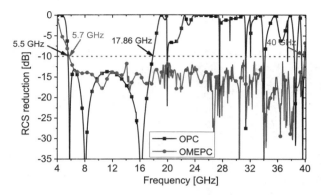

Fig. 10 Comparison of the predicted RCS reduction between OPC and MEPC.

Table IV

The comparison of maximum bandwidth between three cases

	MEPC+SLS	OPC+VLTS	MEPC+VLTS
OFB	8.46-20.88 GHz	5.5-17.86 GHz	5.7-40 GHz
RBW	2.67:1	3.25:1	7.02:1

MSLS: Single layer structure

VLTS: Variable-layer-thickness structure

4. Simulation and Measurement

In this paper, a RCS reducer surface consisting of 20 AMC cells shown in Table III was designed to illustrate the new strategy for the bandwidth breakthrough. Each lattice was a subarray of 6×6 AMC cells. To facilitate manufacturing, the lattices with equal substrate thicknesses was put together. The random distribution of 20 basic AMC lattices and the full structure of the AMC surface with overall dimensions of 300×240 mm2 are depicted in Fig. 11. The monostatic RCS of the proposed AMC surface under normal incidence was simulated by the transient solver of the CST Microwave Studio® based on the Finite Integration Technique (FIT). Numerical simulations run on a workstation with Intel E5-2680 v4 2.4 GHz CPU and 128 GB RAM. The total processing time was approximately 5 hours and 11 minutes. The RCS reduction of the surface normalized to the equal-sized PEC surface is shown in Fig. 12. The AMC surface can achieve 10 dB RCS reduction in an ultrawide band from 5.7 to 40 GHz with a ratio bandwidth of 7.02:1. The simulated results agreed well with the theoretical results in Fig. 10. Both results implied that the combination of variable-layer-thickness lattices, MEPC and the PSO algorithm was able to greatly expand the RCS reduction bandwidth.

Fig. 11   The full structure of the AMC RCS surface.

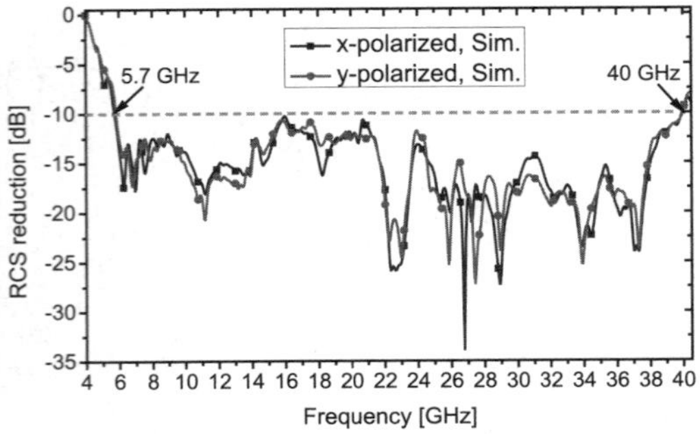

Fig. 12   The simulated RCS reduction versus frequency.

The simulated results of normalized 3-D bistatic scattering patterns at normal incidence are depicted in Fig. 13 at 6 and 24 GHz, respectively. Due to the messy distributions of the reflection phase of 20 lattices, the AMC surface generated diffusion scattering based on array theory. The bistatic RCS of the AMC surface was dramatically decreased compared to that of the equal-sized PEC surface.

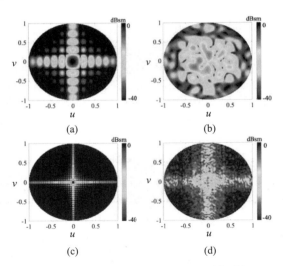

Fig. 13   Simulated bistatic scattering patterns under normal incidences on the normalized wavevector domain ($u = \sin(\theta)\cos(\varphi)$, and $v = \sin(\theta)\sin(\varphi)$). (a) PEC and (b) AMC surface at 6 GHz, (c) PEC and (d) AMC surface at 24 GHz.

Since the phase of the reflection coefficient of the AMC structure depends on the incident angle and polarization of the incident wave; oblique incidence for both transverse-electric (TE) and transverse-magnetic (TM) were investigated. Through simulations, the scattering performances of the metasurface under oblique incidences for both TE and TM polarizations are provided in Fig. 14. As the incident angle increased, the RCS reduction performance of the metasurface slightly deteriorated in the low frequency range. However, it should also be noted that the RCS reduction was more than 6.5 dB at low-frequency region.

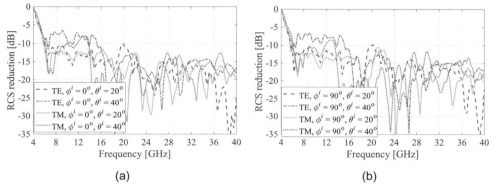

Fig. 14 RCS reduction performances with TE and TM polarizations for various angles of incidence wave when planes of incidence are (a) the xz-plane and (b) the yz-plane, respectively.

A prototype of AMC surface shown in Fig. 11 was fabricated and tested to validate the stealth performance. The monostatic RCS results of the AMC surface and the equal-sized PEC surface were measured by the compact antenna test range (CATR) system of the Science and Technology on Electromagnetic Scattering Laboratory in Beijing. The measured band 4-40 GHz was covered by six pairs of standard linearly-polarized horn antennas operating at C1 (4-6 GHz), C2 (6-8 GHz), X (8-12 GHz), Ku (12-18 GHz), K (18-26.5 GHz) and Ka (26.5-40 GHz) bands, respectively. Figure 15 shows that the measured RCS reductions were larger than 10 dB from 5.76 to 40 GHz with a ratio bandwidth of 7.0:1 under the normal incidence for both polarizations. The average RCS reductions for x- and y-polarizations reached 18.84 dB and 18.26 dB, respectively. We noted that the measurements agreed well with the simulations shown in Fig. 12. The proposed AMC surface demonstrated an ultrawideband low detectable capacity. A comparison between this work and the previous studies of opposite phase cancellation and polarization conversion are provided in Table V. Obviously, our results were far beyond the state-of-the-art of RCS reduction performance.

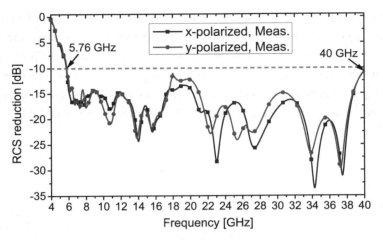

Fig. 15 Measured RCS reduction under normal incidences.

Table V

Comparison of our work and previous studies

Article	Year	σ_R (dB)	OFB (GHz)	FBW (%)	RBW (f_H / f_L)
[12]	2016	10	9.4-23.3	85	2.48:1
[7]	2017	10	3.8-10.7	95	2.82:1
[19]	2017	10	6.1-17.8	98	2.92:1
[8]	2018	10	5.4-14.2	89.8	2.63:1
[13]	2018	10	10-20.7	69.7	2.07:1
[9]	2018	10	6.2-12.5	67	2.02:1
[14]	2019	10	7.98-16.32	68.6	2.05:1
[10]	2019	10	3.77-10.14	91.5	2.69:1
[20]	2019	10	4.29-13.34	102.7	3.11:1
This Work	2020	10	5.7-40	150.1	7.02:1

OFB: Operating frequency band

FBW: The fractional bandwidth

5. Conclusion

In this paper, the root cause of the bandwidth limit of radar cross section (RCS) reduction was discussed and studied by the transmission line theory. A perfect combination of AMC elements with variable substrate thickness, multielement phase cancellation (MEPC), and the PSO algorithm was proposed to effectively overcome the short-circuited problem and extend high-frequency to the band of the grating lobes appearance, which greatly expands the bandwidth of the RCS reduction.

For monolayered structures, the novel approach can avoid the insurmountable bandwidth blind zone and achieve multiband RCS reduction;

For variable-layer-thickness structures, the novel approach can overcome the high-frequency limit and realize ultrawideband RCS reduction.

The designed 20-element AMC surface can achieve 10 dB RCS reduction in an ultrawideband frequency range from 5.7 to 40 GHz under the normal incidence for both polarizations. Compared with the traditional OPC, the proposed MEPC can improve the ratio bandwidth from 3.2:1 to 7.02:1. The theoretical, simulated, and experimental results proved that the proposed approach could break the high-frequency limit and make a significant breakthrough in the bandwidth expansion of RCS reduction.

In this work, a simple structure of the Minkowski fractal patch was chosen as the basic AMC element, and the substrate thickness was discrete. Subsequent studies will further improve the bandwidth and magnitude of RCS reduction in two ways. One approach is to design better AMC elements that can provide a large range phase difference than those previously provided. Another approach is to use a continuously varying substrate thickness, which will make it easier to achieve good destructive interference in multiple blind zones.

Acknowledgments

This work was supported by the National Natural Science Foundation of China (61701448 and 61671415) and Key Laboratory of All Optical Network and Advanced Telecommunication Network, Ministry of Education (Beijing Jiaotong University ZG19002).

Author Information

J. Su, H. Yu, Q. Guo and Z. Li are with the School of Information and Communication Engineering, Communication University of China, Beijing 100024, China, and also with State Key Laboratory of Media Convergence and Communication, Communication University of China.

H. Yin and Z. Xiao are with the Science and Technology on Electromagnetic Scattering Laboratory, Beijing, 100854, China.

References

[1] BECKER R. Electromagnetic field and interactions[M]. New York: Dover Publications, 1982.

[2] PAIN H J, RANKIN P. Introduction to vibrations and waves[M]. 1st ed. New York: Wiley, 2015.

[3] SKOLNIK M I. Radar handbook[M]. 2nd ed. New York: McGrawHill, 1990.

[4] PAQUAY M, IRIARTE J C, EDERRA I, et al. Thin AMC structure for radar cross-section reduction[J]. IEEE Trans. Antennas Propagat., 2007, 55 (12): 3630–3638.

[5] IRIARTE GALARREGUI J C, PEREDA A T, MATINEZ DE FALCON J L, et al. Broadband radar cross-section reduction using AMC technology[J]. IEEE Trans. Antennas Propagat., 2013, 61 (12): 6136–6143.

[6] CHEN W, BALANIS C A, BIRTCHER C R. Checkerboard EBG surfaces for wideband radar cross section reduction[J]. IEEE Trans. Antennas Propag., 2015, 63 (6): 2636-2645.

[7] HAJI-AHMADI M J, NAYYERI V, SOLEIMANI M, et al. Pixelated checkerboard metasurface for ultra-wideband radar cross section reduction[J]. Sci. Rep., 2017, 7: 11437.

[8] XUE J J, JIANG W, GONG S X. Chessboard AMC surface based on quasi-fractal structure for wideband RCS reduction[J]. IEEE Antennas Wireless Propag. Lett., 2018, 17 (2): 201–204.

[9] GHAYEKHLOO A, AFSAHI M, OROUJI A A. An optimized checkerboard structure for cross-section reduction: producing a coating surface for bistatic radar using the equivalent electric circuit model[J]. IEEE Antennas Propag. Mag., 2018, 60 (5): 78–85.

[10] SANG D, GHEN Q, DING L, et al. Design of checkerboard AMC structure for wideband RCS reduction[J]. IEEE Trans. Antennas Propag, 2019, 67 (4): 2604–2612.

[11] SU J, LU Y, LIU J, et al. A novel checkerboard metasurface based on optimized multielement phase cancellation for superwideband RCS reduction[J]. IEEE Trans. Antennas Propag., 2018, 66 (12): 7091–7099.

[12] MODI A Y, BALANIS C A, BIRTCHER C R, et al. New class of RCS-reduction metasurfaces based on scattering cancellation using array theory[J]. IEEE Trans. Antennas Propag., 2019, 67 (1): 298–308.

[13] HAN J F, CAO X Y, GAO J, et al. Broadband radar cross section reduction using dual-circular polarization diffusion metasurface[J]. IEEE Antennas Wireless Propag. Lett., 2018, 17 (6): 969–973.

[14] KIM S H, YOON Y J. Wideband radar cross-section reduction on checkerboard metasurfaces with surface wave suppression[J]. IEEE Antennas Wireless Propag. Lett., 2019, 18 (5): 896–900.

[15] YAN D, GAO Q, WANG C, et al. A novel polarization convert surface based on artificial magnetic conductor[C]. Suzhou, China: Pro. APMC, 2005.

[16] ZHU X, HONG W, WU K, et al. A novel reflective surface with polarization rotation characteristic[J]. IEEE Antennas Wireless Propag. Lett., 2013, 12: 968-971.

[17] GRADY N K, HEYES J E, CHOWDHURY D R, et al. Terahertz metamaterials for linear polarization conversion and anomalous refraction[J]. Science, 2013, 340 (6138): 1304-1307.

[18] CHEN H, WANG J, MA H, et al. Ultra-wideband polarization conversion metasurface based on multiple plasmon resonances[J]. J. Appl. Phys., 2014, 115 : 154504.

[19] JIA Y T, LIU Y, GUO Y J, et al. A dual-patch polarization rotation reflective surface and its application to ultra-wideband RCS reduction[J]. IEEE Trans. Antennas Propagat., 2017, 65 (6): 3291–3295.

[20] SUN S Y, JIANG W, LI X Q, et al. Ultrawideband high-efficiency 2.5-Dimensional polarization conversion metasurface and its application in RCS reduction of antenna[J]. IEEE Antennas Wireless Propag. Lett., 2019, 18: 881-885.

[21] COSTA F, MONORCHIO A, MANARA G. Wideband scattering diffusion by using diffraction of periodic surfaces and optimized unit cell geometries[J]. Sci. Rep., 2016, 6: 25458.

[22] COSTA F, MONORCHIO A, MANARA G. Efficient analysis of frequency selective surfaces by a simple equivalent-circuit model[J]. IEEE Antennas Propag. Mag., 2012, 54 (4): 35–48.

Ultrawideband and High-Efficient Polarization Conversion Metasurface Based on Multi-Resonant Element and Interference Theory[*]

1. Introduction

Metasurface is a kind of two-dimensional metamaterials that have been widely used in various fields such as polarization converter, ultrathin lens, invisible cloak, phase shifter, absorber, and low scattering, because of its low profile characteristics and potential ability to manipulate electromagnetic (EM) waves. The polarization conversion metasurface (PCM) is an important EM structure, which can be widely applied to radar cross section (RCS) reduction, polarization controlled devices, and so on. Generally, PCMs can be classified into transmission and reflection types.

In recent years, many studies have been made on reflective PCMs. In terms of wideband PCMs, an L-shaped structure is proposed in [15], whose polarization conversion ratio (PCR) is higher than 90% from 7.8 to 34.7 GHz under the normal incidence. In [16], a linear-to-linear 2.5-dimensional PCM with a high PCR is presented. Its PCR is higher than 0.96 within a ratio bandwidth (fH/fL) of 3.1:1. A butterfly-shaped reconfigurable and wideband reflective PCM is designed, whose functionalities can be dynamically switched among linear-to-linear, linear-to-elliptical, and linear-to-circular polarization conversions in a wideband. In [21], a single-layer PCM with five resonances is introduced to realize a 4.1:1 ratio bandwidth with the PCR above 90%. Due to their resonant characteristics, it is challenging to control the polarization state of EM waves with a high polarization conversion efficiency over an ultrawide frequency range, which limits their practical applications.

In this paper, a novel ultrawideband and high-efficient PCM composed of two pairs of L-shaped patches is designed. The aim of this work is to expand the bandwidth of the PCM while guaranteeing a high conversion rate and lessening the size of the unit cell. Following this, this PCM can convert linearly polarized waves to their orthogonal direction in an ultra-wideband from 3.37 to 22.07 GHz with a ratio bandwidth of 6.5:1 under the

---

[*] The paper was originally published in *Optics Express*, 2021, 29 (22), and has since been revised with new information. It was co–authored by Hang Yu, Xiaoyu Wang, Jianxun Su, Meijun Qu, Qingxin Guo, Zengrui Li, and Jiming Song.

normal incidence. To the best of our knowledge, the proposed PCM is beyond state-of-the-art of the performance in terms of bandwidth. Besides, the dimensions and thickness of the unit cell in terms of the wavelength at 3.37 GHz is only 0.112 λ0×0.112 λ0×0.095 λ0. The equivalent circuit concept and interference theory are employed to analyze the root cause of the polarization conversion and prove that adding superstrate can widen the bandwidth.

2. Design, Simulation, and Measurement

The proposed PCM's unit cell is composed of two pairs of L-shaped metallic patches covered by a single-layer F4B-2 dielectric superstrate (ε_r =2.65, tanδ =0.001). The metallic ground is placed at the height of h_air away from L-shaped metallic patches. Fig. 1 presents the structure of the unit cell of the proposed PCM. In order to avoid strong magnetic coupling, we set h_air to 5 mm, which is one quarter of the wavelength at 15 GHz. The reason why we choose two pairs of L-shaped patterns is that the length of the L-shaped patch determines the frequency band of resonance. The larger length can achieve the lower frequency band. Therefore, the L-shaped patches with different lengths can obtain the polarization conversion in different frequency bands. From this, two pairs of L-shaped patches with different lengths are combined to expand the bandwidth of the PCM. The unit cell of the PCM was simulated and optimized using the CST Microwave Studio®. The optimized parameters of unit cell are as follows: p = 10 mm, l_1 = 9.57 mm, l_2 = 4.76 mm, w_1 = 1 mm, w_2 = 0.7 mm, d = 1.1 mm, h = 3.5 mm, and h_air = 5 mm.

Because of the same polarization conversion characteristics for x- and y-polarized waves, we supposed that the incident wave is y-polarized. The cross- and co-polarization reflection coefficients can be defined as rxy=|Exr/Eyi| and ryy=|Eyr/Eyi|, in which Eyi presents the electric field of y-polarized incident waves, Exr and Eyr indicate the electric field of x- and y-polarized reflected waves, respectively. To measure the efficiency of the polarization conversion, we introduce the polarization conversion ratio (PCR), which can be written as PCR = /(+). The simulated results of co- and cross-polarization reflection coefficients under y-polarized normal incidence are presented in Fig. 2(a). It can be seen that the reflection coefficient of co-polarization has a wide frequency range from 3.37 to 22.07 GHz below -10 dB, and the cross-polarization reflection coefficient almost reaches 0 dB in this frequency band. There are seven resonant frequencies at 3.65, 5.6, 9.78, 13.69, 16.78, 20.53 and 21.9 GHz, respectively. The PCR at these resonant frequencies can be up to 100%, which indicates the completely polarization conversion. The fractional bandwidth of the PCM is 147% (fH/fL = 6.5:1) with the PCR higher than 90%.

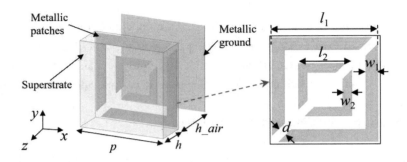

Fig. 1 Geometry structure of the unit cell.

To validate the above-mentioned simulations, a prototype of the PCM with 24×24 unit cells is fabricated, as shown in Fig. 2(b). The sample is fabricated by printed circuit board (PCB) technologies. The metallic bottom ground and the superstructure are connected by nylon columns. High-precision test is performed using the compact antenna test range (CATR) system at Science and Technology on Electromagnetic Scattering Laboratory in Beijing, China. The measured result of co-polarization reflection coefficients is shown in Fig. 2(a), which matches well with the simulated one.

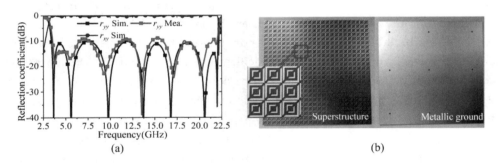

Fig. 2 The ultrawideband PCM (a) Simulated and measured results of co- and cross-polarization reflection coefficients under the normal incidence. (b) The fabricated photograph.

Table I presents a comparison between the proposed PCM and other reported wideband polarization converters. Obviously, the proposed PCM works in a lower frequency band and has better broadband characteristics while ensuring a smaller electrical size. The proposed PCM is far beyond state-of-the-art of the performance in terms of bandwidth.

Table I

Comparison with other wideband PCMs

Year and Ref.	PCR ≥ 90% operating frequency band (GHz)	Number of resonances	Ratio bandwidth (RBW)	Fractional bandwidth (FBW)	Unit cell electrical size / λ_0^3 (width×length×thickness)
2016 [12]	5.71-15.02	5	2.6:1	89.8%	0.190×0.190×0.076
2017 [14]	6.03-17.78	5	2.9:1	98%	0.120×0.120×0.120
2018 [16]	7-19.5	4	2.79:1	94.3%	0.212×0.212×0.0817
2019 [18]	4.29-13.34	4	3.1:1	102.7%	0.197×0.197×0.096
2020 [19]	9.04-20.83	3	2.3:1	79%	0.211×0.211×0.066
2021 [20]	10.2-20.5	5	2.0:1	67.1%	0.408×0.408×0.068
2021 [21]	3.01-13.4	5	4.5:1	126.6%	0.122×0.122×0.081
This work	3.37-22.07	7	6.5:1	147%	0.112×0.112×0.095

3. Analysis and Discussion

3.1 Analysis of Surface Impedance Concept

To understand the principle of polarization conversion, we take the direction of electric field along the y-axis under the normal incidence as an example. We rotate the original coordinate axis by 45° to obtain a u-v coordinate system, as shown in Fig. 3(a). The direction of the electric field can be decomposed into the u- and v-axes, thus the incident waves and the reflected waves can be expressed as

$$\vec{E}_i = \hat{u}E_{ui} + \hat{v}E_{vi} \tag{1}$$

$$\vec{E}_r = \hat{u}\tilde{r}_u E_{ui} + \hat{v}\tilde{r}_v E_{vi} \tag{2}$$

where $\hat{u}$ and $\hat{v}$ are the unit vectors along the u- and v-axes, $\tilde{r}_u$ and $\tilde{r}_v$ are the reflection coefficients under u- and v-polarized normal incidence, respectively. Since the L-shaped patches are symmetrical about the u- and v-axes, no cross-polarization occurs. Additionally, the energy of incident waves can be completely reflected with a metallic ground on the back, so the magnitude of both $\tilde{r}_u$ and $\tilde{r}_v$ would be equal to 1. Due to the anisotropic characteristic of the PCM, there is a phase difference between $\tilde{r}_u$ and [15]. We define the phase difference as $\Delta\varphi$, so $\tilde{r}_v = \tilde{r}_u e^{j\Delta\varphi}$. When $\Delta\varphi = \pm 180°$, the reflected waves can be written as $\vec{E}_r = \hat{u}\tilde{r}_u E_{ui} - \hat{v}\tilde{r}_u E_{vi} = \hat{x}\sqrt{2}E_{ui}$. Obviously, the electric field of the reflected waves is along x-direction, as shown in Fig. 3(a). It means that y-polarized incident wave is reflected and transformed to x-polarized reflected wave. The simulated amplitudes and phase difference under u- and v-polarized normal incidence are shown in Fig. 3(b). It is confirmed that the amplitude of $\tilde{r}_u$ and $\tilde{r}_v$ are both almost equal to 1. Besides, the phase difference (the shaded part) stays around 180° from 3.37 to 22.07 GHz, implying the ability of polarization conversion in this ultra-wide frequency range.

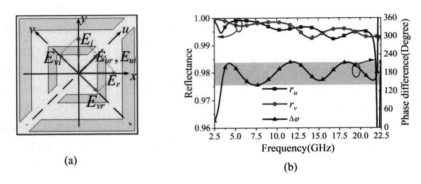

Fig. 3 (a) Electric field vector image of y-polarized incident waves rotated to x-polarized reflection waves. (b) Simulated reflectance and phase difference with the incident electric field along u- and v-axes.

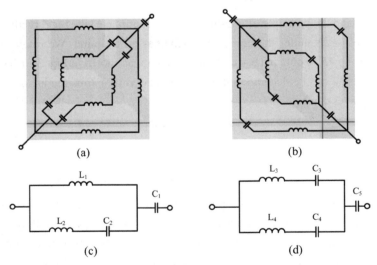

Fig. 4 Analysis diagram of equivalent circuit model for (a) the u-direction and (b) the v-direction and simplified circuit model for (c) the u-direction and (d) the v-direction.

In order to further investigate the cause of the phase difference between $\tilde{r}_u$ and $\tilde{r}_v$, the transmission line theory and equivalent circuits are applied and discussed. Due to the anisotropy of the structure, the L-shaped metallic patches have different equivalent circuit models along the u- and v-directions. The inductor (L) is related to the current distribution in the metallic patches and the capacitor (C) results from the electric field distribution in the gaps between metallic patches. Using the above principles, Fig. 4 shows the analysis diagram by the geometrical parameters and simplified circuit model. The detailed circuit parameters are evaluated by fitting the S-parameters according to the data returned from the full-wave simulation. The parameters are obtained as follows: L1 =2.8 nH, L2 = 4.405 nH, C1 = 0.21 pF, C2 = 0.018 pF, L3 = 0.04 nH, L4 = 3.0 nH, C3 = 0.038 pF, C4 = 0.056 pF, and C5 = 0.12 pF. The surface impedances of the metallic patches are different in these two directions, which results in different input impedances Z_u and Z_v. According to the transmission line theory, the input impedance of u- and v-polarized incident waves under the normal incidence can be expressed as

$$Z_{u,v} = \frac{1+\tilde{r}_{u,v}}{1-\tilde{r}_{u,v}} Z_0 \qquad (3)$$

where Z_0 is the wave impendence of the free space. It is confirmed that $r_u \simeq r_v \simeq 1$, and Zu is different from Zv. Therefore, the reflection phases φ_u and φ_v are different, resulting in a certain phase difference. According to the above theoretical analysis, if the phase difference is kept at around 180° within a frequency range, the y- or x-polarized EM wave will rotate to its cross-polarization in this frequency band.

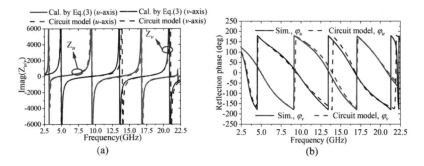

Fig. 5 (a) Imaginary parts of input impedance for an incident field along u- and v-axes. (b) Reflection phase under u- and v-polarized normal incidence.

The imaginary parts of the input impedance for the u- and v-polarized EM components are shown as the solid lines in Fig. 5(a), which are calculated from (3), respectively. Since the metasurface uses a lossless dielectric, the real part of the equivalent impedance is very small and can be ignored. Additionally, in order to accurately describe the input impedance using the equivalent circuit models in Fig. 4, we optimized these two circuit models in the frequency range of 2 to 23 GHz using a commercial software. The air and F4B superstrate were modeled by transmission lines with the characteristic Z_0 and $Z_0/\sqrt{\varepsilon_r}$, respectively. The imaginary parts of the input impedance calculated by the equivalent circuit model in these two directions are shown as the dashed curves in Fig. 5(a). Obviously, the calculation results from (3) are consistent with the results calculated by the equivalent circuit models. It is observed that the imaginary parts of Zu are infinite at frequencies of 3.2, 9.4, 16.6, and 22 GHz, and are almost zero at 5.1, 13.67 and 20.7 GHz; which is opposite to the equivalent impedance in the v-direction. It is consistent with the simulated results when the resonant frequencies are 3.65, 5.6, 9.78, 13.69, 16.78, 20.53, and 21.9 GHz.

At 9.5 GHz, the imaginary part of Zu is infinite, and the reflection phase of u-polarization under the normal incidence is close to 0°, as shown in Fig. 5(b). In this case, in-phase reflection occurs, and the PCM is analogous to a high-impedance surface. Meanwhile, the imaginary part of Zv is close to 0°, and the reflection phase under v-polarized normal incidence is about 180°, so the metasurface is equivalent to a perfect electric conductor. Thus, the phase difference between φ_u and φ_v is about 180°. It is the same as the differences at 3.4 and 16.6 GHz. In addition, the imaginary part of Zv is infinite at about 13.67 GHz and φ_u is about 0°. At the same time, the imaginary part of Zu is close to 0, and φ_v is about 180°. It is the same at 5.1 and 20.7 GHz. Thus, the phase difference between φ_u and φ_v is about -180°. Due to the anisotropy of the input impedance, there will be a 180° phase difference between φ_u and φ_v, which makes the y-polarized incident waves transformed into x-polarized reflected waves.

3.2 Method of Interference Theory

In order to study the polarization conversion characteristics of the PCM, multiple interference theory [22, 23] is applied to model multilayered metasurface. In Fig. 6, the equivalent model of PCM with multilayered media is presented. First, to facilitate the analysis, we regard the part excluding the superstrate (the metal patches and everything to the right) as a reflection surface, then the superstrate and the reflection surface form a three-layer model.

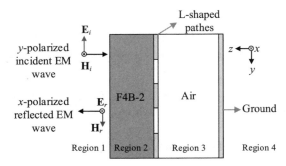

Fig. 6 Schematic of y-polarized incident waves propagating in the PCM described by the reflection and transmission coefficients.

Hence, the total reflection coefficient can be calculated by a consequence of multiple reflections within region 2. Suppose the y-polarized EM waves are incident on from the left, the reflection and transmission will occur on the surface of the superstrate. Due to the isotropy of the superstrate, the reflected and transmitted waves have the same polarization. The Fresnel reflection and transmission coefficients are $\tilde{r}_{yy12}=r_{yy12}e^{j\varphi_{yy12}}$ and $\tilde{t}_{yy12}=t_{yy12}e^{j\phi_{yy12}}$, respectively. The transmitted waves continue to propagate with the propagation phase β_1 in the superstrate until they reach the interface of superstrate and metallic patches, after which they are reflected to the superstrate with the reflection coefficients $\tilde{r}_{xy23}=r_{xy23}e^{j\varphi_{xy23}}$ and $\tilde{r}_{yy23}=r_{yy23}e^{j\varphi_{yy23}}$. These waves are partially reflected back to the superstrate with the reflection coefficients $\tilde{r}_{yy21}=r_{yy21}e^{j\varphi_{yy21}}$ and $\tilde{r}_{xx21}=r_{xx21}e^{j\varphi_{xx21}}$, and partially transmitted into the free space with the transmission coefficients $\tilde{t}_{yy21}=t_{yy21}e^{j\varphi_{yy21}}$ and $\tilde{t}_{xx21}=t_{xx21}e^{j\varphi_{xx21}}$. The overall reflections for y-to-y and y-to-x polarizations consist of superposition of multiple reflections. The propagation process between the EM waves 'touching' the air-superstrate interface two times is defined as a roundtrip. The overall reflection coefficient at the interface between region 1 and region 2 can be written as

$$\begin{aligned}\overline{\mathbf{R}}_{12}^{-} &= \overline{\mathbf{R}}_{12} + \overline{\mathbf{T}}_{12}\overline{\mathbf{R}}_{23}^{-}e^{-j2\beta_1}\overline{\mathbf{T}}_{21} + \overline{\mathbf{T}}_{12}\overline{\mathbf{R}}_{23}^{-}e^{-j2\beta_1}\overline{\mathbf{R}}_{21}\overline{\mathbf{R}}_{23}^{-}e^{-j2\beta_1}\overline{\mathbf{T}}_{21} + \ldots \\ &= \overline{\mathbf{R}}_{12} + \overline{\mathbf{T}}_{12}\overline{\mathbf{R}}_{23}^{-}e^{-j2\beta_1}\left[\overline{\mathbf{I}} - \overline{\mathbf{R}}_{21}\overline{\mathbf{R}}_{23}^{-}e^{-j2\beta_1}\right]^{-1}\overline{\mathbf{T}}_{21}\end{aligned} \quad (4)$$

where $\overline{\mathbf{R}}_{12}^{-}$ is the 2 by 2 reflection matrix which is defined similar to S parameters, and $\beta_1=\sqrt{\varepsilon_r}k_0 h$ is the propagation phase in region 1, $\overline{\mathbf{I}}=\mathrm{diag}(1,1)$, $\overline{\mathbf{R}}_{12}(\overline{\mathbf{R}}_{21})$ and $\overline{\mathbf{T}}_{12}(\overline{\mathbf{T}}_{21})$ are the Fresnel reflection and transmission coefficients. Regions 1 and 2 are isotropic and the reflection matrix at the interface between Region 2 and the reflection surface is symmetrical, we have

$$\overline{\mathbf{R}}_{12} = \tilde{r}_{yy12}\overline{\mathbf{I}} = -\overline{\mathbf{R}}_{21}, \ \overline{\mathbf{T}}_{12} = \tilde{t}_{yy12}\overline{\mathbf{I}} = (1+\tilde{r}_{yy12})\overline{\mathbf{I}}, \ \overline{\mathbf{T}}_{21} = \tilde{t}_{yy21}\overline{\mathbf{I}} = (1-\tilde{r}_{yy12})\overline{\mathbf{I}},$$

$$\overline{\mathbf{R}}_{23}^{-} = \begin{bmatrix} \tilde{r}_{xx23}^{-} & \tilde{r}_{xy23}^{-} \\ \tilde{r}_{yx23}^{-} & \tilde{r}_{yy23}^{-} \end{bmatrix} = \begin{bmatrix} \tilde{r}_{yy23}^{-} & \tilde{r}_{xy23}^{-} \\ \tilde{r}_{xy23}^{-} & \tilde{r}_{yy23}^{-} \end{bmatrix}, \quad (5)$$

In closed form, let

$$e^{-j2\beta_1}\tilde{r}_{yy21}\overline{\mathbf{R}}_{23}^{-} = e^{-j2\beta_1}\tilde{r}_{yy21}\begin{bmatrix} \tilde{r}_{yy23}^{-} & \tilde{r}_{xy23}^{-} \\ \tilde{r}_{xy23}^{-} & \tilde{r}_{yy23}^{-} \end{bmatrix} = \begin{bmatrix} a & b \\ b & a \end{bmatrix} \quad (6)$$

where $a=e^{-j2\beta_1}\tilde{r}_{yy21}\tilde{r}_{yy23}^{-}$, $b=e^{-j2\beta_1}\tilde{r}_{yy21}\tilde{r}_{xy23}^{-}$. Plugging Eqs. (5) and (6) into Eq. (4), the total reflection matrix from region 2 to region 1 can be simplified as

$$\bar{\mathbf{R}}_{12}^{-} = \tilde{r}_{yy12}\bar{\mathbf{I}} + \tilde{t}_{yy12}\tilde{t}_{yy21}e^{-j2\beta_1}\bar{\mathbf{R}}_{23}^{-}\left[\bar{\mathbf{I}} - e^{-j2\beta_1}\tilde{r}_{yy21}\bar{\mathbf{R}}_{23}^{-}\right]^{-1}$$

$$= \tilde{r}_{yy12}\bar{\mathbf{I}} + \frac{\tilde{t}_{yy12}\tilde{t}_{yy21}}{\tilde{r}_{yy21}}\begin{bmatrix} a & b \\ b & a \end{bmatrix}\begin{bmatrix} 1-a & -b \\ -b & 1-a \end{bmatrix}^{-1}$$

$$= \tilde{r}_{yy12}\bar{\mathbf{I}} + \frac{\tilde{t}_{yy12}\tilde{t}_{yy21}}{\tilde{r}_{yy21}\left[(1-a)^2 - b^2\right]}\begin{bmatrix} a(1-a)+b^2 & b \\ b & a(1-a)+b^2 \end{bmatrix} \quad (7)$$

Second, the polarization conversion characteristic between the metallic L-shaped patches and the ground was analyzed. Similar to Eq. (4), the total reflection coefficients from region 3 to region 2 can be written as

$$\bar{\mathbf{R}}_{23}^{-} = \bar{\mathbf{R}}_{23} + \bar{\mathbf{T}}_{23}\bar{\mathbf{R}}_{34}^{-}e^{-j2\beta_0}\left[\bar{\mathbf{I}} - \bar{\mathbf{R}}_{32}\bar{\mathbf{R}}_{34}^{-}e^{-j2\beta_0}\right]^{-1}\bar{\mathbf{T}}_{32} \quad (8)$$

where $\beta_0 = k_0 h_{\_air}$ is the propagation phase in the air. Because the substrate at region 3 is isotropic and the metasurface is symmetrical, we have

$$\bar{\mathbf{R}}_{34}^{-} = \begin{bmatrix} \tilde{r}_{xx34}^{-} & \tilde{r}_{xy34}^{-} \\ \tilde{r}_{yx34}^{-} & \tilde{r}_{yy34}^{-} \end{bmatrix} = -\bar{\mathbf{I}},$$

$$\bar{\mathbf{R}}_{23} = \begin{bmatrix} \tilde{r}_{xx23} & \tilde{r}_{xy23} \\ \tilde{r}_{yx23} & \tilde{r}_{yy23} \end{bmatrix} = \begin{bmatrix} \tilde{r}_{yy23} & \tilde{r}_{xy23} \\ \tilde{r}_{xy23} & \tilde{r}_{yy23} \end{bmatrix},$$

$$\bar{\mathbf{R}}_{32} = \begin{bmatrix} \tilde{r}_{xx32} & \tilde{r}_{xy32} \\ \tilde{r}_{yx32} & \tilde{r}_{yy32} \end{bmatrix} = \begin{bmatrix} \tilde{r}_{yy32} & \tilde{r}_{xy32} \\ \tilde{r}_{xy32} & \tilde{r}_{yy32} \end{bmatrix}, \quad (9)$$

$$\bar{\mathbf{T}}_{23} = \begin{bmatrix} \tilde{t}_{xx23} & \tilde{t}_{xy23} \\ \tilde{t}_{yx23} & \tilde{t}_{yy23} \end{bmatrix} = \begin{bmatrix} \tilde{t}_{yy23} & \tilde{t}_{xy23} \\ \tilde{t}_{xy23} & \tilde{t}_{yy23} \end{bmatrix},$$

$$\bar{\mathbf{T}}_{32} = \begin{bmatrix} \tilde{t}_{xx32} & \tilde{t}_{xy32} \\ \tilde{t}_{yx32} & \tilde{t}_{yy32} \end{bmatrix} = \begin{bmatrix} \tilde{t}_{yy32} & \tilde{t}_{xy32} \\ \tilde{t}_{xy32} & \tilde{t}_{yy32} \end{bmatrix}.$$

Plugging Eq. (9) into Eq. (8), the total reflection matrix can be simplified as

$$\bar{\mathbf{R}}_{23}^{-} = \bar{\mathbf{R}}_{23} - \bar{\mathbf{T}}_{23}e^{-j2\beta_0}\left[\bar{\mathbf{I}} + e^{-j2\beta_0}\bar{\mathbf{R}}_{32}\right]^{-1}\bar{\mathbf{T}}_{32}$$

$$= \begin{bmatrix} \tilde{r}_{yy23} & \tilde{r}_{xy23} \\ \tilde{r}_{xy23} & \tilde{r}_{yy23} \end{bmatrix} - \frac{e^{-j2\beta_0}}{(1+e^{-j2\beta_0}\tilde{r}_{yy32})^2 - (e^{-j2\beta_0}\tilde{r}_{xy32})^2}\begin{bmatrix} c & d \\ d & c \end{bmatrix} \quad (10)$$

where $c = (\tilde{t}_{yy23}\tilde{t}_{yy32} + \tilde{t}_{xy23}\tilde{t}_{xy32})(1+e^{-j2\beta_0}\tilde{r}_{yy32}) - e^{-j2\beta_0}\tilde{r}_{xy32}(\tilde{t}_{yy23}\tilde{t}_{xy32} + \tilde{t}_{xy23}\tilde{t}_{yy32})$ and $d = (\tilde{t}_{yy23}\tilde{t}_{xy32} + \tilde{t}_{xy23}\tilde{t}_{yy32})(1+e^{-j2\beta_0}\tilde{r}_{yy32}) - e^{-j2\beta_0}\tilde{r}_{xy32}(\tilde{t}_{yy23}\tilde{t}_{yy32} + \tilde{t}_{xy23}\tilde{t}_{xy32})$.

Using Eqs. (7) and (10), the overall co- and cross-polarized reflection coefficients at the different interfaces are written as

$$\tilde{r}_{yy12}^{-} = \tilde{r}_{yy12} + \tilde{t}_{yy12}\tilde{t}_{yy21} \cdot \frac{[e^{-j2\beta_1}\tilde{r}_{yy23}(1-e^{-j2\beta_1}\tilde{r}_{yy21}\tilde{r}_{yy23}^{-}) + \tilde{r}_{yy21}(e^{-j2\beta_1}\tilde{r}_{xy23}^{-})^2]}{\left[(1-e^{-j2\beta_1}\tilde{r}_{yy21}\tilde{r}_{yy23}^{-})^2 - (e^{-j2\beta_1}\tilde{r}_{yy21}\tilde{r}_{xy23}^{-})^2\right]} \quad (11)$$

$$\tilde{r}_{xy12}^{-} = \frac{e^{-j2\beta_1}\tilde{t}_{yy12}\tilde{t}_{yy21}\tilde{r}_{xy23}^{-}}{(1-e^{-j2\beta_1}\tilde{r}_{yy21}\tilde{r}_{yy23}^{-})^2 - (e^{-j2\beta_1}\tilde{r}_{yy21}\tilde{r}_{xy23}^{-})^2} \quad (12)$$

$$\tilde{r}_{yy23}^{-} = \tilde{r}_{yy23} - \frac{e^{-j2\beta_0}}{(1+e^{-j2\beta_0}\tilde{r}_{yy32})^2 - (e^{-j2\beta_0}\tilde{r}_{xy32})^2}[(\tilde{t}_{yy23}\tilde{t}_{yy32} + \tilde{t}_{xy23}\tilde{t}_{xy32}) \cdot \\ (1+e^{-j2\beta_0}\tilde{r}_{yy32}) - e^{-j2\beta_0}\tilde{r}_{xy32}(\tilde{t}_{yy23}\tilde{t}_{xy32} + \tilde{t}_{xy23}\tilde{t}_{yy32})] \quad (13)$$

$$\tilde{r}_{xy23}^{-} = \tilde{r}_{xy23} - \frac{e^{-j2\beta_0}}{(1+e^{-j2\beta_0}\tilde{r}_{yy32})^2 - (e^{-j2\beta_0}\tilde{r}_{xy32})^2}[(\tilde{t}_{yy23}\tilde{t}_{xy32} + \tilde{t}_{xy23}\tilde{t}_{yy32}) \cdot \\ (1+e^{-j2\beta_0}\tilde{r}_{yy32}) - e^{-j2\beta_0}\tilde{r}_{xy32}(\tilde{t}_{yy23}\tilde{t}_{yy32} + \tilde{t}_{xy23}\tilde{t}_{xy32})] \quad (14)$$

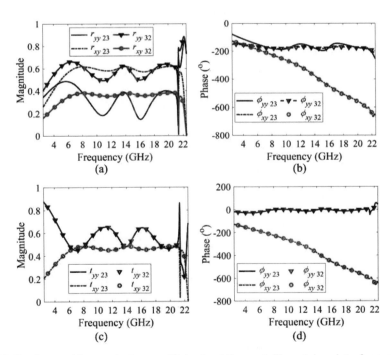

Fig. 7 Reflection and transmission coefficients at the metallic patches interface obtained by the full-wave simulations. (a) Magnitudes and (b) phases of the reflection coefficient. (c) Magnitudes and (d) phases of the transmission coefficient.

Reflection and transmission coefficients at the metallic patches interface obtained by simulations are shown in Fig. 7. Bring those parameters into the above formula (11)-(14), the overall co- and cross-polarized reflectance at the different interfaces obtained by simulations and theory calculations are shown in Fig. 8. It can be seen that the minimum operating frequency without substrate is higher than that with substrate. Besides, the introduction of the substrate with an optimized thickness h generates more resonances and improves the polarization conversion efficiency. Since the operational wavelength is much larger than the dimensions of the unit cell at low-frequency region, a good approximation is to keep only zero-order Bragg modes when analyzing the interference process, as inter-order couplings are expected to be weak. The neglect of near-field interactions is the reason for the difference at high-frequency region between theoretical calculation and simulation. In general, the calculated and simulated results are similar to each other, which provides a theoretical basis for the performance of the double-layer PCM better than the single layer one.

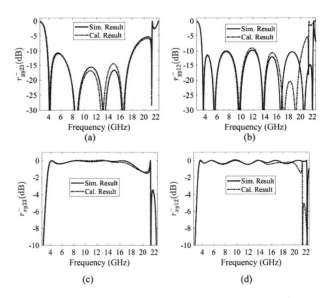

Fig. 8 Reflection coefficients at the different interfaces obtained by the simulations and the theory calculations: (a) r^-_{yy23}, (b) r^-_{yy12}, (c) r^-_{xy23}, and (d) r^-_{xy12}.

3.3 Surface Current Distributions

In order to get a better physical insight, Fig. 9 illustrates the surface current distributions at resonant frequencies of 3.65, 5.6, 9.78, 13.69, 16.78, 20.53, and 21.9 GHz. It is quite helpful to show the features of both resonances at the seven frequencies, and then, to determine the contributions of electric and magnetic modes. Magnetic resonances are generated by anti-parallel currents coupling between the metallic patches and the ground plane due to the circulating current flow. Electric resonances are generated by parallel currents between them. At the lowest resonant frequency of 3.65 GHz [see Fig. 9(a)], it can be observed that surface currents flow along the outer L-shaped patches without changing direction, which makes the outer patches equivalent to a cut-wire resonator in the fundamental resonant mode. The currents on the ground plane are antiparallel to the induced currents on the patches layer generating magnetic resonance. For Fig. 9 (b), the currents on the patches layer are perpendicular to the induced currents on the ground plane. According to the principle of vector decomposition, the black arrows can be decomposed into two components including the green and blue arrows. For the currents indicated by blue arrows, they indicate electric resonance due to the parallel direction; while for the currents represented by green arrows, they form magnetic resonance duo to the directions are antiparallel. Therefore, the mode pattern of 5.6 GHz is the combination of both electric and magnetic resonances.

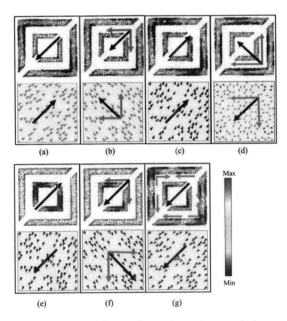

Fig. 9 Surface current distributions of metallic parts and ground plane of the proposed PCM for normal y-direction incident wave at the seven resonant frequencies: (a) 3.65 GHz, (b) 5.6 GHz, (c) 9.78 GHz, (d) 13.69 GHz, (e) 16.78 GHz, (f) 20.53 GHz, and (g) 21.9 GHz.

Parallel surface currents are induced by the metallic resonator and the ground plane, forming electric dipole resonance in 9.78 GHz, as shown in Fig. 9(c). From Fig. 9(d), it can be observed that surface current currents flow along the outer L-shaped patches and change direction in the half section, which is the characteristic of the second-order resonant mode. Similar to Fig. 9(b), due to the combination of both electric and magnetic resonances, the currents on the patches layer are perpendicular to the induced currents on the ground plane. Additionally, the resonance patterns of Figs. 9(e) and (f) are similar to Figs. 9(a) and (b), respectively. The difference is the surface currents are transferred from the outer L-shaped patches to the inner L-shaped patches. The third-order mode resonant mode of outer L-shaped patches and the fundamental resonant mode of the inner L-shaped patches are excited in Fig. 9(g). According to the vector synthesis principle, all arrows can be combined into a black arrow, the resultant currents on the metallic patches and ground plane are parallel to each other generating electric resonance. In summary, the types of resonance modes in Figs. 9(a) and (e) are magnetic modes; the resonant frequencies of Figs. 9(c) and (g) are electric resonance; the resonance modes of Figs. 9(b), (d), and (f) are induced by the combination of both electric and magnetic resonances. Consequently, the proposed PCM works in an ultrawideband frequency domain due to the resonances mentioned above.

3.4 Oblique Incidence Performance and Parametric Analysis

It is important to investigate the polarization conversion of the proposed PCM for oblique incidence. Fig. 10 shows simulated PCRs of the proposed PCM for different oblique incident angles. It is clearly shown that the incident angles have a great influence on the bandwidth of polarization conversion. The main reason is the propagation phase changes between normal and oblique incidence, which create a destructive interference condition at some frequencies. Due to the additional propagation phase changes more drastically at the higher frequencies, PCR rapidly decreases at an increment of incident angle.

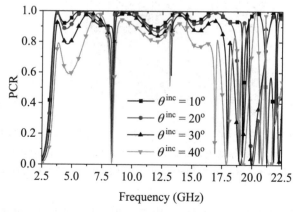

Fig. 10 Simulated PCRs of the proposed PCM for different oblique incident angle.

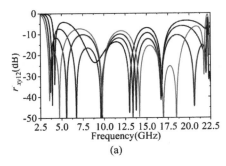
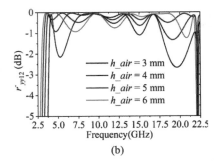

Fig. 11 Simulated coefficients of the proposed PCM for different thickness (h_air) of the air layer: (a) r^-_{yy12}, and (b) r^-_{xy12}.

In order to investigate the influence of the height (h_air) of the air layer on the polarization conversion characteristics, we simulated the reflection coefficients of co- and cross-polarization with different thicknesses, as displayed in Figs. 11(a) and (b), respectively. It can be seen that increasing the height of air can generate more resonant frequencies, i.e., when the thicknesses of air space are 3 and 6 mm, the proposed PCM has four and seven resonant frequencies, respectively; and shifts the operating frequency band to lower frequency. The near-equal ripple reflection coefficients with an optimized thickness of 5 mm are selected to balance the bandwidth and polarization conversion efficiency.

4. Conclusion

In this paper, an ultrawideband and high-efficient L-shaped PCM was designed, simulated, and fabricated. The bandwidth of the PCM was broadened to 6.5:1 with PCR more than 0.9, which is wider than other PCMs. Besides, The PCR can achieve 100% at seven resonant frequencies. The root cause of multi-resonance and polarization conversion behaviors was analyzed through the transmission line theory and equivalent circuit model. We showed that the fundamental reason for the polarization conversion is a certain reflection phase difference in u- and v-directions, which is caused by the anisotropy of the equivalent impedance. Besides, we proved that adding dielectric superstrate can widen the bandwidth using simulations and the interference theory. Simulation, theoretical calculation and measurement results validate the capability of the proposed PCM to convert linearly polarized waves to its orthogonal direction in an ultra-wide frequency range.

Funding. National Natural Science Foundation of China (61701448 and 62071436).

Disclosures. The authors declare no conflicts of interest.

Data availability. Data underlying the results presented in this paper are not publicly available at this time but may be obtained from the authors upon reasonable request.

References

[1] GAO X, HAN X, CAO W, et al. Ultrawideband and high-efficiency linear polarization converter based on double V-shaped metasurface[J]. IEEE Trans. Antennas Propag., 2015, 63 (8): 3522-3530.

[2] ALETA F, GENEVET P, KATS M A, et al. Aberration-free ultrathin flat lenses and axicons at telecom wavelengths based on plasmonic metasurfaces[J]. Nano Lett, 2012, 12: 4932-4936.

[3] MA H F, Cui T J. Three-dimensional broadband ground-plane cloak made of metamaterials[J]. Nat. Commun., 2010, 1 (3): 211-216.

[4] ZHANG J, CHEUNG S W, YUK T I. Design of n-bit phase shifters with high power-handling capability inspired by composite right-left-handed transmission line unit cells[J]. IET Microw. Antennas Propag., 2010, 4 (8): 991-999.

[5] NGUYEN A B, LEE J W. A K-band CMOS phase shifter MMIC based on a tunable composite metamaterial[J]. IEEE Microw. Wireless Compon. Lett., 2011, 21 (6): 311-313.

[6] SUN L K, CHENG H F, ZHOU Y J, et al. Broadband metamaterial absorber based on coupling resistive frequency selective surface[J]. Opt. Express, 2012, 20 (4): 4675-4680.

[7] YANY M, ZHOU X Y, QIANG C, et al. Diffuse reflections by randomly gradient index metamaterials[J]. Opt. Lett., 2010, 35 (6): 808-810.

[8] XI Y, JIANG W, HONG T, et al. Wideband and wide-angle radar cross section reduction using a hybrid mechanism metasurface[J]. Opt. Express, 2021, 29 (14): 22427-22441.

[9] ZHAO Y, ALU A. Manipulating light polarization with ultrathin plasmonic metasurfaces[J]. Phys. Rev. B, 2011, 84: 205428.

[10] MENZEL C, HELGERT C, ROCKSTUHL C, et al. Asymmetric transmission of linearly polarized light at optical metamaterials[J]. Phys. Rev. Lett., 2010, 104: 253902.

[11] AMIT K B, SHASHANK S K, SISIR K N. Linear-to-cross-polarization transmission converter using ultrathin and smaller periodicity metasurface[J]. IEEE antennas wireless propag. lett., 2019, 18 (7): 1433-1437.

[12] LI S J, CAO X Y, XU L M, et al. Ultra-broadband reflective metamaterial with RCS reduction based on polarization convertor, information entropy theory and genetic optimization algorithm[J]. Sci. Rep., 2016, 6: 37409.

[13] ZHAO Y, CAO X, GAO J, et al. Jigsaw puzzle metasurface for multiple functions: polarization conversion, anomalous reflection and diffusion[J]. Opt. Express, 2016, 24 (10): 11208-11217.

[14] JIA Y, LIU Y, GUO Y J, et al. A dual-patch polarization rotation reflective surface and its application to ultra-wideband RCS reduction[J]. IEEE Trans. Antennas Propag., 2017, 66 (6): 3291-3295.

[15] JIA Y, LIU Y, ZHANG W, et al. Ultra-wideband and high-efficiency polarization rotator based on metasurface[J]. Appl. Phys. Lett., 2016, 109 (5): 051901.

[16] XU, LI R, QIN J, et al. Ultra-broadband wide-angle linear polarization converter based on H-shaped metasurface[J]. Opt. Express, 2018, 26 (16): 20913-20919.

[17] GAO X, YANG W L, MA H F, et al. A reconfigurable broadband polarization converter based on an active metasurface[J]. IEEE Trans. Antennas Propag., 2018, 66 (11): 6086-6095.

[18] SUN S, JIANG W, LI X, et al. Ultrawideband high-efficiency 2.5-dimensional polarization conversion metasurface and its application in RCS reduction of antenna[J]. IEEE Antennas Wireless Propag. Lett, 2019, 18 (5): 881-885.

[19] LIU C, GAO R, WANG Q, et al. Design of ultra-wideband linear cross-polarization conversion metasurface with high efficiency and ultra-thin thickness[J]. J. Appl. Phys., 2020, 127 (15): 153103.

[20] KARAMIRAD M, GHOBADI C, NOURINIA J. Metasurfaces for wideband and efficient polarization rotation[J]. IEEE Trans. Antennas Propag., 2021 69 (3): 1799-1804.

[21] ZHANG Z, WANG J, FU X, et al. Single-layer metasurface for ultra-wideband polarization conversion: bandwidth extension via Fano resonance[J]. Sci. Rep., 2021, 11: 585.

[22] BAO Y, SONG J M. Effective medium model for multi-layered anisotropic media with different orientations[J]. Appl. Comput. Electrom., 2017, 32 (6): 491-497.

[23] Chew W C. Waves and fields in inhomogeneous media[M]. Oxford: Oxford University, 1995.

[24] GRADY N K, HEYES J E, CHOWDHURY D R, et al. Terahertz metamaterials for linear polarization conversion and anomalous refraction[J]. Science, 2013, 340 (6138): 1304–1307.

[25] TUNG N T, THUY V T T, PARK J W, et al. Left-handed transmission in a simple cut-wire pair structure[J]. J. Appl. Phys., 2010, 107 (2): 023530.

On the Use of Hybrid CFIE-EFIE for Objects Containing Closed-Open Surface Junctions[*]

1. Introduction

In the analysis of electromagnetic (EM) scattering or radiation properties, the integral equations in conjunction of the method of moments (MoM) are competitive approaches. During the numerical modeling of perfect electric conductor (PEC) objects which often contain both open and closed surfaces, traditionally, the electric field integral equation (EFIE) is formulated due to its independence of the surface type. Unfortunately, for the closed PEC part, using the EFIE alone may encounter the interior resonance problem. Moreover, discretizing the EFIE which is a first-kind Fredholm integral equation usually yields an ill-conditioned matrix equation that is difficult to converge during the iterative solution. To avoid the interior resonance problem as well as to improve the matrix condition, some articles proposed the so-called hybrid combined field integral equation-electric field integral equation (CFIE-EFIE). That is, on the closed PEC parts of the objects establish the second-kind CFIE, which is derived from the linear combination of the EFIE and the magnetic field integral equation (MFIE), while the open parts still keep the EFIE. When the major part of the object is closed, the hybrid CFIE-EFIE can improve the solving efficiency substantially. Nevertheless, the existing articles only presented the discussions when the closed and open parts are separate, or dealt with the surface-wire junctions. For the objects containing closed-open surface junctions, the derivation of a rational CFIE-EFIE is more complicated, which will be shown in Section II of this letter.

In the process of MoM solution, the induced surface current is expanded with a series of basis functions. Because of the convenience of discretizing arbitrary surfaces and the quality of being free of pseudo line charges, the divergence-conforming RWG basis functions based on triangular patches are being widely used. However, with the RWG basis functions, the numerical results from the MFIE are usually not as accurate as that from the EFIE. This phenomenon will be more obvious when the calculated objects contain sharp edges or tips. Some researchers focused on looking for the reasons of this inaccuracy, such

---

[*] The paper was originally published in *IEEE Antennas and Wireless Propagation Letters*, 2021, 20 (7), and has since been revised with new information. It was co-authored by Jinbo Liu, Jin Yuan, Wen Luo, Zengrui Li, and Jiming Song.

as the singularities arising in the outer integrals, the improper expression of solid angle, and so on. Among them, the most likely one is that for the MFIE, the solution accuracy strongly depends on the quality of the current expression, while the "normal-linear" and "tangential-constant" RWG basis functions cannot properly represent an arbitrarily continuous current distribution. To overcome this problem, lots of novel basis functions were proposed. The set of curl-conforming $\hat{n} \times$ RWG basis functions was used to improve the MFIE accuracy, while it is not suitable for the solution of CFIE. The monopolar-RWG basis functions were proposed in [18] for the sharp-edged objects accurately solved by the MFIE, and were then combined with the RWG basis functions to form a hybrid discretization scheme for the CFIE implementation by setting the monopolar-RWG for those edges between non-coplanar triangles and the RWG for the others. Besides, the linear-linear (LL) basis functions, also called as Trintinalia-Ling (TL) functions, which are "normal-linear" and "tangential-linear" and capable of expressing any linear current distribution, are attractive. The LL basis functions were first employed for the accurate solution of EFIE, and then extended to the MFIE and CFIE. It was shown that with the use of the LL functions, the results from the MFIE and CFIE can be significantly improved compared with the RWG functions.

However, since one LL basis function concurrently contains two linear vector functions associated with each common edge shared by two adjacent triangles, for a same object, the number of unknowns using the LL basis functions is the double of that using the RWG functions. As a result, the use of LL basis functions sacrifices the computational efficiency. As is well known, the distribution of induced current usually changes rapidly over the fine structures, while the change is slow for the smooth surfaces. Based on this fact, in this letter, the LL and RWG basis functions are simultaneously used to expand the surface current for the solution of the CFIE-EFIE, which will be established in Section II for the objects containing closed-open surface junctions. To be more specific, the LL basis functions are used to express the current on the fine structures, while the RWG basis functions are on other relatively smooth surfaces. The validity of this strategy is verified in Section III.

2. CFIE-EFIE Formulations and LL Basis Functions

Consider a PEC object in the free space that contains both closed surface Sc and open surface So, illuminated by an incident EM wave $\vec{E}^i$, $\vec{H}^i$ from an arbitrary direction. By vanishing the tangential component of total electric field, the EFIE is formed on all the open and closed surfaces . Imposing the boundary condition on the magnetic field over the closed surface Sc, the MFIE can be obtained and linearly added to the EFIE to form the so-called hybrid CFIE-EFIE as [4-6].

$$\text{CFIE} = \alpha(\vec{r})\text{EFIE} + \eta_0 \beta(\vec{r})\text{MFIE} \tag{1}$$

where both α and β are $\vec{r}$-dependent real combined coefficients, and η_0 is the intrinsic impedance of the free space. In [5], it was stated that $\beta(\vec{r}) = 1 - \alpha(\vec{r})$, and $0 < \alpha(\vec{r}) < 1$ when $\vec{r} \in S_c$ while $\alpha(\vec{r}) = 1$ for $\vec{r} \in S_o$. If the closed and open surfaces are totally separate, we can take the values of α and β like this without any doubt. On the contrary, however, if the object contains closed-open surface junctions where the closed and open surfaces have conjunct boundary, how to set up the values of α and β is worth further discussion.

Using the Galerkin's MoM, (1) is transformed into a generalized impedance matrix equation. During the current discretization, at the closed-open junctions, the basis functions are defined using the rule established in [23] to ensure no line charge accumulation. The matrix entry Zji, which denotes the interaction between the ith basis function $\vec{f}_i$ whose domain is Si and the jth testing function $\vec{f}_j$ with domain Sj, is obtained by

$$\begin{aligned} Z_{ji} = & j\omega\mu_0 \int_{S_j} \alpha(\vec{r})\vec{f}_j(\vec{r}) \cdot \int_{S_i} \vec{f}_i(\vec{r}')G dS' dS \\ & + \frac{j}{\omega\varepsilon_0} \int_{S_j} \alpha(\vec{r})\vec{f}_j(\vec{r}) \cdot \nabla \int_{S_i} \nabla'_S \cdot \vec{f}_i(\vec{r}')G dS' dS \\ & + \frac{\eta_0}{2} \int_{S_j} \beta(\vec{r})\vec{f}_j(\vec{r}) \cdot \vec{f}_i(\vec{r}) dS \\ & + \eta_0 \int_{S_j} \beta(\vec{r})\vec{f}_j(\vec{r}) \cdot \hat{n}(\vec{r}) \times P.V. \int_{S_i} \vec{f}_i(\vec{r}') \times \nabla G dS' dS \end{aligned} \tag{2}$$

where $j = \sqrt{-1}$, ε_0 and μ_0 are the permittivity and permeability of the free space, P.V. means the principal value integral, $\nabla'_S \cdot$ denotes the surface divergence operation, and $G = G(\vec{r}, \vec{r}')$ is the Green's function in the free space. The jth element of the excitation vector is

$$V_j = \int_{S_j} \left[\alpha(\vec{r})\vec{f}_j(\vec{r}) \cdot \vec{E}^i(\vec{r}) + \eta_0 \beta(\vec{r})\vec{f}_j(\vec{r}) \cdot \hat{n}(\vec{r}) \times \vec{H}^i(\vec{r}) \right] dS \tag{3}$$

In (2), it is noticed that for the second term, the gradient operator is placed on the observation point $\vec{r}$, leading to a two-order singularity during $\vec{r} \to \vec{r}'$. To reduce the order of singularity, taking the surface Gauss theorem, the second term of (2) is usually transformed into

$$\int_{S_j} \alpha(\vec{r})\vec{f}_j(\vec{r}) \cdot \nabla \int_{S_i} \nabla'_S \cdot \vec{f}_i(\vec{r}')GdS'dS$$

$$= \left\{ \begin{array}{l} \int_{S_j} \nabla_S \cdot \left[\alpha(\vec{r})\vec{f}_j(\vec{r}) \int_{S_i} \nabla'_S \cdot \vec{f}_i(\vec{r}')GdS' \right] dS \\ -\int_{S_j} \nabla_S \cdot \left[\alpha(\vec{r})\vec{f}_j(\vec{r}) \right] \int_{S_i} \nabla'_S \cdot \vec{f}_i(\vec{r}')GdS'dS \end{array} \right\} \quad (4)$$

$$= \left\{ \begin{array}{l} \int_{\partial S_j} \alpha(\vec{r}) \left[\hat{n}_{\partial S_j} \cdot \vec{f}_j(\vec{r}) \right] \int_{S_i} \nabla'_S \cdot \vec{f}_i(\vec{r}')GdS'dl \\ -\int_{S_j} \nabla_S \cdot \left[\alpha(\vec{r})\vec{f}_j(\vec{r}) \right] \int_{S_i} \nabla'_S \cdot \vec{f}_i(\vec{r}')GdS'dS \end{array} \right\}$$

where $\hat{n}_{\partial S_j}$ denotes the outer-normal direction of ∂S_j, the boundary of S_j. Through this transformation, the singularity order is degraded to one. On the other hand, it is observed that the surface divergence operators are placed on not only the single $\vec{f}_i(\vec{r}')$ but also the product $\alpha(\vec{r})\vec{f}_j(\vec{r})$, both of which are then restricted to be divergence conforming. Under this restriction, if $\vec{r} \in S_j$ which belongs to a junctional region contains both the part of closed surface Sc and the part of open surface So, the value of $\alpha(\vec{r})$ for $\vec{r} \in S_c$ and that for $\vec{r} \in S_o$ must be the same. In other words, $\alpha(\vec{r})$ should be constant everywhere. Therefore, in the implementation, we set

$$\alpha(\vec{r}) = \alpha_0 \qquad \forall \vec{r} \in S_c + S_o$$
$$\beta(\vec{r}) = \begin{cases} 1 - \alpha(\vec{r}) & \forall \vec{r} \in S_c \\ 0 & \forall \vec{r} \in S_o \end{cases} \quad (5)$$

while α_0 is constant and $0 < \alpha_0 < 1$ ($\alpha_0 = 0.5$ in all numerical examples presented later). Please note that mathematically, the values of α and β in depend on the position of observation point, but not the row number of the matrix equation as [4-6].

In the choice of basis functions to solve the CFIE-EFIE, because there is no accumulation of pseudo line charges, the divergence-conforming RWG basis functions are widely used. However, as mentioned above, the set of RWG basis functions cannot express arbitrary current distribution, while the MFIE is sensitive to the accuracy of current expression. Therefore, if the CFIE-EFIE that contains MFIE is used to model the object, the use of RWG basis functions may lead to inaccurate results. To express the surface current more accurately, a set of LL basis functions has been developed. Similar to the RWG basis functions, the LL basis functions are also defined on pairs of adjacent triangles. The LL basis function shares the following two properties with the RWG basis function: 1) Its normal component on the common edge is continuous when across the common edge, while that on the non-common edges is strictly equal to zero. 2) Its surface divergence is piecewise uniform which is inversely proportional to the corresponding triangle area, accomplishing charge neutrality over the pair of adjacent triangular patches. Actually, adding the two linear functions of the ith LL function can obtain the ith RWG function. Due to this property, the computational code using the LL basis functions can be obtained by modifying the conventional one using the RWG basis functions easily, while the LL and RWG basis functions can be simultaneously used to discretize objects without worrying about the mesh boundary continuity.

3. Numerical Validations

In the following calculations, the GMRES with a restart number 100 is used as the iterative solver to reach 0.001 residual error. All calculations are executed serially on a workstation with 3.2 GHz CPU and 16 GB RAM.

In the first case, using the RWG basis functions, the bistatic radar cross section (RCS) at xoz plane of a PEC semisphere of radius 1.5λ clung to a square plate of side length 3.1λ, which is illuminated by an x-polarized plane wave propagating in -z-axis, is calculated. After discretization with an average 0.1λ mesh size, the numbers of triangles on the closed semisphere part and the other open part are 5,304 and 576, respectively, which results in 8,805 unknowns. The numerical results from the EFIE and CFIE-EFIE (CFIE for the closed semisphere and EFIE for the remaining open part) are shown in Fig. 1. In addition, the CFIE-EFIE result from the wrong choice of $\alpha(\vec{r})$, i.e., $\alpha(\vec{r})$ is 0.5 for $\vec{r} \in S_c$ and 1 for $\vec{r} \in S_o$, is also given. It is observed that the results from the EFIE and CFIE-EFIE are almost in excellent agreement everywhere, while the result with wrong α shows a totally unacceptable difference. It states that for simple objects, the rational use of the CFIE-EFIE companied with the RWG basis functions can give reliable results. During the iterative solution, the CFIE-EFIE reaches the convergence with 81 iterations and 1.6 sec CPU time, about four times faster than the EFIE converged after 329 iterations with 6.1 sec CPU time. On the other hand, if the area of the bottom plate becomes larger, the advantage of the CFIE-EFIE on the convergence speed will be weaker. That is to say, the CFIE-EFIE is actually effective only when the closed part occupies a main proportion of the calculated object.

In the second case, the radiation patterns and the input impedances of a monopole mounted on the center of a PEC box are calculated at 300 MHz. The size of the box is 1 m × 1 m × 0.1 m, and the length and width of the strip-shaped monopole are 0.25 m and 0.01 m, respectively. After discretizing, the number of triangles is 546. In the CFIE-

EFIE implementation, the EFIE and CFIE are applied to the open PEC monopole and the six faces of the closed PEC box, respectively. The monopole is fed with a delta-function voltage source associated with the common edge that belongs to the closed-open surface junction. According to [25, (4.2)], the incident electric field within the edge can be expressed as $\vec{E}^i = -\nabla\varphi$ with the electric potential φ. From the Maxwell's equation $\nabla \times \vec{E}^i = -j\omega\mu_0\vec{H}^i$, we have $\vec{H}^i \equiv 0$. Therefore, when the integral in (3) is executed over the closed triangular patches that contains the feed edge, the second term of the kernel related to the magnetic field is zero. The calculated radiation patterns are shown in Fig. 2, while the input impedances as well as the computational details such as the numbers of unknowns and iterations are listed in Table I. For comparison, the result from the EM simulation software Altair FEKO is also shown as the baseline. It is observed that compared with the FEKO result, the EFIE one shows a good agreement, while the CFIE-EFIE result has a clear difference. The maximum difference of the radiation patterns between the FEKO and CFIE-EFIE results occurring over the peak range (2.62 dBi vs. 1.62 dBi at about 56°) is about 1 dB. Physically, besides the monopole part, the top face of the box also has a big influence on the numerical results, while the influence of other five faces is believed to be slight. Modeled by the CFIE-EFIE and expanded by the RWG basis functions, the normalized magnitude of the current density on the top face of the PEC box is shown in Fig. 3 (a). It is evident that the center current nearby the fine feed port mightily changes and abruptly varies on both sides of the common edge. As the CFIE-EFIE solution accuracy involving the MFIE strongly depends on the quality of current expression, the result difference in Fig. 2 between RWG&EFIE and RWG&CFIE-EFIE implementations is obvious.

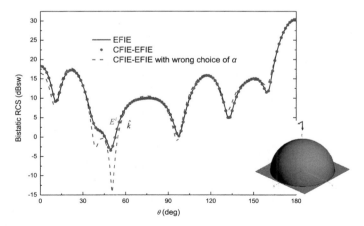

Fig. 1 Bistatic RCS of a PEC semisphere of radius 1.5λ clung to a square plate of side length 3.1λ.

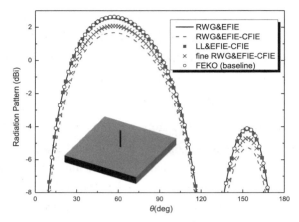

Fig. 2 Radiation patterns of a 0.25m×0.01m strip-shaped monopole mounted on a 1m×1m×0.1m PEC box at 300 MHz using different implementations.

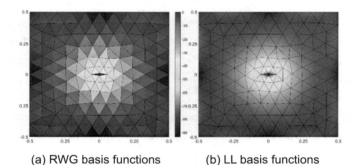

(a) RWG basis functions   (b) LL basis functions

Fig. 3 Normalized magnitude (in dB) of the current density solved by the CFIE-EFIE on the top face of the PEC cube, on whose center the striped monopole is mounted.

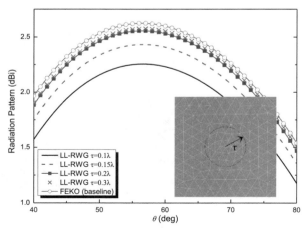

Fig. 4 Radiation patterns nearby the peak range using LL-RWG strategy with different values of τ to determine the LL region.

Table I

Computational Details for Different Implementations, Input Impedance from FEKO is 42.8+j27.0

Strategy	UN	IE	IN	Time (sec)	MD (dB)	Input impedance
RWG	814	EFIE	93	2.31	0.02	42.9+j26.9
RWG	814	CE	28	0.84	1.01	57.7+j18.6
LL	1,628	CE	57	3.92	0.03	43.6+j25.7
fine RWG	3,208	CE	121	10.1	0.54	51.9+j23.1
LR τ=0.1λ	832	CE	30	0.87	0.37	46.4+j24.6
LR τ=0.15λ	848	CE	32	0.90	0.19	45.0+j25.2
LR τ=0.2λ	874	CE	35	1.02	0.06	44.1+j25.4
LR τ=0.3λ	922	CE	49	1.49	0.05	43.9+j25.5

Note - CE: CFIE-EFIE, LR: LL-RWG, UN: unknowns number, IE: integral equation, IN: iterations number, Time: total CPU time, MD: maximal difference over the peak range between the computational and the FEKO results.

To express the surface current more accurately, using the LL basis functions, the monopole object is recalculated, while the numerical results and computational details are also shown in Fig. 2 and Table I, respectively. For comparison, we remesh the whole object with a fine mesh size and use the RWG basis functions alone to expand the current (denoted by fine RWG). It is seen that when the LL basis functions are used, the result from the CFIE-EFIE is in quite agreement with the FEKO result, which demonstrates that the accuracy problem of the CFIE-EFIE arising from the RWG basis functions can be mitigated by employing the LL basis functions. The normalized magnitude of the current density on the top face is shown in Fig. 3 (b). As expected, the current behavior obtained from the LL basis functions is clearly smoother than that from the RWG basis functions, which illustrates that the LL basis functions provide a much better representation of the current distribution. This is also the main reason why the result accuracy from the CFIE-EFIE can be improved. On the other hand, the fine meshes with the RWG basis functions have a very limited role in improving the numerical accuracy. In other words, when the RWG basis functions are used, the numerical accuracy cannot be significantly improved just through a fine-mesh scheme.

However, the cost of the improvement using the LL is that the number of unknowns is doubled, resulting in more memory usage, relatively slow convergence during the iterative solution, and long total CPU time, as shown in Table I. To alleviate this problem, the necessity of the LL basis functions is analyzed carefully. Logically, because of the location of the fine feed port, the magnitude of the current density on the center of the top face of the PEC cube is distinctly larger than other regions and drastically changes, where the discontinuity of the current distribution is then obvious. The phenomenon shown in Fig. 3 conforms to this anticipation. Due to this fact, we discretize the closed box parts modeled by the CFIE using different kinds of basis functions simultaneously, i.e., the LL basis functions are used to express the center region of the top face which contains the fine feed port, while the RWG basis functions are to other parts (denoted by LL-RWG). Since the striped monopole is modeled by the EFIE, it is still expressed by the RWG basis functions. On the choice of the LL and the RWG regions, we use a flexible mesh information-based strategy with the following filtering criterion

$$\begin{cases} \left|\vec{r}_m^c - \vec{r}_{\text{feed}}^c\right| \leq \tau & T_m \in \text{LL region} \\ otherelse & T_m \in \text{RWG region} \end{cases} \quad (6)$$

where $\vec{r}_m^c$ and $\vec{r}_{\text{feed}}^c$ are the centers of the triangle Tm on the top face and the feed port, respectively, and τ is a turning parameter. In the LL-RWG strategy, the computational accuracy and efficiency can be conveniently controlled by setting different τ. Please note that for this monopole object, the distance between some triangles on the bottom face and the feed port center may be also smaller than τ, but because the bottom face is not illuminated by the monopole directly, these triangles also belong to the RWG region. Under this criterion, with different values of τ, the numerical results and computational details are shown in Fig. 4 and Table I, respectively. It is found that the LL-RWG strategy does an excellent job on improving the calculation accuracy with slight more unknowns and an acceptable convergence rate. When $\tau=0.2\lambda$, the maximal difference over the peak range between the result from FEKO and that from the CFIE-EFIE with the LL-RWG (2.62 dBi vs. 2.56 dBi) is about 0.06 dB. Through massive numerical experiments, it is found that $\tau=0.2\lambda$ can give an acceptable accuracy for most radiation problems.

4. Conclusion

In this letter, the hybrid CFIE-EFIE is presented to model the objects that contain closed-open surface junctions. In the MoM solution, when the RWG basis functions are used to expand the current distribution, because the RWG basis functions cannot properly expand the current distribution, the CFIE-EFIE may result in less accurate solutions. The LL basis functions are introduced to solve this problem. Nevertheless, the number of unknowns using the LL basis functions is twice of that using the RWG basis functions, leading to a better accuracy but less efficient. To break this limitation, we use the RWG and LL basis functions to discretize different parts of the objects simultaneously according to their structural characters, while a criterion with a tuning parameter to determine how to choose the LL region is proposed. Numerical results show that this strategy has an acceptable accuracy with a high efficiency.

Acknowledgments

This work was supported in part by the National Natural Science Foundation of China under Grant 61971384 and Grant 62071436, and in part by the Fundamental Research Funds for the Central Universities under Grant CUC210B013 and Grant CUC19ZD001. *(Corresponding authors: Jinbo Liu and Zengrui Li.)*

Author Information

Jinbo Liu, Jin Yuan, Wen Luo, and Zengrui Li are with the State Key Laboratory of Media Convergence and Communication and the School of Information and Communication Engineering, Communication University of China, Beijing 100024, China (e-mail: liuj@cuc.edu.cn; zrli@cuc.edu.cn).

Jiming Song is with the Electrical and Computer Engineering, Iowa State University, Ames, Iowa 50011, USA (e-mail: jisong@iastate.edu).

References

[1] HARRINGTON R F. Field computation by moment methods[M]. New York: Macmillan, 1968.

[2] CHEW W C, JIN J M, MICHIELSSEN E, et al. Fast and efficient algorithms in computational electromagnetics[M]. London: Artech House, 2001.

[3] S RAO, WILTON, et al. Electromagnetic scattering by surfaces of arbitrary shape[J]. IEEE transactions on antennas and propagation, 1982, 30(3): 409-418.

[4] GÜREL L, ERGÜ Ö. Extending the applicability of the combined-field integral equation to geometries containing open surfaces[J]. IEEE antennas and wireless propagation letters, 2006, 5(1): 515-516.

[5] ÖZGÜR ERGUL, GUREL L. Iterative solutions of hybrid integral equations for coexisting open and closed surfaces[J]. IEEE transactions on antennas and propagation, 2009, 57(6): 1751-1758.

[6] FAN Z, DING D Z, CHEN R S. The efficient analysis of electromagnetic scattering from composite structures using hybrid CFIE-IEFIE[J]. Progress in electromagnetics research B, 2008, 10:131-143.

[7] KARAOSMANOĞLU B, ÖNOL C, ERGÜL Ö. Optimizations of EFIE and MFIE combinations in hybrid formulations of conducting bodies[C]. Turin, Italy: International Conference

on Electromagnetics in Advanced Applications. IEEE, 2015.

[8] YLA-OIJALA P, TASKINEN M. Calculation of CFIE impedance matrix elements with RWG and n × RWG functions[J]. IEEE transactions on antennas and propagation, 2003, (8): 51.

[9] CHAO H Y, ZHAO J S, CHEW W C. Application of curvilinear basis functions and MLFMA for radiation and scattering problems involving curved PEC structures[J]. IEEE transactions on antennas and propagation, 2003, 51(2):331-336.

[10] EWE W B, LI L W, CHANG C S, et al. AIM analysis of scattering and radiation by arbitrary surface-wire configurations[J]. IEEE transactions on antennas and propagation, 2007, 55: 162-166.

[11] ERGÜL Ö, GÜREL L. Investigation of the inaccuracy of the MFIE discretized with the RWG basis functions[C]. Monterey, CA, the USA: IEEE Antennas and Propagation Society Symposium, 2004.

[12] GÜREL L, ERGÜL Ö. Singularity of the magnetic-field Integral equation and its extraction[J]. IEEE antennas and wireless propagation letters, 2005, 4: 229-232.

[13] RIUS J M, UBEDA E, PARRON J. On the testing of the magnetic field integral equation with RWG basis functions in method of moments[J]. IEEE transactions on antennas and propagation, 2001, 49(11): 1866-1868.

[14] GRAGLIA R D, WILTON D R, PETERSON A F. Higher order interpolatory vector bases for computational electromagnetics[J]. IEEE transactions on antennas and propagation, 2002, 45(3): 329-342.

[15] WANG J, WEBB J P. Hierarchal vector boundary elements and p-adaption for 3-D electromagnetic scattering[J]. IEEE transactions on antennas and propagation, 1997, 45(12): 1869-1879.

[16] EDUARD, UBEDA, JUAN, et al. MFIE MoM-formulation with curl-conforming basis functions and accurate kernel integration in the analysis of perfectly conducting sharp-edged objects[J]. Microwave and optical technology letters, 2005, 44(4): 354-358.

[17] ERGÜL, ÖZGÜR, GÜREL. et al. The use of curl-conforming basis functions for the magnetic-field integral equation[J].IEEE transactions on antennas and propagation, 2006,54(7): 1917-1926.

[18] UBEDA E, RIUS J M. Novel monopolar MFIE MoM-discretization for the scattering analysis of small objects[J]. IEEE transactions on antennas and propagation, 2006, 54(1): 50-57.

[19] UBEDA, EDUARDSEKULIC, IVANRIUS, et al. Accurate, grid-robust and versatile combined-field discretization for the electromagnetic scattering analysis of perfectly conducting targets[J]. Journal of computational physics, 2020, 407(1).

[20] TRINTINALIA L C, LING H. First order triangular patch basis functions for electromagnetic scattering analysis[J]. Journal of electromagnetic waves & applications, 2001, 15(11): 1521-1537.

[21] ZGÜR ERGÜL, LEVENT GÜREL. Improving the accuracy of the magnetic field integral

equation with the linear-linear basis functions[J]. Radio science, 2006, 41(4): 1-15.

[22] ERGUL Z, GUREL L. Linear-linear basis functions for MLFMA solutions of magnetic-field and combined-field integral equations[J]. IEEE transactions on antennas and propagation, 2007, 55(4): 1103-1110.

[23] KOLUNDZIJA B M. Electromagnetic modeling of composite metallic and dielectric structures[J]. IEEE transactions on microwave theory and techniques, 2002, 47(7): 1021-1032.

[24] SAAD Y. A generalized minimum residual algorithm for solving nonsymmetric linear systems[J]. SIAM J. Stat. Comput, 1986, 7(3): 856-869.

[25] MAKAROV S N. Dipole and monopole antennas: the radiation algorithm[M]. Houston, USA: Princeton University Press, 2002: 57-88.

[26] https://www.altair.com/feko

Ultrawideband Frequency-Selective Absorber Designed with an Adjustable and Highly Selective Notch[*]

1. Introduction

Absorbers are widely used in stealth technology, RCS reduction, radomes, electromagnetic shielding, and many other applications that may reduce or selectively reduce reflection or transmission of electromagnetic waves. Research on absorbers began in 1952 when the first absorber was proposed by Salisbury with a quarter-wavelength thickness, which utilizes the $\lambda/2$ path difference between the incident wave and the return wave to achieve mutual cancellation. A Salisbury absorber is characterized by its simple principle, whereas it suffers from limited bandwidth. Multilayered Salisbury screen (Jaumann absorber) is one solution to improve bandwidth; however, stacked layers end with a large thickness. The wedge-tapered absorber proposed in 1971 showed the best absorption while also having a cumbersome structure.

Circuit analog absorber, which was proposed around the turn of this century, achieved a wider frequency response while maintaining a lightweight. Instead of homogeneous resistive sheets, band-stop FSSs were applied to absorbers, which were modeled by series *RLC* components. FSSs can be divided into conductive FSSs (lossless FSSs) and resistive FSSs (lossy FSSs). When applied to absorbers, lossy FSSs are either made using resistive sheets or loading lumped resistors. Employing multiple resonances, single-layer wideband absorbers were implemented with a rather thin structure. Recently, there have also been some novel absorber designs, such as adopting the combination of plasma and resistive FSS, fractal FSS, and absorbers on the base of a magnetic substrate and FSS. As we can see, however, no matter how innovative the designs, they are most often combined with FSSs.

At present, the demand for absorbers mainly lies in slimmer and lighter structures, broader bands, and better absorption effects. At the same time, for different application scenarios, such as radomes, in addition to the absorption performance, there are also needs for allowing electromagnetic waves to transmit through the radomes in the

---

[*] The paper was originally published in *IEEE Transactions on Antennas and Propagation*, 2021, 69 (3), and has since been revised with new information. It was co-authored by Yuxuan Ding, Mengyao Li, Jianxun Su, Qingxin Guo, Hongcheng Yin, Zengrui Li, and Jiming Song.

operation band without distorting the radiation performance of the antennas. These are the so-called frequency-selective absorbers (FSAs) that generate a transmission band based on a wideband absorber. In addition, notched absorbers that can be used as antennas' ground plane, serving as an FSS reflector, are designed to realize low-RCS antennas, which are another type of FSA, or more specifically, absorptive frequency-selective reflector (AFSR).

The design of a reflection band, either narrow or wideband, between two absorptive bands is not challenging work, if both the lossy and lossless FSSs are tuned simultaneously. However, realizing a reconfigurable and narrow reflective band with low insertion loss by merely tuning the lossless FSS is a considerable challenge.

In this article, we first derived a theoretical formula based on transmission line theory to indicate that the notch can be flexibly tuned by only adjusting the lossless layer. The absorption and reflection mechanism of the proposed FSS absorbers is shown in Fig. 1. Our research showed that the absorption band ($f_{L1}\ f_{H2}$) and the notch f_N can be controlled independently by the lossy layer and lossless layer, respectively. The independent design of the lossy layer and the lossless layer greatly simplifies the design process of the FSA with a dynamically tunable and highly selective notch. High selectivity (narrow reflection band) can guarantee two wide absorptive bands ($f_{L1}\ f_{H1}$ and $f_{L2}\ f_{H2}$), which can ensure good stealth performance. In most previous publications, the bandwidth of two absorption bands is limited. The dynamically tunable notch can better meet wideband antenna application. Both cases for electrically and geometrically tunable notches are provided for different application scenarios.

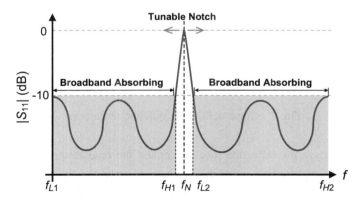

Fig. 1 Absorption and reflection mechanism of an FSA with a tunable notch.

The remainder of this paper is organized as follows. In Section II, the rationale for the independently regulated notch-band is analyzed. The modeling and performance of two absorbers with geometrically and electrically adjustable notches are detailed in Section

III separately. In Section IV, taking the geometrically controlled notched absorber as an example, the numerical solution of impedance conditions for the metal-backed band-notched absorber is derived based on the general equivalent circuit model (ECM). Strict ECMs are calculated in Section V. The fabricated and measured prototypes are described in Section VI, and one application scenario of our proposed FSA serving as a monopole's ground plane is given with a time-domain simulation and actual measurement results. Section VII provides concluding remarks.

2. Derivation of the Notch-Band Control Principle

As we know, an infinite periodic structure in free space can be seen as a space filter of electromagnetic waves, which is equivalent to a two-port network. The free space and substrates are equivalent to transmission lines of corresponding electrical length with characteristic impedance $Z_0 = 120\pi$ Ω and $Z_1 = Z_0/\sqrt{\varepsilon_r}$, respectively. The metal-backed substrate is equivalent to short-circuited transmission lines. The general ECM for a metal-backed notched absorber is shown in Fig. 2. Each FSS layer is equivalent to a shunt impedance in the circuit, where $Z_R = R_R + jX_R$ and $Z_F = jX_F$ denote the equivalent impedance of the lossy FSS and lossless FSS, respectively. The coupling effect between the lossless FSS and the metal ground is represented by C_T.

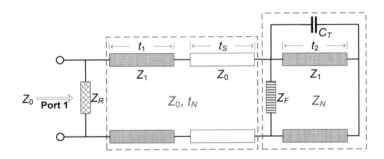

Fig. 2 General ECM for notched absorbers.

To further analyze the reflection mechanism and the regulating principle of the notch band, we derive a simplified circuit of the notched absorber. For the sake of simplicity, the equivalent electrical length from Port 1 to the lossless FSS layer is approximately expressed as a total $t_N \approx t_1\sqrt{\varepsilon_r} + t_S$, and the lossless FSS as well as the supporting substrate and the metal ground are considered together as a reactive load of $Z_N = jX_N$ (see the red dotted boxes in Fig. 2). Therefore, the transfer matrix can be simplified as follows:

$$\begin{bmatrix} A & B \\ C & D \end{bmatrix} = \begin{bmatrix} 1 & 0 \\ \dfrac{1}{Z_R} & 1 \end{bmatrix} \begin{bmatrix} \cos\delta_N & jZ_0\sin\delta_N \\ j\dfrac{\sin\delta_N}{Z_0} & \cos\delta_N \end{bmatrix}$$

$$= \begin{bmatrix} \cos\delta_N & jZ_0\sin\delta_N \\ \dfrac{\cos\delta_N}{R_R+jX_R} + \dfrac{j\sin\delta_N}{Z_0} & \cos\delta_N + \dfrac{jZ_0\sin\delta_N}{R_R+jX_R} \end{bmatrix}, \quad (1)$$

where $\delta_N = \beta t_N = (2\pi t_N / c)f$ is the total phase path between lossy FSS and lossless FSS. The reflection coefficient at the notch frequency is calculated as follows:

$$|S_{11}| = \left|\frac{AZ_N + B - CZ_0Z_N - DZ_0}{AZ_N + B + CZ_0Z_N + DZ_0}\right| = 1. \quad (2)$$

It should be mentioned that the absorption rate is calculated as $1-|S_{11}|^2$ with $S_{21}=0$ for the metal-backed absorbers throughout the whole discussion.

The complete expression of (2) is shown at the bottom of this page. The analytical solution to X_N is

$$X_N = -Z_0 \tan\delta_N. \quad (3)$$

Thus, the notch frequency f_N can be expressed by

$$f_N = \frac{c}{2\pi t_N} \arctan\left(-\frac{X_N}{Z_0}\right), \quad (4)$$

that is, when the thickness of dielectric slabs is fixed, the notch frequency has a function relationship with the reactance X_N of the lossless layer and is independent of the lossy layer. We thereby prove that the notch frequency can be fully controlled by adjusting the lossless layer, while the absorption band remains stable. The independent design of the lossy layer and the lossless layer greatly simplifies the design process of an FSA with a dynamically tunable and highly selective notch.

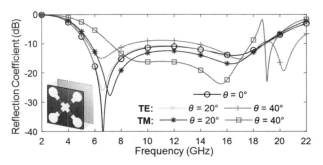

Fig. 3 Reflection coefficient of a crossed dipole-shaped wideband absorber under different incident angles.

3. Modeling and Performance of Two Absorbers

3.1 FSA with a Geometrically Controlled Notch-Band

For the first absorber designed with a geometrically controlled notch-band, a crossed dipole-shaped wideband absorber loaded with four lumped resistors is used as a baseline, which is composed of two perpendicular dipole-shaped patches in each unit and has a period of 10.2 mm. More than 10-dB RCS reduction for two polarizations is realized with a ratio bandwidth over 3.55:1 within the incident angle of 40°.

The lumped resistors are used to realize wideband impedance matching between the structured absorber and the free space. The absorption performance is mainly attributed to the lumped resistors due to the low dielectric loss. The absorption characteristic of the proposed wideband absorber (without a notch) is simulated and analyzed by the frequency-domain solver of CST, as shown in Fig. 3.

To generate a notch within the absorption band, a lossless FSS and a substrate supporting it are added to the wideband absorber. The complete model of the notched absorber is shown in Fig. 4. A deformation exists at the end of dipoles to increase the terminal capacitance. The sensitivity to incident angles and polarization has been decreased due to its rotationally symmetric and compact structure.

$$|S_{11}|^2 = \frac{\left(jX_N + Z_0 - X_N\tan\delta_N + jZ_0\tan\delta_N - \dfrac{jZ_0^2\tan\delta_N}{R_R - jX_R} - \dfrac{jX_N Z_0}{R_R - jX_R}\right)}{\left(jX_N + Z_0 - X_N\tan\delta_N + jZ_0\tan\delta_N + \dfrac{jZ_0^2\tan\delta_N}{R_R - jX_R} + \dfrac{jX_N Z_0}{R_R - jX_R}\right)}$$

$$\frac{\left(jX_N - Z_0 + X_N\tan\delta_N + jZ_0\tan\delta_N - \dfrac{jZ_0^2\tan\delta_N}{R_R + jX_R} - \dfrac{jX_N Z_0}{R_R + jX_R}\right)}{\left(jX_N - Z_0 + X_N\tan\delta_N + jZ_0\tan\delta_N + \dfrac{jZ_0^2\tan\delta_N}{R_R + jX_R} + \dfrac{jX_N Z_0}{R_R + jX_R}\right)} = 1$$

Ultrawideband Frequency-Selective Absorber Designed with an Adjustable and Highly Selective Notch

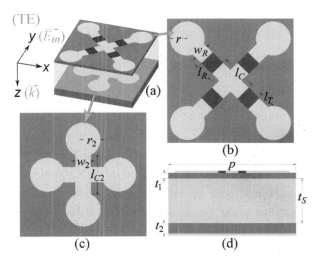

Fig. 4 Unit cell of the first absorber with a geometrically controlled notch band. The four chip resistors with a resistance value R = 125 Ω are indicated in red. (a) Perspective view. (b) Upper lossy layer. (c) Lower lossless layer. (d) Side view. (l_R = 1.2 mm, w_R = 1.25 mm, r = 1.4 mm, l_C = 2.9 mm, l_T = 1 mm, t_1 = 0.25 mm, t_S = 3.8 mm, t_2 = 1 mm, r_2 = 1.4 mm, w_2 = 1.25 mm, l_{C2} = 3.25 mm, p = 10.2 mm, ε_r = 2.65, and tan δ = 0.002)

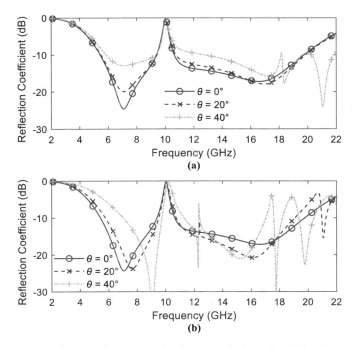

Fig. 5 Reflection coefficient of a geometrically controlled notched absorber under different incident angles. (a) TE polarization. (b) TM polarization.

It is well-known that the free space wave impedance at oblique incidence varies with polarization, and the reflection coefficient varies accordingly. Any polarization can be decomposed into TE and TM waves, while $Z_0^{TE}=Z_0/cos\theta$ and $Z_0^{TM}=Z_0 cos\theta$. The impedance variation trends with incident angle are opposite for the two polarizations, which means it is impossible to achieve perfect impedance matching for two polarizations simultaneously under oblique incidences; there must be compromises between these two polarizations. As shown in Fig. 5, the absorption bands are well-maintained when the incident angle is less than 40° for TE polarization or is less than 30° for TM polarization. With an increasing incident angle, the lowest frequency of the grating lobes decreases [4], i.e., more grating lobes appear in the operating frequency band, leading to deterioration of angular stability. In addition to the grating lobes, for TM only, we can see spikes at 12.35 GHz under oblique incidences, which are the so-called Fano resonance. These polarization-dependent Fano peaks result from the asymmetry of the lossless layer at oblique incidences. As we know, for TE polarization, the incident electric field is always in the y-direction, but for TM polarization, when the FSS is incident with an angle θ, there is a pitch angle between the incident electric field and the FSS plane; thus, the symmetry of the system is destroyed, and as a result, Fano resonance is enhanced.

One advantage of our design is that the position of the sharp notch is completely controlled by the lossless FSS whose shape is determined by three parameters. By adjusting these three parameters, the center frequency of the notch-band can be adjusted across the whole band. The frequency-domain simulation results are shown in Fig. 6.

In fact, with the other two parameters fixed, simply changing r_2 can adjust the notch position throughout the entire frequency band, as shown in Fig. 6(a). As r_2 increases, the center of the notch moves to the lower frequencies.

With r_2 and w_2 fixed, changing l_{C2} can adjust the notch position over a relatively narrow band, as shown in Fig. 6(b). As l_{C2} increases, the center of the notch moves to lower frequencies. When l_{C2} increases from 0.4 to 0.5 mm, there is a hop in the position of the notch and a loss in bandwidth; therefore, the value of l_{C2} should not be too large.

With r_2 and l_{C2} fixed, changing w_2 can fine-tune the notch position, which is shown in Fig. 6(c). As w_2 increases, the center of the notch moves to higher frequencies. Meanwhile, w_2 affects the width of the notch-band. As w_2 increases, the notch becomes wider, resulting in poor selectivity; therefore, the value of w_2 should not be too large.

3.2 FSA with an Electrically Controlled Notch-Band

The second notched absorber has a similar structure to the first but has different FSS patterns. This absorber is designed based on an octagonal ring-shaped wideband absorber (see inset in Fig. 7). The lossy FSS of this absorber is composed of an octagonal ring

Ultrawideband Frequency-Selective Absorber Designed with an Adjustable and Highly Selective Notch

and four lumped resistors in each unit and is rotationally symmetric. Good absorption performance is achieved within the incident angle of 30° for both TE and TM polarizations, as shown in Fig. 7. The operating frequency band ranges from 4.63 to 19.82 GHz for normal incidence.

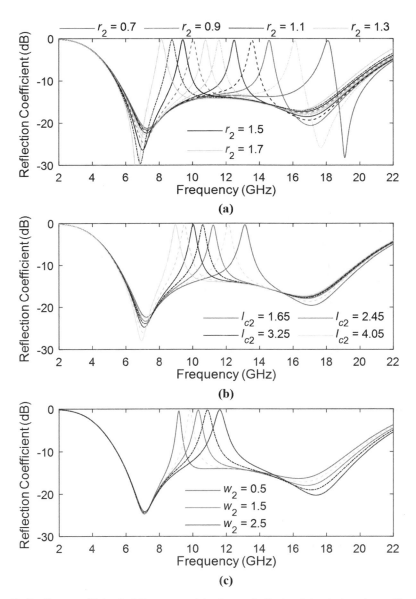

Fig. 6 Reflection coefficient of the geometrically controlled notched absorber with different values of (a) r_2, (b) l_{c2}, and (c) w_2 (mm).

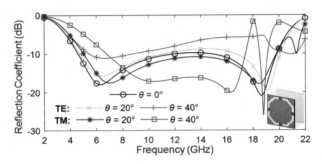

Fig. 7 Reflection coefficient of an octagonal ring-shaped wideband absorber under different incident angles.

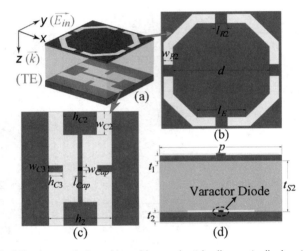

Fig. 8 Unit cell of the second absorber with an electrically controlled notch band. The four chip resistors with a resistance value $R_2 = 340\ \Omega$ are indicated in red. (a) Perspective view. (b) Upper lossy layer. (c) Lower lossless layer. (d) Side view. (l_{R2} = 1 mm, w_{R2} = 0.8 mm, l_E = 4.1 mm, d = 8.2 mm, t_1 = 0.5 mm, t_{S2} = 4.4 mm, t_2 = 1 mm, l_{Cap} = 0.4 mm, w_{Cap} = 0.3 mm, h_2 = 5.5 mm, h_{C2} = 2.9 mm, w_{C2} = 2.1 mm, h_{C3} = 1.3 mm, w_{C3} = 0.6 mm, p = 10.2 mm, ε_r = 2.65, and $\tan \delta$ = 0.002)

Fig. 8 shows the unit cell structure of the second notched absorber. A major difference from the geometrically controlled absorber is the use of a varactor diode on its lossless layer, which makes it possible to shift the notch position electrically. Hence, one can implement a real-time adjustment of the notch band by simply changing the bias voltage of the varactor. The lossless layer is based on an H-shaped pattern with a narrow slot on the edge, which permits a y-polarized notch, as seen in Fig. 9(a). When the incident electric field is perpendicular to the direction of the varactor diode, which means the varactor diode is inactive and the induced current on the lossless layer is very low; therefore, there will be no notch band. The H-shaped structure also makes it easy to apply a bias voltage to the varactor diodes. To narrow the notch-band and achieve high selectivity, some deformation is adopted with no more tautology. The simulation results of oblique incidences are shown in Fig. 9.

The type of varactor diode we chose is a MA46H120 with low parasitic capacitance and high Q, and the varactor has a linear tuning range from 0.14 to 1.1 pF. As shown in Fig. 10, the notch frequency decreases with increasing capacitance C under y-polarization. For x-polarization, the absorber retains wideband absorption characteristics and will not be affected by the capacitance C of the varactor. Therefore, when applied to antennas, the antenna polarization should be consistent with the polarization of the notch, which can effectively ensure antenna gain and achieve good stealth performance for both polarizations.

4. Impedance Condition for Notched Absorbers

In this section, all numerical calculations are based on the geometrically controlled notched absorber. The electrically controlled design can be solved similarly. To simplify the calculation of the numerical solution, only the impedance values of FSSs are set as dependent variables.

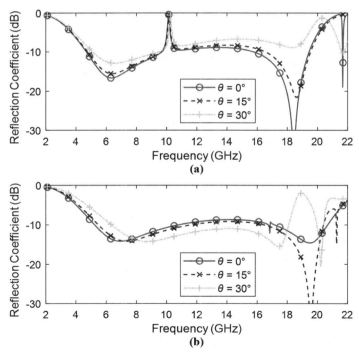

Fig. 9 Reflection coefficient of an electrically-controlled notched absorber under different incident angles when the capacitance value C of the varactor diode is 0.4 pF. (a) TE polarization. (b) TM polarization.

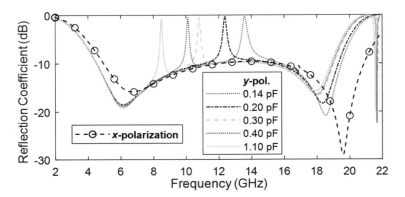

Fig. 10 Reflection coefficient of the electrically controlled notched absorber with different capacitance values C (pF).

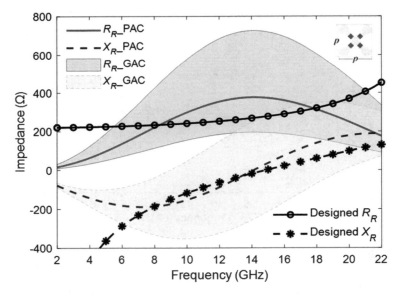

Fig. 11 PAC curves and GAC value range of lossy FSS for wideband absorber compared with impedance values of the freestanding lossy FSS in our design (t1 = 0.25 mm and tS = 4.9 mm).

First, we plug in the thickness values of the wideband absorber and calculate the values of ZR that meet the absorption conditions. The equation $|S_{11}|=0$ is solved numerically at each frequency point with the help of MATLAB, enabling us to obtain RR and XR curves satisfying the perfect absorption condition (PAC). For a practical design, we usually choose $|S_{11}| < -10dB$ (with a linear value of 0.316) as a criterion to evaluate the absorption performance, which was referred to as a general absorption condition (GAC). We successively get the value ranges of RR and XR satisfying GAC when XR and RR are constrained to PAC, respectively. When both the real part and the imaginary part of ZR come within the areas defined by GAC, less than −10 dB reflection can be realized. As shown in Fig. 11, the impedance values of the freestanding lossy FSS in our design lie in the absorption areas at the operating band.

Then comes the generation of the notch-band. From this point forward, we take into account Z_F, C_T, and t_2. As before, the thickness t_2 is set to a constant.

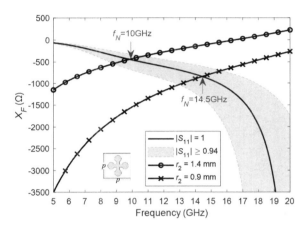

Fig. 12 Impedance conditions of freestanding lossless FSS at notch frequencies compared with impedance values of our designed lossless FSSs.
(t_1 = 0.25 mm, t_s = 3.8 mm and t_2 = 1 mm)

It should be noted that the coupling capacitance C_T is considerable when calculating the notch frequency f_N, which can be confirmed by observing the distribution of the electric field. The value of C_T directly depends on the dimensions of the lossless FSS, and because the dimensions of lossless FSS determine the notch frequency f_N, we can deduce that C_T has a function relationship with f_N. Using simulation software Ansys Electronics Desktop, we can get impedance curves of the freestanding lossless FSS and the lossless FSS backed with a grounded substrate, namely Z_F and Z_N (see Fig. 2), respectively. With the help of the optimization tool, we determined the values of C_T for different radius values of r_2, and then, the relationship between C_T and f_N is obtained by curve fitting.

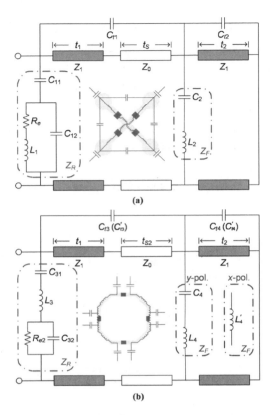

Fig. 13 Integrated ECM for (a) geometrically controlled notched absorber of two polarizations and (b) electrically controlled notched absorber of y- and x-polarization, respectively.

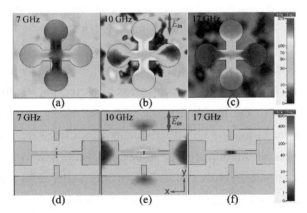

Fig. 14 Surface current distribution of lossless FSS layers at a lower absorption band of 7 GHz, notch frequency of 10 GHz, and a higher absorption band of 17 GHz. (a)-(c) Geometrically controlled lossless layer with r2 = 1.4 mm. (d)-(f) Electrically controlled lossless layer under y-polarization with C = 0.4 pF.

$$C_T = \frac{1}{0.0421 f_N^2 - 0.2142 f_N + 1.0366} \tag{5}$$

Plugging in the designed ZR of the upper lossy FSS, the reflection condition is calculated in an ideal world with $|S_{11}|=1$ and in a practical situation with less than 0.5 dB insertion loss, i.e., $|S_{11}| \geqslant -0.5$ dB (linear value of 0.94). The result is shown in Fig. 12. Through comparison with the designed impedance values of the freestanding lossless FSS, the notch frequency can be recognized and corresponds to the simulation results of Ansys Electronics Desktop. Moreover, the light blue ($|S_{11}| \geqslant 0.94$) can be seen to widen, which means the selectivity of notch band worsens as fN increases, and this trend is consistent with full-wave simulation result (see Fig. 6). The slight discrepancies should be attributed to the coupling effect between lossy FSS and lossless FSS, which is relatively weak and is not taken into consideration in the general ECM.

5. Equivalent Circuit Extraction and Analysis

By observing the distribution of electric field and magnetic field, the distribution of equivalent capacitance and inductance can be roughly inferred (see Fig. 13). For the geometrically controlled notched absorber, we see that the magnetic field concentrates on the metal cross in the center. The coupling between units is primarily electric coupling, which exists between the ends of adjacent dipoles, and the coupling is stronger at lower frequencies. As frequency increases, the coupling between units weakens, and the electric coupling is gradually distributed between the intersecting dipole ends within a unit. In general, we can deduce that the equivalent capacitance generated by the interaction between units is larger than that from within a unit, i.e., C_{11} representing the equivalent capacitance between units is much larger than C_{12} representing the terminal capacitance within a unit. The metal cross in the center contributes to the equivalent inductance L_1.

Similarly, for the octagonal ring-shaped lossy FSS of the electrically controlled notched absorber, it can be seen from the electric field distribution that the electric coupling between the units is stronger at lower frequencies and mainly spreads over the gaps in the direction of the incident electric field. Meanwhile, the magnetic field concentrates on the metallic stripes along the incident electric field. Two arms perpendicular to the incident electric field have almost no current flowing through. The equivalent capacitance in the direction of the incident electric field is much larger, that is, C_{31} is much larger than C_{32}. In this context, the octagonal rings loaded with resistors are modeled as distributed RLC components.

The band-stop FSS in free space, which can be modeled by a series LC circuit, resonates to be zero impedance when it presents reflection characteristics, and it is inductive below the resonant frequency and capacitive above the resonant frequency. In addition to the rigorous calculations in Section IV, there is a simple method to estimate the resonance frequency of the band-stop lossless FSS in the geometrically controlled FSA.

The traditional cross-shaped FSS is composed of two orthogonal microstrip lines, which allow it to work with an electric field in either direction. In free space, the first resonance occurs at a frequency point where the length of the microstrip line equals half-wavelength. In this instance, the lossless FSS is almost sandwiched by two F4B substrates, which makes the resonance frequency of the freestanding lossless FSS go down by about $\sqrt{\varepsilon_r}$; that is, the resonance frequency of the lossless FSS can be roughly estimated as $c/(2l_e \times \sqrt{\varepsilon_r})$, where le is the length of the microstrip line. However, while employing a traditional cross-shaped FSS to produce a notch, the adjustable range of notch-band is limited with a large insertion loss. To overcome these shortcomings, the crosswise dipole-shaped FSS with $l_e \approx l_{C2} + 4r_2$ is designed to increase terminal capacitance, and as a result, a sharp notch with an insertion loss less than 0.5 dB is obtained, which can be adjusted across the whole band.

To further validate the circuit model, the surface current distribution of the lower band-stop FSS is shown in Fig. 14. Strong surface current is excited only at the notch frequency and mainly along the direction of the incident electric field. At other frequencies, a very low surface current is induced on the surface of the lossless FSS, which means that the tangential electric field is very weak, and consequently, the band-stop FSS is transparent to EM waves and the FSA works as a wideband absorber.

The lossless FSS of the electrically controlled absorber, which is loaded with a varactor, is designed with several gaps to increase the distributed capacitance. After the simulation, it can be seen that the electric field concentrates on the meandering edges of the shape. Strong surface current is induced around the varactor at notch frequency, and a notch band can only be generated under a y-polarized wave (see Fig. 14(e)) when the incident electric field is parallel to the varactor. For x-polarization, because the incident electric field is perpendicular to the varactor, very low surface current is induced on the lossless layer, and as a result, there will be no notch band. Moreover, changing C cannot influence Z_F; thus, the S_{11} curve is not affected by the varactor and remains unchanged under x-polarization.

A general ECM for metal-backed notched absorbers has been discussed in Section II, as is shown in Fig. 2, where variables are functions of both frequency and dimension. For the variables in ECM to be independent of frequency, the upper lossy FSS is modeled by an RLC network, and the lower band-stop FSS is modeled by a series LC circuit. To accurately model the absorber under specific dimensions and values, Ct1 is introduced to ECM, which represents the coupling effect between lossy FSS and lossless FSS and is much smaller than Ct2. The final circuit schematics of our proposed FSAs are presented in Fig. 13. The first FSA is rotationally symmetric, so its ECM is the same for two polarizations under normal incidence.

Table I

Optimized values of ecm components for two FSAs

C_{11}/pF	C_{12}/pF	L_1/nH	R_e/Ω	C_2/pF
0.08	0.01	1.86	211.42	0.03
L_2/nF	C_{t1}/pF	C_{t2}/pF	C_{31}/pF	C_{32}/pF
1.92	0.01	0.3	0.13	0.03
L_3/nH	R_{e2}/Ω	C_4/pF	L_4/nH	C_{t3}/pF
3	269.74	0.04	1.49	0.03
C_{t4}/pF	L_4'/nH	C_{t3}'/pF	C_{t4}'/pF	
0.2	0.5	0.03	0.25	

As shown in Fig. 13(a), chip resistors in each unit are in parallel, so we have modified the formula in [21] and the total equivalent resistance can be estimated as

$$R_e \approx \frac{R_S}{N}\frac{S}{S_R}, \quad (6)$$

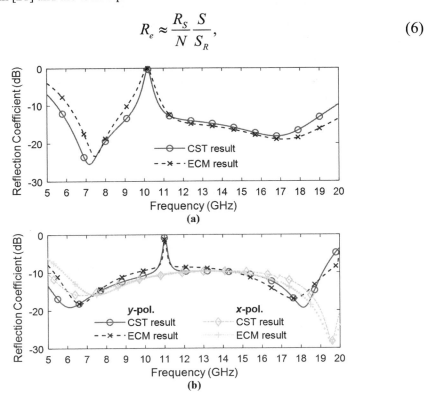

Fig. 15 Reflection coefficients obtained via full-wave simulation and equivalent circuit calculation. (a) Geometrically controlled notched absorber with r2 = 1.4 mm. (b) Electrically controlled notched absorber with C = 0.4 pF.

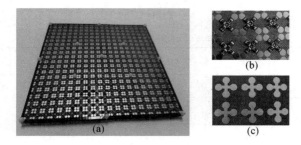

Fig. 16 Photos of the fabricated geometrically controlled FSA. (a) Assembled absorber. (b) Upper lossy layer. (c) Lower lossless layer.

where S is the patch area corresponding to the resistance branch, SR is the surface area of a chip resistor, $R_S = R \times w_R / l_R$ is the surface resistance of chip resistors, and N is the number of branches in parallel. The above estimated Re from observations is not accurate, so the optimization tool of Ansys Electronics Desktop is used to approximate the S_{11} results of the full-wave simulation when the impedance values of the freestanding FSSs have been taken as a reference primarily. The optimized parametric values of equivalent circuits are listed in Table I. The S_{11} result of our proposed ECM is consistent with the S_{11} curve simulated by CST (see Fig. 15). In the ECM, the dielectric loss is not taken into account, and only major distributed components are introduced, which are the main causes of deviation from the simulation results.

6. Experimental Verification

In this section, samples of the proposed FSAs with geometrically controlled and electrically controlled notch band have been manufactured and tested.

6.1 Measurement of a Geometrically Controlled FSA

For the geometrically controlled FSA, which is composed of 19×19 unit cells and is 200×200 mm² in size, we fabricated one lossy layer and one lossless layer and assembled them successively with nylon screws. As explained before, for the geometrically controlled FSA, we can obtain different notch frequencies by replacing lossless layers of different dimensions. In this experiment, we choose two instances with the only difference of r_2, which has a length of 1.4 mm and 0.9 mm, respectively, and their time-domain simulation results from CST are provided to validate the adjustability of the notch band.

As shown in Fig. 16, the metallic patterns were printed on F4B substrates (ε_r = 2.65, $tan\delta$ = 0.002) using the printed circuit board (PCB) technology, and four chip resistors (0805, 120 Ω) were soldered on the copper strips of each resistive element using the surface mount technology. To avoid warping of the upper thin substrate, 13 isolated columns were

Ultrawideband Frequency-Selective Absorber Designed with an Adjustable and Highly Selective Notch

sandwiched between the upper substrate and the lower substrate to keep them parallel.

The absorbing/reflecting performance of the geometrically controlled FSA was measured by the compact antenna test range system at the Science and Technology on Electromagnetic Scattering Laboratory, Beijing, China. To improve measurement precision, five pairs of standard linearly polarized horn antennas working at 4–6 GHz, 6–8 GHz, 8–12 GHz, 12–18 GHz, and 18–24 GHz were used for transmitting and receiving electromagnetic waves.

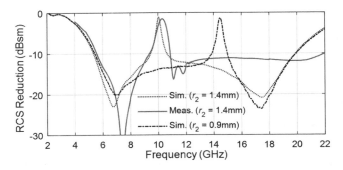

Fig. 17 Measured and time-domain simulated RCS reduction of the geometrically controlled FSA with different radius values of r_2.

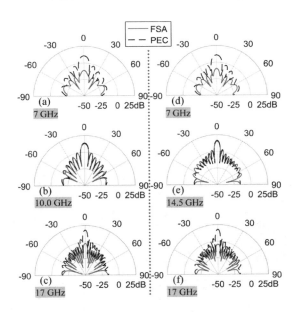

Fig. 18 Comparison of the bistatic scattering RCS between the proposed geometrically controlled FSA and equal-sized PEC surface under normal incidence at the lower absorption band of 7 GHz, the notch frequencies of 10.0 and 14.5 GHz, and the higher absorption band of 17 GHz, respectively. (a)-(c) Geometrically controlled FSA with r_2 = 1.4 mm. (d)-(f) Geometrically controlled FSA with r_2 = 0.9 mm.

The RCS is measured using a Keysight N5234A vector network analyzer. To eliminate the interferences from multiple reflections and the coupling effect between two horns, the time-domain gating function of the network analyzer was used to accurately measure the reflected wave. The RCS reduction results of the prototype were obtained after calibrating with a metal ground plane test.

Fig. 17 shows a comparison of the measured and simulated RCS reduction results of the geometrically controlled FSA with $r_2 = 1.4$ mm. The measured data are largely in keeping with the time-domain simulation results. Except for the fabrication error and resistor value tolerances, the slight discrepancy between time-domain simulation and measurement results should be attributed to the unevenness of the handmade air layer and the parasitic effect of lumped resistors, which was not considered in the full-wave simulation. Compared with the frequency-domain simulation results for infinite periodic structures, the higher insertion losses result from the edge effects of finite arrays. The time-domain simulation result of $r_2 = 0.9$ mm is also provided. Compared with the result for $r_2 = 1.4$ mm, the notch frequency shifts from 10.0 GHz to 14.5 GHz. The adjustability of the notch is proved in this way.

To reflect the applicability of the geometrically controlled FSA with bistatic scenarios, the time-domain simulation results of bistatic scattering RCS in the E-plane are shown in Fig. 18. At the absorption frequencies of 7 GHz and 17 GHz, the bistatic RCS of the FSA in the entire space is much lower than that of the equal-sized PEC ground surface. The bistatic stealth performance is better than diffuse scattering based on the phase cancellation method, whose sidelobe level will increase according to the law of conservation of energy. Moreover, the FSA serves as a PEC surface where almost all incident energy is reflected at notch frequencies of 10 GHz and 14.5 GHz for $r_2 = 1.4$ mm and $r_2 = 0.9$ mm, respectively.

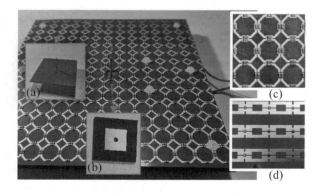

Fig. 19　Photos of the electrically controlled FSA and an FSA-backed monopole. (a) Photo of the fabricated monopole antenna. (b) Ground plane of the monopole. (c) Top lossy layer of the FSA. (d) Lower lossless layer of the FSA.

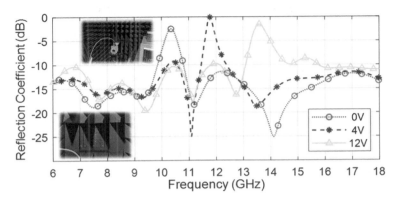

Fig. 20   Measured reflection of the electrically controlled FSA with different bias voltages.

6.2 Measurement of an Electrically Controlled FSA

For the electrically controlled FSA, we fabricated a prototype with 14×14 unit cells, which is shown in Fig. 19. The upper lossy layer is mounted with four chip resistors (0603, 340 Ω) in each unit, and the lower lossless layer is welded with one varactor diode (MA46H120) in each unit. The size of the F4B substrates is 147 mm × 143 mm due to the addition of bias lines printed on two sides to feed the varactor diodes. For subsequent antenna-loaded tests, one unit cell has been removed. The upper lossy layer is held up with ten nylon screws and gaskets to keep a 4.4 mm distance from the lower ossless layer.

The measurement setup is shown in the insets of Fig. 20. The electrically-controlled FSA was measured in an anechoic chamber using Keysight E5071A with time-domain gating, which was used to reduce multipath interferences during the measurement of the reflection coefficient. The aperture of the horn antenna for testing is 75 mm × 75 mm, and the horn antenna is 1 m away from the FSA. The FSA was surrounded with absorbent foam to reduce the effect of edge diffraction. Limited by this measurement setup, the far-field conditions were hard to meet for the entire frequency range.

In testing of the electrically controlled FSA, the notch frequency is regulated by changing the bias voltage of the varactor diodes. With an increasing bias voltage, the capacitance of the varactors decreases, and thereby, the notch moves to higher frequencies. As shown in Fig. 20, the variation trend of the notch band is consistent with previous conclusions.

To illustrate the application scenario of these band-notched FSAs, we conducted experiments of a monopole antenna backed with the electrically controlled FSA. As shown in Fig. 19, we fabricated a monopole antenna, and then made the monopole perpendicular to the surface of the FSA. The length of the monopole is 18 mm.

The monopole is fixed to an F4B substrate, which is 20 mm × 20 mm in size. A metal ground patch with a side length of 10 mm is printed on the back, as shown in Fig. 19(b).

It is learned that a well-structured FSA with more unit cells produces a better ground plane for the antenna, but there is a trade-off between efficiency and RCS. FSA-backed antennas suffer from low-efficiency, which is a compromise to low-RCS. From Fig. 21, we can see that modification to the central unit of FSA has little impact on the performance of RCS reduction; the FSA serving as a ground plane of the monopole effectively lowers the out-of-band RCS level of the monopole.

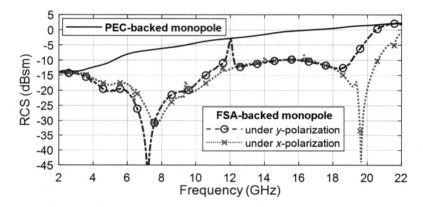

Fig. 21 Simulated RCS of an FSA-backed monopole under y-polarized and x-polarized normal incidences compared with a PEC-backed monopole.

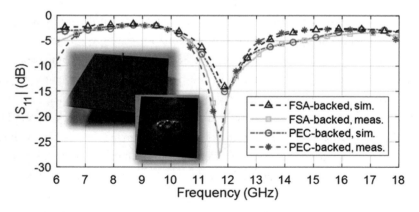

Fig. 22 Simulated and measured reflection of an FSA-backed monopole compared with a PEC-backed monopole. The insets are photos of the PEC-backed monopole.

Ultrawideband Frequency-Selective Absorber Designed with an Adjustable and Highly Selective Notch

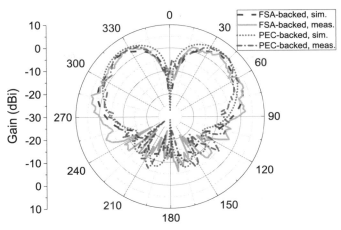

Fig. 23 Gain of an FSA-backed monopole compared with a PEC-backed monopole at a notch frequency of 11.8 GHz.

For comparison, we also measured a PEC-backed monopole antenna. The ground plane of the single PEC-backed monopole is 50 mm in diameter printed on the underside of a perforated F4B substrate, as shown in the insets of Fig. 22. The substrate for the monopole is 1 mm in thickness. According to Fig. 22, the center frequency of the FSA-backed monopole shifts slightly toward higher frequencies, and the bandwidth narrows, which is a consequence of near-field coupling. In Fig. 23, the scattering and radiation performance of the single PEC-backed monopole is also provided as a reference. From Figs. 22 and Fig. 23, it is clear that our designed band-notched FSA allows the monopole antenna to work properly at the notch band and maintains its radiation pattern, whereas at other frequencies, the FSA still works as an absorber, thereby reducing the RCS of the antenna.

For an electrically controlled FSA loaded with varactors, in actual use, we recommend using an analog-to-digital converter (DAC) to control the bias voltage of the varactor diodes and a microcontroller to control the DAC. In this way, the notch frequency of the FSA can be instantly manipulated by a computer via communication with the microcontroller. The operating frequency of the antenna is correlated with the notch frequency; thus, the RCS of a frequency-sweep antenna can be reduced in real time.

Table II provides a comparison of our two designs with two other reported band-notched FSAs. Only our designs have adjustable and highly selective notches. The tunable range of the electrically controlled FSA is limited by the capacitance range of the varactor diode (MA46H120).

Table II
Performance comparison of band-notched FSAs

Ref.	Notch Band			Absorption Band			Other Attributes	
	Notch Adjustability	Tunable Range (GHz)	Insertion Loss and f_N	Absorptive Bands (GHz)	Relative Absorption Bandwidth	aPolarization	Cell Size (mm)	TotalThickness(mm)
[16]	No	8.2–9.8	−0.2 dB @7.9 GHz	4.8–8 & 10.2–16	86.5%	Single	16 (0.26λ_L) × 11.4 (0.18λ_L)	6.5 (0.1λL1)
[28]	No	4.15–4.56	−0.12 dB @3.54 GHz	2.05–4.02 & 4.78–7.24	95.4%	Dual	25 (0.17λ_L) × 25 (0.17λ_L)	13 (0.09λL1)
Our work (r_2 = 1.4 mm)	Yes(Geometrical)	5.35–19.9	−0.45 dB @10 GHz	5.35–9.46 & 10.63–19.9	106%	Dual	10.2 (0.18λ_L) × 10.2 (0.18λ_L)	5.05 (0.09λ_L)
Our work (C = 0.4 pF)	Yes(Electrical)	8–14	−0.4 dB @10.14 GHz	4.77–10.86 & 11.96–19.08	110.8%	Single	10.2 (0.16λ_L) × 10.2 (0.16λ_L)	5.9 (0.09λL1)λ_L

7. Conclusion

In this paper, we presented two ultrawideband FSA designs with a flexibly adjustable and highly selective notch. By cascading a band-stop FSS layer after the resistive layer, a notch is inserted into the wide absorption band, and the notch frequency is fully controlled by the lossless layer, which can be shifted across the entire absorption band by simply adjusting the lossless layer. Less than 0.5 dB insertion loss is realized at notch frequencies. It is expected that the notched absorber can be combined with an antenna or an antenna array and serve as a ground plane or FSS reflector to realize out-of-band RCS reduction.

The notch frequency of the first design is geometrically controlled, which means the notch is adjusted by replacing the lossless layer and can be applied to narrowband antennas. Due to the smaller cell size compared to wavelength and the rotationally symmetric structure, this absorber has good angular stability, and is equally effective for both y- and x-polarization within the incident angle of 30°.

Loading a varactor in the lossless FSS, the notch frequency of the second design is electrically controllable under y-polarization, which is better for real-time control of notch frequency, thereby realizing efficient and cost-effective regulation by simply changing the applied voltage of the varactor. Due to its real-time control of the notch band, this band-notched absorber can be used not only in narrowband antennas, but also in wideband antennas to create low RCS antennas. A ratio bandwidth (f_{H2} / f_{L1}) of 4:1 is realized for this FSA. Under x-polarization, the FSA retains wideband absorption characteristics.

In the final stages of this work, prototypes of the notched absorbers were fabricated and measured. The theoretical analysis, full-wave simulation, ECM, and experimental results were found to be in reasonable agreement, demonstrating the validity of the proposed design strategies. Moreover, an FSA-backed monopole is measured and vividly shows that the two FSAs can serve as antenna's ground plane and reduce RCS of the antenna.

Acknowledgments

This work was supported by the National Natural Science Foundation of China (61701448 and 61671415) and the Fundamental Research Funds for the Central Universities (18CUCTJ076).

Author Information

Y. X. Ding, M. Y. Li, Q. X. Guo, and Z. R. Li are with the School of Information and Communication Engineering, Communication University of China, Beijing 100024, China (e-mail: nxdingyx@163.com, lmy_88515@sina.com, qxguo@cuc.edu.cn, zrli@cuc.edu.cn).

J. X. Su and H. C. Yin are with the School of Information and Communication Engineering, Communication University of China, Beijing 100024, China and the Science and Technology on Electromagnetic Scattering Laboratory, Beijing, 100854, China (e-mail: sujianxun_jlgx@163.com, yinhc207@126.com).

J. M. Song is with the Department of Electrical and Computer Engineering, Iowa State University, Ames, IA 50011, USA (e-mail: jisong@iastate.edu).

References

[1] SALISBURY W W. Absorbent body for electromagnetic waves[J]. U.S. patent, 1952: No.2599944 A.

[2] CHAMBERS B, TENNANT A. Design of wideband Jaumann radar absorbers with optimum oblique incidence performance[J]. Electron. Lett., 1994, 30(18): 1530-1532.

[3] BUCCI O, FRANCESCHETTI G. Scattering from wedge-tapered absorbers[J]. IEEE transactions on antennas and propagation, 1971, 19(1): 96-104.

[4] MUNK B A. Frequency selective surfaces: theory and design[M]. New York, NY, USA: Wiley, 2000.

[5] COSTA F, MONORCHIO A. A frequency selective radome with wideband absorbing properties[J]. IEEE transactions on antennas and propagation, 2012, 60(6): 2740-2747.

[6] CHEN, QIANG, YANG. Design of absorptive/transmissive frequency-selective surface based on parallel resonance[J]. IEEE transactions on antennas and propagation, 2017. 65(9): 4897-4902.

[7] ZAHIR JOOZDANI M, KHALAJ AMIRHOSSEINI M. Wideband absorber with combination of plasma and resistive frequency selective surface[J]. IEEE transactions on plasma ence, 2016, (12): 1-8.

[8] PANWAR, RAVI, PUTHUCHERI, et al. Fractal frequency-selective surface embedded thin broadband microwave absorber coatings using heterogeneous composites[J]. IEEE transactions on microwave theory and techniques, 2015, 63(8): 2438-2448.

[9] ZHANG L, ZHOU P, ZHANG H, et al. A broadband radar absorber based on perforated magnetic polymer composites embedded with FSS[J]. IEEE transactions on magnetics, 2014, 50(5): 1-5.

[10] MUNK B. Metamaterials: critique and alternatives[M]. Hoboken: John Wiley, 2009.

[11] ZHENG S, YIN Y, FAN J, et al. Analysis of miniature frequency selective surfaces based on fractal antenna–filter–antenna arrays[J]. IEEE antennas and wireless propagation letters, 2012, 11: 240-243.

[12] GUO Q, LI Z, SU J, et al. Dual-polarization absorptive/transmissive frequency selective surface based on tripole elements[J]. IEEE antennas and wireless propagation letters, 2019: 961-965.

[13] HUANG H, SHEN Z, OMAR A A. 3-D Absorptive frequency selective reflector for antenna

radar cross section reduction[J]. IEEE transactions on antennas and propagation, 2017, (99): 1-1.

[14] MEI P, LIN X Q, YU J W, et al. Development of a low radar cross section antenna with band-notched absorber[J]. IEEE transactions on antennas and propagation, 2017: 1-1.

[15] GAO J, ZHOU C, HUA B, et al. Frequency-Selective Rasorber with a Transparent Window Between Absorption Bands[C]. Chengdu, China: 2018 International Conference on Microwave and Millimeter Wave Technology (ICMMT).2018.

[16] MEI P, LIN X Q, YU J W, et al. A band-notched absorber designed with high notch-band-edge selectivity[J]. IEEE transactions on antennas and propagation, 2017: 1-1.

[17] LIN X, MEI P, ZHANG P, et al. Development of a resistor-loaded ultrawideband absorber with antenna reciprocity[J]. IEEE transactions on antennas and propagation, 2016: 4910-4913.

[18] OGAWA S, TAKAGAWA Y, KIMATA M. Errata: Fano resonance in asymmetric-period two-dimensional plasmonic absorbers for dual-band uncooled infrared sensors[J]. Optical engineering, 2016, 55(11): 117105.

[19] OMAR, AHMED A, SHEN. Double-sided parallel-strip line resonator for dual-polarized 3-D frequency-selective structure and absorber[J]. IEEE transactions on microwave theory and technigues, 2017, 65 (10): 3744-3752.

[20] MUNK B A, MUNK P, PRYOR J. On designing jaumann and circuit analog absorbers (CA absorbers) for oblique angle of incidence[J]. IEEE transactions on antennas and propagation, 2007, 55: 186-193.

[21] COSTA F, MONORCHIO A, MANARA G. Analysis and design of ultra thin electromagnetic absorbers comprising resistively loaded high impedance surfaces[J]. IEEE transactions on antennas and propagation, 2010, 58(5): 1551-1558.

[22] KNOTT E F. Radar cross section measurements[M]. 2nd ed. New York: Van Nostrand Reinhold, 1993.

[23] JENN D C. Radar and laser cross section engineering[M]. Reston, VA, USA: AIAA, 2005.

[24] GAO L, et al. Broadband diffusion of terahertz waves by multi-bit coding metasurfaces[J]. L. Sci. Appl., 2015, 4(9): e324.

[25] CHEN W, BALANIS C A, BIRTCHER C R. Checkerboard EBG surfaces for wideband radar cross section reduction [J]. IEEE transactions on antennas and propagation, 2015, 63(6): 2636-2645.

[26] LI S, et al. Ultra-broadband reflective metamaterial with RCS reduction based on polarization convertor, information entropy theory and genetic optimization algorithm[J]. Sci. Rep., 6(37409), 2016: 1-12.

[27] HUM S V, OKONIEWSKI M, DAVIES R J. Realizing an electronically tunable reflectarray using varactor diode-tuned elements [J]. IEEE microwave and wireless components letters: a publication of the IEEE microwave theory and techniques society. 2005, 15(6): 422-424.

[28] HAN Y, ZHU L, CHANG Y, et al. Dual-polarized bandpass and band-notched frequency-selective absorbers under multimode resonance[J]. IEEE transactions on antennas and propagation, 2018, 66(12): 7449-7454.

A Well-Conditioned Integral Equation for Electromagnetic Scattering from Composite Inhomogeneous Bi-Anisotropic Material and Closed PEC Objects[*]

1. Introduction

With the rapid development of materials science, the applications of the bi-anisotropic materials become various, e.g. in the designs of antennas, waveguide mode converters, radar absorbing materials, antireflection coatings, microwave devices, and so on. Consequently, in the field of computational electromagnetics, the EM scattering from composite objects containing anisotropic or bi-anisotropic materials and perfect electric conductors (PECs) has inspired quantities of researches. However, because the constitutive relations of bi-anisotropy are enforced as an additional coupling between the electric and magnetic fields as well as their constitutive parameters are all tensors, it is quite a challenge to accurately analyze such composite objects. Among numerous numerical methods, the integral equation method, in conjunction of the method of moments (MoM), is one of the most competitive choices to analyze the composite complex material-PEC objects and discussed in several articles. In [9], the volume integral equation (VIE) was presented to analyze the EM scattering from inhomogeneous anisotropic objects, which was further extended to analyze the electric anisotropy-PEC objects. In , scattering from bi-anisotropic objects were analyzed by the hybrid finite element-boundary integral (FEBI) method. A PEC object coated with homogenous bi-isotropic materials was modelled using the surface integral equation (SIE) method in [12], whose authors later proposed a VIE method with solenoidal basis functions for objects with inhomogeneous bi-isotropy. Article [14] used the adaptive integral method (AIM) to accelerate the MoM solution of the VIE for inhomogeneous bi-anisotropic objects. Using the volume-surface integral equation (VSIE), the scattering problem from composite electric anisotropy-PEC objects was solved in [15]. With the equivalent currents expanded by piecewise constant basis functions, article [16] used the VIE to calculate extremely anisotropic objects. In [17], the VIE was applied to analyze bi-anisotropic objects, while the eigenvalue spectrum was derived and analyzed.

---

[*] The paper was originally published in *IET Microwaves, Antennas & Propagation*, 2021, 15 (4), and has since been revised with new information. It was co-authored by Jinbo Liu, Zengrui Li, and Jiming Song.

A Well-Conditioned Integral Equation for Electromagnetic Scattering from Composite Inhomogeneous Bi-Anisotropic Material and Closed PEC Objects

On the choice of integral equations, compared with the pure SIE-based methods, the VSIE is more robust and generalized in modelling composite objects containing thin inhomogeneous materials because only the Green's function of the background medium is needed. This generality owes to the fact that according to the equivalence principle, the VSIE implementation simultaneously retains two kinds of generally applicable integral equations, i.e. the VIE to model the field superposition in the material regions, and the SIE to impose the boundary condition on the PECs. For the SIE part, the electric field integral equation (EFIE) is widely adopted because it can be used to model both the open and closed PECs. However, the EFIE is a first-kind Fredholm integral equation, usually resulting in an ill-conditioned impedance matrix. For the closed PEC part, using the EFIE alone will also encounter the interior resonance problem. On the other hand, in practical applications, metal objects having finite thickness can be modelled as closed PECs. To avoid the interior resonance problem as well as to improve the matrix condition, one can linearly combine the magnetic field integral equation (MFIE) with the EFIE to form the combined field integral equation (CFIE) which is the second-kind. For a closed PEC object which has a finite thickness, the CFIE can give an acceptable accuracy at a not very low frequency with a faster convergence. Furthermore, the CFIE can be combined with the VIE to yield a new generalized VSIE, called VIE-CFIE, which is expected to make the matrix equation easier to iteratively solve than the conventional VIE-EFIE. In the authors' previous work, we have successfully adopted the VIE-CFIE to analyze the EM scattering from composite inhomogeneous bi-isotropy and closed PEC objects, and accelerated the MoM solution of the VIE from inhomogeneous bi-anisotropic objects by the spherical harmonics expansion-based multilevel fast multipole algorithm (SE-MLFMA).

To the authors' best knowledge, so far no article shows the detail that how to use the MoM to solve the VSIE with the composite bi-anisotropy-PEC objects, especially when the VIE-CFIE is conditionally applied. In this paper, the VIE-CFIE is presented to analyze the EM scattering from arbitrarily shaped composite objects comprised of both inhomogeneous bi-anisotropic materials and closed PECs. By discretizing the equivalent surface and volume currents using the commonly used RWG and SWG basis functions which are defined on the triangular and tetrahedral cells, the VIE-CFIE yields a well-conditioned matrix equation. Due to the introduction of MFIE during modelling the closed PEC surface, compared with the existing articles such as [10] and [15], some new singularities appear in the process of matrix filling, which will be properly handled and elaborated.

2. Derivation of VSIE for Bi-Anisotropy-PEC Objects

Consider a composite object which contains both inhomogeneous bi-anisotropic material occupying a region V and PEC surface S in the free space. Assume this object is illuminated by an incident EM plane wave $(\vec{E}^i, \vec{H}^i)$ from an arbitrary direction, radiating the scattered fields $(\vec{E}^s, \vec{H}^s)$ into space. In the bi-anisotropic region V, the coupled constitutive relations between the electric flux density $\vec{D}$, magnetic flux density $\vec{B}$, and electric field $\vec{E}$, magnetic field $\vec{H}$ are written as

$$\begin{bmatrix} \vec{D}(\vec{r}) \\ \vec{B}(\vec{r}) \end{bmatrix} = \begin{bmatrix} \bar{\bar{\varepsilon}}(\vec{r}) & \bar{\bar{\xi}}(\vec{r}) \\ \bar{\bar{\varsigma}}(\vec{r}) & \bar{\bar{\mu}}(\vec{r}) \end{bmatrix} \begin{bmatrix} \vec{E}(\vec{r}) \\ \vec{H}(\vec{r}) \end{bmatrix} \forall \vec{r} \in V \qquad (1)$$

where all the constitutive parameters (permittivity $\bar{\bar{\varepsilon}}$, permeability $\bar{\bar{\mu}}$, and coupling parameters $\bar{\bar{\xi}}$, $\bar{\bar{\varsigma}}$) are $\vec{r}$-dependent tensors. Equation can also be rewritten as

$$\begin{bmatrix} \vec{E} \\ \vec{H} \end{bmatrix} = \begin{bmatrix} \bar{\bar{\varepsilon}} & \bar{\bar{\xi}} \\ \bar{\bar{\zeta}} & \bar{\bar{\mu}} \end{bmatrix}^{-1} \begin{bmatrix} \vec{D} \\ \vec{B} \end{bmatrix} = \begin{bmatrix} \bar{\bar{\alpha}}_{11} & \bar{\bar{\alpha}}_{12} \\ \bar{\bar{\alpha}}_{21} & \bar{\bar{\alpha}}_{22} \end{bmatrix} \begin{bmatrix} \vec{D} \\ \vec{B} \end{bmatrix} \qquad (2)$$

while the variable $\vec{r}$ is omitted for the purpose of easy reading. According to the volume equivalence principle, the scattered fields from the bi-anisotropic material can be seen as producing by both the equivalent volume electric currents $\vec{J}_V$ and magnetic currents $\vec{M}_V$. From the two curl equations of Maxwell's equations, $\vec{J}_V$ and $\vec{M}_V$ for bi-anisotropy are further derived as [20].

$$\begin{bmatrix} \vec{J}_V \\ \vec{M}_V \end{bmatrix} = j\omega \begin{bmatrix} \bar{\bar{I}} - \varepsilon_0 \bar{\bar{\alpha}}_{11} & -\varepsilon_0 \bar{\bar{\alpha}}_{12} \\ -\mu_0 \bar{\bar{\alpha}}_{21} & \bar{\bar{I}} - \mu_0 \bar{\bar{\alpha}}_{22} \end{bmatrix} \begin{bmatrix} \vec{D} \\ \vec{B} \end{bmatrix} = j\omega \begin{bmatrix} \bar{\bar{\beta}}_{11} & \bar{\bar{\beta}}_{12} \\ \bar{\bar{\beta}}_{21} & \bar{\bar{\beta}}_{22} \end{bmatrix} \begin{bmatrix} \vec{D} \\ \vec{B} \end{bmatrix} \qquad (3)$$

where $\bar{\bar{I}}$ denotes the identity tensor, the time-harmonic factor is $e^{j\omega t}$ with the imaginary unit $j = \sqrt{-1}$.

Due to $\vec{J}_V$ and $\vec{M}_V$ in V as well as the equivalent surface electric current $\vec{J}_S$ on S, the scattered fields are cast in terms of auxiliary potentials as

$$\begin{cases} \vec{E}^s = -j\omega(\vec{A}_S^J + \vec{A}_V^J) - \nabla(\varphi_S^J + \varphi_V^J) - \dfrac{1}{\varepsilon_0} \nabla \times \vec{A}_V^M \\ \vec{H}^s = \dfrac{1}{\mu_0} \nabla \times (\vec{A}_S^J + \vec{A}_V^J) - j\omega \vec{A}_V^M - \nabla \varphi_V^M \end{cases} \qquad (4)$$

The vector and scalar potentials are expressed as the convolutions of currents or their divergences and the Green's function as

A Well-Conditioned Integral Equation for Electromagnetic Scattering from Composite Inhomogeneous Bi-Anisotropic Material and Closed PEC Objects

$$\begin{cases} \vec{A}_T^J(\vec{r}) = \mu_0 \int_T \vec{J}_T(\vec{r}')G(\vec{r},\vec{r}')dT' \\ \varphi_T^J(\vec{r}) + \dfrac{j}{\omega\varepsilon_0}\int_T \nabla' \cdot \vec{J}_T(\vec{r}')G(\vec{r},\vec{r}')dT' \\ \vec{A}_V^M(\vec{r}) = \varepsilon_0 \int_V \vec{M}_V(\vec{r}')G(\vec{r},\vec{r}')dV' \\ \varphi_V^M(\vec{r}) + \dfrac{j}{\omega\mu_0}\int_V \nabla' \cdot \vec{M}_V(\vec{r}')G(\vec{r},\vec{r}')dV' \end{cases} \quad T = S, V \tag{5}$$

The Green's function of free space is expressed as

$$G(\vec{r},\vec{r}') = G = \frac{e^{-jk_0|\vec{r},\vec{r}'|}}{4\pi|\vec{r},\vec{r}'|} \tag{6}$$

with free space wavenumber k_0.

In the region V, the VIE is formed by making the incident fields equal to the total fields $(\vec{E},\vec{H})$ minus the scattered fields as

$$\left[\vec{E}(\vec{r}),\vec{H}(\vec{r})\right] - \left[\vec{E}^s(\vec{r}),\vec{H}^s(\vec{r})\right] = \left[\vec{E}^i(\vec{r}),\vec{H}^i(\vec{r})\right] \quad \forall \vec{r} \in V \tag{7}$$

On the PEC surface S, by vanishing the tangential (tan) component of total electric field, the EFIE is formed as

$$\vec{E}(\vec{r})\big|_{\tan} = \left[\vec{E}^i(\vec{r}) + \vec{E}^s(\vec{r})\right]_{\tan} = 0 \quad \forall \vec{r} \in S \tag{8}$$

The VIE can be combined with the EFIE to form the commonly used VIE-EFIE, a first-kind integral equation which is usually ill-conditioned. Furthermore, on the closed PECs, the MFIE

$$\frac{1}{2}\vec{J}_s(\vec{r}) - \hat{n}(\vec{r}) \times \vec{H}^s(\vec{r}) = \hat{n}(\vec{r}) \times \vec{H}^i(\vec{r}), \quad \forall \vec{r} \in S \tag{9}$$

can be modelled and linearly added to the EFIE to form the well-conditioned CFIE as [18].

$$\text{CFIE} = \alpha\text{EFIE} + (1-\alpha)\eta_0\text{MFIE} \tag{10}$$

where $\hat{n}$ is the outwardly directed normal, α ($0 \leq \alpha \leq 1$) is a real constant, and η_0 is the intrinsic impedance of free space. We can combine the VIE and CFIE together to build the VIE-CFIE to solve the EM scattering from composite objects comprised of bi-anisotropic materials and closed PECs. Generally, when the constitutive parameter tensors have the same order of magnitude, the VIE-CFIE is well-conditioned.

3. MoM Solution

3.1 Discretization of the VIE-CFIE

By dispersing the equivalent currents or flux densities, the MoM discretizes the VSIE into a matrix equation. In the implementation, $\vec{J}_S$ on S and $\vec{D}$ and $\vec{B}$ in V are respectively expanded using the set of RWG basis functions $\vec{f}_i^S$ [21] defined in the domain S_i and SWG basis functions $\vec{f}_i^V$ [22] defined in the domain V_i as

$$\begin{cases} \vec{J}_S = \sum_{i=1}^{N_S} I_i^S \vec{f}_i^S \\ j\omega \vec{D} = \sum_{i=1}^{N_V} I_i^D \vec{f}_i^V \\ j\omega \vec{B} = \eta_0 \sum_{i=1}^{N_V} I_i^B \vec{f}_i^V \end{cases} \quad (11)$$

In (11), N_S and N_V are numbers of RWG and SWG basis functions, while the total number of unknowns is $N_S + 2N_V$. I_i^S, I_i^D and I_i^B are the corresponding unknown expansion coefficients, respectively. To hold the continuity of the normal component that is consistent with the boundary condition of material interface, we disperse $\vec{D}$ and $\vec{B}$ instead of $\vec{J}_V$ and $\vec{M}_V$. It is further assumed that $\bar{\bar{\varepsilon}}$, $\bar{\bar{\mu}}$, $\bar{\bar{\xi}}$, $\bar{\bar{\varsigma}}$ are approximately constant tensors inside each tetrahedron, which is a generalization of that presented in [22]. As a consequence, the tensors $\bar{\bar{\alpha}}_{pq}$ and $\bar{\bar{\beta}}_{pq}$ with p/q=1 or 2 defined in (2) and (3) over a single tetrahedron are also considered as constant.

Substituting (11) into (3)-(10) and combining with the Galerkin's testing result in a generalized impedance matrix equation, which can succinctly be represented as

$$\begin{bmatrix} Z_{SS} & Z_{SD} & Z_{SB} \\ Z_{DS} & Z_{DD} & Z_{DB} \\ Z_{BS} & Z_{BD} & Z_{BB} \end{bmatrix} \begin{Bmatrix} I_S \\ I_D \\ I_B \end{Bmatrix} = \begin{Bmatrix} V_S \\ V_D \\ V_B \end{Bmatrix} \quad (12)$$

with

$$\begin{aligned} \left[Z_{SS} \right] &= \alpha \left[Z_{SS}^E \right] + (1-\alpha)\eta_0 \left[Z_{SS}^M \right] \\ \left[Z_{SD} \right] &= \alpha \left[Z_{SD}^E \right] + (1-\alpha)\eta_0 \left[Z_{SD}^M \right] \\ \left[Z_{SB} \right] &= \alpha \left[Z_{SB}^E \right] + (1-\alpha)\eta_0 \left[Z_{SB}^M \right] \\ \{V_S\} &= \alpha \{V_S^E\} + (1-\alpha)\eta_0 \{V_S^M\} \end{aligned} \quad (13)$$

where I_S, I_D and I_B are the vectors of unknown expansion coefficients, V_S, V_D and V_B are the excitation vectors, $[Z_{PQ}]$ (P, $Q = S$, D, or B) denotes the impedance submatrix representing the interactions between various types of testing and basis functions, and the superscript E or M means the corresponding submatrix or subvector is from EFIE or MFIE, respectively. For convenience, beforehand, three linear vector operators are defined as

$$\begin{cases} \vec{P}_{T_1}(\vec{X}) = \int_{T_1} \vec{X}(\vec{r}') G dT' \\ \vec{Q}_{T_1}(\vec{X}) = \nabla' \cdot \vec{X}(\vec{r}') G dT' \quad T = S \text{ or } V \\ \vec{K}_{T_1}(\vec{X}) = \int_{T_1} \vec{X}(\vec{r}') \nabla G dT' \end{cases} \quad (14)$$

Each submatrix entry denoting the interaction between the jth testing function and ith basis function in (12) and (13) is then given by

$$\left[Z_{SS}^E\right]_{ji} = j\omega\mu_0 \left\langle \vec{f}_j^S, \vec{P}_{S_i}\left(\vec{f}_i^S\right)\right\rangle + \frac{j}{\omega\varepsilon_0}\left\langle \vec{f}_j^S, \vec{Q}_{S_i}\left(\vec{f}_i^S\right)\right\rangle \quad (15)$$

$$\left[Z_{SS}^M\right]_{ji} = \frac{1}{2}\left\langle \vec{f}_j^S, \vec{f}_i^S\right\rangle + \left\langle \vec{f}_j^S, \hat{n}_j, \vec{K}_{S_i}\left(\vec{f}_i^S\right)\right\rangle \quad (16)$$

$$\left[Z_{DS}\right]_{ji} = j\omega\mu_0 \left\langle \vec{f}_j^V, \vec{P}_{S_i}\left(\vec{f}_i^S\right)\right\rangle + \frac{j}{\omega\varepsilon_0}\left\langle \vec{f}_j^V, \vec{Q}_{S_i}\left(\vec{f}_i^S\right)\right\rangle \quad (17)$$

$$\left[Z_{DS}\right]_{ji} = \left\langle \vec{f}_j^V, \vec{K}_{S_i}\left(\vec{f}_i^S\right)\right\rangle \quad (18)$$

and

$$\begin{cases} \left[Z_{SD}^E\right]_{ji} = Z_{ji}^E\left(\bar{\bar{\beta}}_{i11}, \bar{\bar{\beta}}_{i21}\right) \\ \left[Z_{SB}^E\right]_{ji} = Z_{ji}^E\left(\bar{\bar{\beta}}_{i12}, \bar{\bar{\beta}}_{i22}\right) \\ \left[Z_{SD}^M\right]_{ji} = Z_{ji}^M\left(\bar{\bar{\beta}}_{i11}, \bar{\bar{\beta}}_{i21}\right) \\ \left[Z_{SB}^M\right]_{ji} = Z_{ji}^M\left(\bar{\bar{\beta}}_{i12}, \bar{\bar{\beta}}_{i22}\right) \\ \left[Z_{DD}\right]_{ji} = Z_{ji}^V\left(\bar{\bar{\alpha}}_{i11}, \bar{\bar{\beta}}_{i21}, \bar{\bar{\beta}}_{i21}, \varepsilon_0, \mu_0\right) \\ \left[Z_{DB}\right]_{ji} = Z_{ji}^V\left(\bar{\bar{\alpha}}_{i11}, \bar{\bar{\beta}}_{i21}, \bar{\bar{\beta}}_{i22}, \varepsilon_0, \mu_0\right) \\ \left[Z_{BD}\right]_{ji} = Z_{ji}^V\left(\bar{\bar{\alpha}}_{i21}, \bar{\bar{\beta}}_{i22}, -\bar{\bar{\beta}}_{i11}, \mu_0, \varepsilon_0\right) \\ \left[Z_{BB}\right]_{ji} = Z_{ji}^V\left(\bar{\bar{\alpha}}_{i22}, \bar{\bar{\beta}}_{i22}, -\bar{\bar{\beta}}_{i12}, \mu_0, \varepsilon_0\right) \end{cases} \quad (19)$$

with

$$Z_{ji}^{R}\left(\overline{\overline{\beta}}_{ip_1q},\overline{\overline{\beta}}_{ip_2q}\right) = j\omega\mu_0\left\langle \vec{f}_j^s, \vec{P}_{V_i}\left(\overline{\overline{\beta}}_{ip_1q}\cdot\vec{f}_i^V\right)\right\rangle$$
$$+\frac{j}{\omega\varepsilon_0}\left\langle \vec{f}_j^s, \vec{Q}_{V_i}\left(\overline{\overline{\beta}}_{ip_1q}\cdot\vec{f}_i^V\right)\right\rangle - \left\langle \vec{f}_j^s, \vec{K}_{V_i}\left(\overline{\overline{\beta}}_{ip_2q}\cdot\vec{f}_i^V\right)\right\rangle \quad (20)$$

$$Z_{ji}^{M}\left(\overline{\overline{\beta}}_{ip_1q},\overline{\overline{\beta}}_{ip_2q}\right) = \left\langle \vec{f}_j^s\times\hat{n}_j, \vec{K}_{V_i}\left(\overline{\overline{\beta}}_{ip_1q}\cdot\vec{f}_i^V\right)\right\rangle$$
$$+j\omega\varepsilon_0\left\langle \vec{f}_j^s\times\hat{n}_j, \vec{F}_{V_i}\left(\overline{\overline{\beta}}_{ip_2q}\cdot\vec{f}_i^V\right)\right\rangle \quad (21)$$
$$+\frac{j}{\omega\mu_0}\left\langle \vec{f}_j^s\times\hat{n}_j, \vec{Q}_{V_i}\left(\overline{\overline{\beta}}_{ip_2q}\cdot\vec{f}_i^V\right)\right\rangle$$

$$Z_{ji}^{V}\left(\overline{\overline{\alpha}}_{ip_1q},\overline{\overline{\beta}}_{ip_1q},\overline{\overline{\beta}}_{ip_2q},\chi,\gamma\right)$$
$$=\frac{1}{j\omega}\left\langle \vec{f}_j^V,\overline{\overline{\alpha}}_{ip_1q}\cdot\vec{f}_i^V\right\rangle + j\omega\gamma\left\langle \vec{f}_j^V(\vec{r}),\vec{P}_{V_i}\left(\overline{\overline{\beta}}_{ip_1q}\cdot\vec{f}_i^V\right)\right\rangle + \quad (22)$$
$$\frac{j}{\omega\chi}\left\langle \vec{f}_j^V,\vec{Q}_{V_i}\left(\overline{\overline{\beta}}_{ip_1q}\cdot\vec{f}_i^V\right)\right\rangle - \left\langle \vec{f}_j^V(\vec{r}),\vec{K}_{V_i}\left(\overline{\overline{\beta}}_{ip_2q}\cdot\vec{f}_i^V\right)\right\rangle$$

where $\langle\cdot,\cdot\rangle$ denotes the L^2-inner product, $\hat{n}_j$ denotes the outer-normal direction of the triangle containing the jth RWG test function, $\overline{\overline{\alpha}}_{ipq}$ and $\overline{\overline{\beta}}_{ipq}$ are constant tensors over the tetrahedron containing the ith SWG basis function, and $p_1/p_2/q = 1$ or 2 with $p_1 + p_2 \equiv 3$.

Furthermore, the continuity condition (CC) of electric flux can be explicitly enforced on the material-PEC interfaces to reduce the number of volumetric electric unknowns. The CC establishes the relation between $\vec{D}$ and $\vec{J}_S$, which can be written as [19].

$$\hat{n}(\vec{r})\cdot\vec{D}(\vec{r}) = \frac{\nabla\cdot\vec{J}_S(\vec{r})}{j\omega} \quad (23)$$

Because the CC comes from the current continuity equation which is independent of the material type, it can be safely adopted to the bi-anisotropy-PEC interfaces theoretically.

3.2 Matrix Filling

In the following, the details of matrix filling process especially how to handle the singularities are described. The ith RWG basis function is defined over a common side of length l_i shared by two triangles $S_i^{\pm}$ of areas $s_i^{\pm}$ as [21].

$$\vec{f}_i^S(\vec{r}') = \pm\frac{a_i}{3v_i^{\pm}}(\vec{r}'-\vec{r}_i^{\pm}) \quad \forall \vec{r}' \in S_i^{\pm} \quad (24)$$

Similarly, the ith SWG basis function is defined over a common face of area ai shared by two tetrahedrons $V_i^\pm$ of volumes $v_i^\pm$ as [22].

$$\vec{f}_j^V(\vec{r}') = \pm \frac{a_i}{3v_i^\pm}(\vec{r}' - \vec{r}_i^\pm) \quad \forall \vec{r}' \in V_i^\pm \tag{25}$$

In (24) and (25), the sign ± means the current flowing direction of the ith basis function is outward or inward relative to $T_i^\pm$ ($T = S$ for RWG or V for SWG), and $\vec{r}_i^\pm$ is the free vertex of the ith basis function in $T_i^\pm$. If the field point $\vec{r}$ is far from the source point $\vec{r}'$, all of the matrix entries in (12) can be easily evaluated using a universal quadrature rule. The Gaussian quadrature rule with 4/5 sampling points is recommended for integrations over the triangle/tetrahedron domain during calculating the interactions between the testing and basis functions, which ensures accurate integration of up to 3rd order of polynomials. On the contrary, when $\vec{r}$ approaches $\vec{r}'$, because of the Green's function (6) or its gradient, the singularities or near singularities will appear and need special attention. How to evaluate the values of $\left[Z_{SS}^E\right]_{ji}$ and $\left[Z_{SS}^M\right]_{ji}$ generated by the SIE part can be employed as proposed in [21] or [18], while that of $\left[Z_{PQ}\right]_{ji}$ ($P, Q = B$ or D) generated by the VIE part can be found in [9]. The evaluation of $\left[Z_{SD}\right]_{ji}$ and $\left[Z_{BS}\right]_{ji}$ are shown in [19], both of which are independent of the material type contained by the calculated object.

For $\left[Z_{SQ}^E\right]_{ji}$, three types of integrals involving the linear vector operators $\vec{P}_{V_i}$, $\vec{Q}_{V_i}$ and $\vec{K}_{V_i}$ need to be handled. During $\vec{r} \to \vec{r}'$, the inner product calculation relating $\vec{P}_{V_i}$ or $\vec{Q}_{V_i}$ can be easily transformed into a specific form with one order singularity, which can be expediently handled using either the singularity extraction or Duffy transform method. For the calculation relating $\vec{K}_{V_i}$, one can exchange the integral order as

$$\left\langle \vec{f}_j^S, \vec{K}_{V_i}\left(\bar{\bar{\upsilon}}_i \cdot \vec{f}_i^V\right)\right\rangle = \int_{S_j^\pm} \vec{f}_j^S(\vec{r}) \cdot \int_{V_i^\pm} \bar{\bar{\upsilon}}_i \cdot \vec{f}_i^V(\vec{r}') \times \nabla G dV' dS$$

$$= -\frac{l_j a_i}{6s_j^\pm v_i^\pm} \int_{V_i^\pm} \bar{\bar{\upsilon}}_i \cdot (\vec{r}_i^\pm - \vec{r}') \cdot \int_{S_j^\pm} \left[(\vec{r}' - \vec{r}_j^\pm) + (\vec{r} - \vec{r}')\right] \times \nabla G dS dV' \tag{26}$$

$$= -\frac{l_j a_i}{6s_j^\pm v_i^\pm} \int_{V_i^\pm} \bar{\bar{\upsilon}}_i \cdot (\vec{r}_i^\pm - \vec{r}') \cdot \left[(\vec{r}' - \vec{r}_j^\pm) \times \int_{S_j^\pm} \nabla G dS\right] dV'$$

where $\bar{\bar{\upsilon}}_i$ denotes an arbitrary tensor which is constant in single tetrahedron. The above derivation takes the nature of $(\vec{r} - \vec{r}') \times \nabla G \equiv 0$. In this way, the singularity in the gradient of Green's function over a plane triangle can be handled as [27].

For $\left[Z_{SQ}^{M}\right]_{ji}$ which is introduced by the MFIE, there are also three types of integrals. The singularity involving $\vec{P}_{V_i}$ is easy to handle. For the inner product involving $\vec{Q}_{V_i}$, according to the two-dimensional Gauss theorem and the identity $\nabla \cdot (\bar{\bar{a}} \cdot \vec{b}) = \left[\nabla \cdot (\bar{\bar{a}})\right] \cdot \vec{b} + \text{Tr}(\bar{\bar{a}}^T \cdot \nabla \vec{b})$, we can translate it as

$$\begin{aligned}
&\left\langle \vec{f}_j^S \times \hat{n}_j, \vec{Q}_{V_i}\left(\bar{\bar{\upsilon}}_i \cdot \vec{f}_i^V\right) \right\rangle \\
&= \int_{S_j^\pm} \vec{f}_j^S(\vec{r}) \times \hat{n}_j^\pm \cdot \nabla_S \int_{V_i^\pm} \nabla' \cdot \left[\bar{\bar{\upsilon}}_i \cdot \vec{f}_i^V(\vec{r}')\right] G dV' dS \\
&= \int_{S_j^\pm} \nabla_S \cdot \left\{ \left[\vec{f}_j^S(\vec{r}) \times \hat{n}_j^\pm\right] \int_{V_i^\pm} \nabla' \cdot \left[\bar{\bar{\upsilon}}_i \cdot \vec{f}_i^V(\vec{r}')\right] G dV' \right\} dS \\
&= \oint_{\partial S_j^\pm} \hat{n}_{\partial S_j^\pm} \cdot \left[\vec{f}_j^S(\vec{r}) \times \hat{n}_j^\pm\right] \int_{V_i^\pm} \nabla' \cdot \left[\bar{\bar{\upsilon}}_i \cdot \vec{f}_i^V(\vec{r}')\right] G dV' dl \\
&= \pm \text{Tr}(\bar{\bar{\upsilon}}_i) \frac{a_i}{3 v_i^\pm} \oint_{\partial S_j^\pm} \vec{f}_j^S(\vec{r}) \times \hat{n}_j^\pm \cdot \hat{n}_{\partial S_j^\pm} \int_{V_i^\pm} G dV' dl \\
&\quad - \oint_{\partial S_j^\pm} \left[\vec{f}_j^S(\vec{r}) \times \hat{n}_j^\pm \cdot \hat{n}_{\partial S_j^\pm}\right] \left[\hat{n}_{\partial V_i^\pm} \cdot \bar{\bar{\upsilon}}_i \cdot \oint_{\partial V_i^\pm} \vec{f}_i^V(\vec{r}') G dS'\right] dl
\end{aligned} \quad (27)$$

where $\text{Tr}(\bar{\bar{\upsilon}}_i)$ denotes the trace of $\bar{\bar{\upsilon}}_i$, $\hat{n}_j^\pm$ and $\hat{n}_{\partial V_i^\pm}$ denote the outer-normal direction of triangle $S_j^\pm$ and that of the four triangular faces of tetrahedron $V_i^\pm$, respectively. Then the order of singularities occurring in is reduced to one.

For the inner product involving $\vec{K}_{V_i}$, we transform it as

$$\begin{aligned}
&\left\langle \vec{f}_j^S \times \hat{n}_j, \vec{K}_{V_i}\left(\bar{\bar{\upsilon}}_i \cdot \vec{f}_i^V\right) \right\rangle \\
&= \int_{S_j^\pm} \vec{f}_j^S(\vec{r}) \times \hat{n}_j^\pm \cdot \int_{V_i^\pm} \bar{\bar{\upsilon}}_i \cdot \vec{f}_i^V(\vec{r}') \times \nabla G dV' dS \\
&= -\frac{l_j a_i}{6 s_j^\pm v_i^\pm} \int_{V_i^\pm} \bar{\bar{\upsilon}}_i \cdot \left(\vec{r}' - \vec{r}_i^\pm\right) \cdot \int_{S_j^\pm} \hat{n}_j^\pm \times \left(\vec{r} - \vec{r}_j^\pm\right) \times \nabla' G dS dV' \\
&= -\frac{l_j a_i}{6 s_j^\pm v_i^\pm} \hat{n}_j^\pm \times \vec{r}_j^\pm \cdot \int_{V_i^\pm} \bar{\bar{\upsilon}}_i \cdot \left(\vec{r}' - \vec{r}_i^\pm\right) \times \int_{S_j^\pm} \nabla' G dS dV' \\
&\quad - \frac{l_j a_i}{6 s_j^\pm v_i^\pm} \int_{V_i^\pm} \bar{\bar{\upsilon}}_i \cdot \left(\vec{r}' - \vec{r}_i^\pm\right) \cdot \int_{S_j^\pm} \hat{n}_j^\pm \times \vec{r} \times \nabla' G dS dV'
\end{aligned} \quad (28)$$

Therefore, the two-order singularity appeared in the first term of the last right-hand side (RHS) can be handled. However, how to deal with the second one, which is also order two and first encountered, is not obvious. Here we deal with it using the singularity extraction method and simply summarize the key steps. We translate the inner integral of the second term and extract the singularity as

$$\int_{S_j^{\pm}} \hat{n}_j^{\pm} \times \vec{r} \times \nabla' G dS = \int_{S_j^{\pm}} \vec{r} \left(\hat{n}_j^{\pm} \cdot \nabla' G \right) dS - \hat{n}_j^{\pm} \int_{S_j^{\pm}} \vec{r} \cdot \nabla' G dS$$
$$= \int_{S_j^{\pm}} \vec{r} \left[\hat{n}_j^{\pm} \cdot \nabla' \left(G - \frac{1}{4\pi R} \right) \right] dS - \hat{n}_j^{\pm} \int_{S_j^{\pm}} \vec{r} \cdot \nabla' \left(G - \frac{1}{4\pi R} \right) dS \quad (29)$$
$$+ \frac{1}{4\pi} \int_{S_j^{\pm}} \vec{r} \left(\hat{n}_j^{\pm} \cdot \nabla' \frac{1}{R} \right) dS - \frac{1}{4\pi} \hat{n}_j^{\pm} \int_{S_j^{\pm}} \vec{r} \cdot \nabla' \frac{1}{R} dS$$

where $R = |\vec{r} - \vec{r}'|$. Thus, the first two terms of RHS in the last equation can be evaluated numerically using a Gaussian quadrature rule, while the last two terms need to be further analyzed.

Table I

Different mesh sizes with respect to the numbers of unknowns, triangles, and tetrahedrons, RMS and MAX errors of different implementations compared with benchmark results (from the VIE-EFIE with 0.017λ mesh size), and the number of iterations

Mesh size (λ)	0.017 λ		0.03 λ		0.055 λ		0.1 λ	
Number of unknowns	159,896		53,961		20,410		4,916	
Number of triangles	3,716		1,702		896		272	
Number of tetrahedrons	36,382		11,846		4,236		996	
Equation type	VIE-EFIE	VIE-CFIE	VIE-EFIE	VIE-CFIE	VIE-EFIE	VIE-CFIE	VIE-EFIE	VIE-CFIE
RMS error (dB)	—	0.018	0.023	0.037	0.19	0.23	0.67	0.75
MAX error (dB)	—	0.024	0.095	0.11	0.76	0.87	2.19	2.53
Number of iterations	103	33	105	33	102	33	102	31

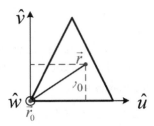

Fig. 1 Notations for integrals on a triangle

Assume that a coordinate transformation is applied so that the triangle locates on the *u-v* plane having a normal in the direction *w*, where $\hat{u} \cdot (\vec{r} - \vec{r}_0) = u_0$, $\hat{v} \cdot (\vec{r} - \vec{r}_0) = v_0$ and $\hat{w} \cdot (\vec{r} - \vec{r}_0) = 0$, as shown in Fig. 1. Using the matrix notation and operation, $\vec{r}$ can be expressed as

$$\vec{r} = \begin{bmatrix} \hat{u}^T \\ \hat{v}^T \\ \hat{w}^T \end{bmatrix}^{-1} \begin{Bmatrix} u_0 \\ v_0 \\ 0 \end{Bmatrix} + \vec{r}_0 = \begin{bmatrix} m_{11} & m_{12} & m_{13} \\ m_{21} & m_{22} & m_{23} \\ m_{31} & m_{32} & m_{33} \end{bmatrix} \begin{Bmatrix} u_0 \\ v_0 \\ 0 \end{Bmatrix} + \vec{r}_0 \quad (30)$$

where $\hat{u}^T$, $\hat{v}^T$ and $\hat{w}^T$ are the row vector expressions of $\hat{u}$, $\hat{v}$ and $\hat{w}$. Further, according to [25], three different types of inner integrals can be analytically evaluated as

$$\int_{S_j^{\pm}} \begin{pmatrix} 1 \\ u_0 \\ v_0 \end{pmatrix} \nabla' \frac{1}{R} dS = \begin{pmatrix} \vec{I}_1 \\ \vec{I}_u \\ \vec{I}_v \end{pmatrix} \quad (31)$$

Then according to (30) and (31), the third term of (29) is handled as

$$\int_{S_j^{\pm}} \vec{r} \left(\hat{n}_j^{\pm} \cdot \nabla' \frac{1}{R} \right) dS$$

$$= \int_{S_j^{\pm}} \begin{bmatrix} m_{11} & m_{12} & m_{13} \\ m_{21} & m_{22} & m_{23} \\ m_{31} & m_{32} & m_{33} \end{bmatrix} \begin{Bmatrix} u_0 \\ v_0 \\ 0 \end{Bmatrix} \left(\hat{n}_j^{\pm} \cdot \nabla' \frac{1}{R} \right) dS + \vec{r}_0 \hat{n}_j^{\pm} \cdot \int_{S_j^{\pm}} \nabla' \frac{1}{R} dS \quad (32)$$

$$= \begin{bmatrix} m_{11} & m_{12} & m_{13} \\ m_{21} & m_{22} & m_{23} \\ m_{31} & m_{32} & m_{33} \end{bmatrix} \begin{Bmatrix} \hat{n}_j^{\pm} \cdot \vec{I}_u \\ \hat{n}_j^{\pm} \cdot \vec{I}_v \\ 0 \end{Bmatrix} + \vec{r}_0 \hat{n}_j^{\pm} \cdot \vec{I}_1$$

Substituting (30) into the fourth term of (29), we have [25].

A Well-Conditioned Integral Equation for Electromagnetic Scattering from Composite Inhomogeneous Bi-Anisotropic Material and Closed PEC Objects

$$\int_{S_j^\pm} \vec{r} \cdot \nabla' \frac{1}{R} dS$$

$$= \int_{S_j^\pm} \begin{bmatrix} m_{11} & m_{12} & m_{13} \\ m_{21} & m_{22} & m_{23} \\ m_{31} & m_{32} & m_{33} \end{bmatrix} \begin{Bmatrix} u_0 \\ v_0 \\ 0 \end{Bmatrix} \cdot \nabla' \frac{1}{R} dS + \vec{r}_0 \cdot \int_{S_j^\pm} \nabla' \frac{1}{R} dS \quad (33)$$

$$= \begin{Bmatrix} m_{11} \\ m_{21} \\ m_{31} \end{Bmatrix} \cdot \vec{I}_u + \begin{Bmatrix} m_{21} \\ m_{22} \\ m_{32} \end{Bmatrix} \cdot \vec{I}_v + \vec{r}_0 \cdot \vec{I}_1$$

According to (30)-(33), all kinds of the singularity in (29) can be handled analytically.

4. Numerical Results

In this section, the bistatic or monostatic radar cross section (RCS) of several composite objects are calculated. When the VIE-CFIE is adopted to model the objects, $\alpha=0.5$. GMRES with a restart number 100 is used as the iterative solver to reach the convergence with a relative residual error of 0.001. A simple diagonal preconditioner is used to accelerate the iterative solving process. A zero vector is taken as the initial guess for all calculations, which are serially carried out on a workstation with 2.4 GHz CPU and 384 GB RAM in single precision.

The first object is a bi-anisotropic cylinder, whose radius and height are 0.5λ and 0.2λ, respectively. The constitutive parameters of the bi-anisotropy are

$$\bar{\bar{\varepsilon}}_r = \begin{bmatrix} 2 & 0 & 0 \\ 0 & 3 & 0 \\ 0 & 0 & 2 \end{bmatrix} \quad \bar{\bar{u}}_r = \begin{bmatrix} 1.2 & 0 & 0 \\ 0 & 1.2 & 0 \\ 0 & 0 & 1 \end{bmatrix}$$
$$\bar{\bar{\xi}}_r = \begin{bmatrix} 0 & 0 & 0 \\ -j\Omega & 0 & 0 \\ 0 & 0 & 0 \end{bmatrix} \quad \bar{\bar{\zeta}}_r = \begin{bmatrix} 0 & j\Omega & 0 \\ 0 & 0 & 0 \\ 0 & 0 & 0 \end{bmatrix} \quad (34)$$

where Ω is variable. The cylinder is meshed into 7,490 tetrahedrons with respect to 31,872 unknowns. Illuminated by an EM plane wave from +z-axis, the $\theta\theta$-and $\varphi\varphi$-polarized bistatic RCS at $\varphi = 0°$ plane with different values of Ω are calculated and given in Fig. 2 (denoted by cal.). For comparison, the results using FEBI from [11, Fig. 5] are also shown (ref.). Good agreements are observed for each Ω.

The second object is a coated PEC sphere. The radius of the inner PEC sphere and the thickness of the coating electric anisotropy are 1 m and 0.42 m, respectively. The relative permittivity tensors of the coated electric anisotropy are defined under the spherical coordinate as $\bar{\bar{\varepsilon}}_r = \text{diag}(\varepsilon_r, \varepsilon_t, \varepsilon_t)$ and $\bar{\bar{\mu}}_r = \text{diag}(\mu_r, \mu_t, \mu_t)$. In this case, $\varepsilon_r = 4$, $\varepsilon_t = 2$ and $\bar{\bar{\mu}}_r = \bar{\bar{I}}$. When the object is illuminated by a θ-polarized plane wave from-z-axis at 22 MHz, the bistatic RCS at $\varphi=0°$ plane is calculated using the VIE-CFIE, while the numbers of triangles and tetrahedrons are 112 and 587 with respect to 1,422 unknowns after discretizing. During the calculation, the CC is alternatively enforced and combined with the VIE-CFIE (denoted by CC-VIE-CFIE) to reduce 112 volumetric unknowns. The numerical results are shown in Fig. 3, while for comparison, the exact result from Mie series [10, Fig. 1] is also given. It is seen that the numerical results with or without the CC agree well with the exact result everywhere, which states that the VIE-CFIE has a high calculation accuracy. Because this object is small, the computational details are not reported.

The third object is also a coated PEC sphere. The radius of the inner PEC sphere is 0.3λ, and the thickness of coating material is 0.05λ. The coating material is bi-anisotropic, whose relative permittivity and permeability tensors are defined under the spherical coordinate while $\varepsilon_r = \mu_r = 4$ and $\varepsilon_t = \mu_t = 2$. Another two coupling parameters are defined as $\bar{\bar{\xi}}_r = \bar{\bar{\zeta}}_r^* = (0.5 - j0.5)\bar{\bar{I}}$, while the asterisk denotes the complex conjugate. This object is illuminated by a plane wave from +z-axis. It is impossible to analyze this object using the SIE-based schemes but the VSIE can do. Both the VIE-EFIE and VIE-CFIE are used to calculate the co-polarization of bistatic RCS, while the observation range is $0°\leq\theta\leq180°$ and $\varphi=0°$. In this calculation, four types of mesh size are adopted to discrete the object as shown in Table I, where the numbers of triangles, tetrahedrons and unknowns are listed, respectively. To investigate the numerical accuracy, we set the VIE-EFIE result from the mesh size of 0.017λ as the benchmark, while the root-mean-square (RMS) error is applied to analyze the accuracy, defined as

$$\text{RMS} = \sqrt{\frac{1}{M}\sum_{i=1}^{M}\left|\sigma_i^{\text{cal}} - \sigma_i^{\text{ben}}\right|^2} \tag{35}$$

where M is the number of observation angles, σ_i^{cal} and σ_i^{ben} denote the calculated and the benchmark RCS measured in dB in the ith observation angle, respectively. Table I also shows when the incident plane wave is θ-polarized, the RMS and Maximal (MAX) errors from different implementations, as well as the number of iterations during the iterative solution. It is observed that when the mesh size is fixed, the RMS and MAX errors from the VIE-EFIE and VIE-CFIE are quite close, which means the new matrix entries as well as singularities introduced by the MFIE have been handled properly, and the VIE-CFIE can also give reliable results. On the other hand, due to the improvement of matrix condition, the convergence speed of the VIE-CFIE is always several times faster than that of the VIE-EFIE. For different mesh sizes, the errors are tolerable if the mesh size is smaller than 0.055λ. However, if the mesh size is set as 0.1λ, the errors are unacceptable. The reason is that for this coating material, 0.1λ is roughly equivalent to 0.4 times of the wavelength in the coating material, which is too large. The numerical results from 0.017λ and 0.1λ are shown in Fig. 4, while the maximal errors arise over the valley range (in this case, about 45° for the $\theta\theta$-polarization and 75° for the $\varphi\varphi$-polarization).

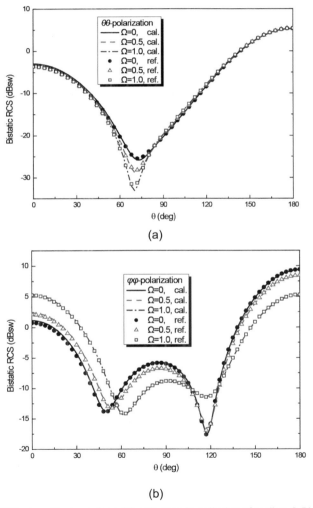

Fig. 2 Bistatic RCS at $\varphi=0°$ plane for a bi-anisotropic cylinder of radius 0.5λ and height 0.2λ, illuminated by an EM plane wave from +z-axis

(a) $\theta\theta$-polarization, (b) $\varphi\varphi$-polarization

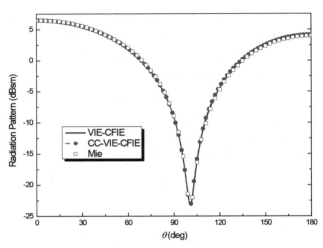

Fig. 3 Bistatic RCS for a PEC sphere of radius 0.1 m coated with 0.42 m thick electric anisotropic material with the constitutive parameters ε_r=4 and ε_t=2 (defined in spherical coordinate), illuminated by a θ-polarized *EM* plane wave from -z-axis at 22 MHz

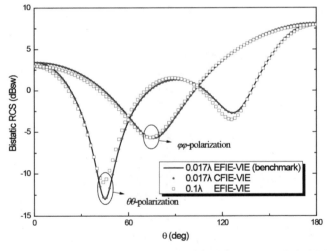

Fig. 4 Bistatic RCS for a PEC sphere of radius 0.3λ coated with 0.05λ thick bi-anisotropic material with the constitutive parameters $\varepsilon_r = \mu_r = 4$, $\varepsilon_t = \mu_t = 2$ (defined in spherical coordinate), and $\bar{\bar{\xi}}_r = \bar{\bar{\zeta}}_r^* = (0.5 - j0.5)\bar{\bar{I}}$, illuminated by a θ-polarized *EM* plane wave from +z-axis

A Well-Conditioned Integral Equation for Electromagnetic Scattering from Composite Inhomogeneous Bi-Anisotropic Material and Closed PEC Objects

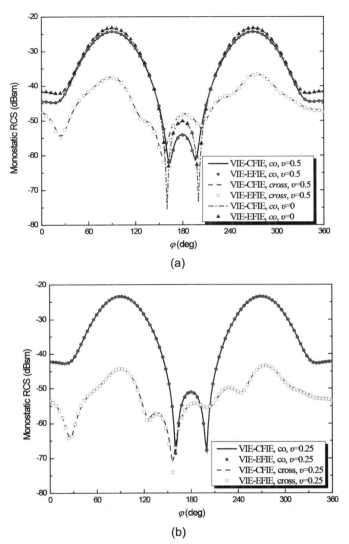

Fig. 5 Monostatic RCS for a 252.4 mm length PEC almond coated with 10 mm thick bi-anisotropic material, illuminated by a θ-polarized EM plane wave at 1 GHz
(a) $v=0$ and $v=0.5$, (b) $v=0.25$

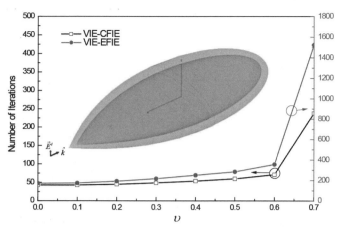

Fig. 6 Numbers of iterations with respect to different values of υ, illuminated by a θ-polarized EM plane wave from +x-axis

The fourth object is a coated PEC almond containing sharp tips, shown as an inset in Fig. 6. The length of the PEC almond is 252.4 mm, and the coating thickness is 10 mm. The frequency of the incident θ-polarized EM wave is 1 GHz. A moderate mesh size is chosen to generate totally 25,611 unknowns with respect to 1,983 triangles and 5,188 tetrahedrons. The constitutive parameters of the coating material are

$$\bar{\bar{\varepsilon}}_r = \bar{\bar{u}}_r = \begin{bmatrix} 2 & j & 0 \\ -j & 2 & 0 \\ 0 & 0 & 1.5 \end{bmatrix} \quad \bar{\bar{\xi}}_r = \bar{\bar{\zeta}}_r^* = \upsilon(1-j)\bar{\bar{I}} \tag{36}$$

where υ is a variable. Figure 5 shows the monostatic RCS with $\upsilon=0$, 0.25, or 0.5 and the observation range is $0°\leq\varphi\leq360°$ and $\theta=90°$ with 181 observation angles, while *co* and *cross* denote the co- and cross-polarization, respectively. Besides, the *cross* RCS of the coated almond with $\upsilon=0$ (in this case, the coating material degrades into anisotropy) is below -80 dBsm everywhere and not shown. Excellent agreements are observed between the results from the VIE-CFIE and those from the VIE-EFIE, indicating that despite the sharp structures are contained, the results from the VIE-CFIE are also dependable. Meanwhile, with different values of υ, the magnitude of the co-polarization RCS is almost the same in most angles, while that of the cross-polarization one is largely varied. Besides, the larger the value of υ, the larger the cross-polarization RCS. This phenomenon demonstrates the effect of the coupled parameters on the electric and magnetic fields inside the bi-anisotropic coating.

In order to investigate how υ influences the condition of the impedance matrix, Fig. 6 shows the numbers of iterations for different values of υ, while the relative residual error is fixed to 0.001. In this process, the incident wave is from +x-axis, illuminating the tip of the coated almond. It is observed that with different values of v, the convergence of the VIE-CFIE is always several times faster than that of the VIE-EFIE, illustrating the robustness and efficiency of the proposed scheme. Another finding is that when v is small, both the VIE-EFIE and VIE-CFIE can reach the target convergence after dozens of iterations. Along with the increase of v, the iterations will slightly increase, followed by a sharp increase. When v is larger than about 0.7, none can reach the target convergence after 2000 iterations. The reason is that under the fixed values of $\bar{\bar{\varepsilon}}_r$ and $\bar{\bar{u}}_r$, if $v>0.7$, the main diagonal elements of $\bar{\bar{\alpha}}_{i11}$ and $\bar{\bar{\alpha}}_{i22}$ in will be negative, which leads to a particularly ill-conditioned impedance matrix.

The fifth object is a multi-tablet containing five layers of different materials, as shown inside Fig. 7 (a). The size of each tablet is $0.5\lambda \times 0.5\lambda \times 0.05\lambda$. Besides, both the second and fourth layers are PEC, while the others are penetrable material. The first layer is bi-anisotropic, whose constitutive parameters are

$$\bar{\bar{\varepsilon}}_{1r} = \bar{\bar{u}}_{1r} = \begin{bmatrix} 2 & j & 0 \\ -j & 2 & 0 \\ 0 & 0 & 1.5 \end{bmatrix} \quad \bar{\bar{\xi}}_{1r} = \bar{\bar{\zeta}}_{1r}^* = (0.5 - j0.5)\bar{\bar{I}} \qquad (37)$$

The third layer is isotropic with scalar relative permittivity and permeability as $\varepsilon_{3r} = u_{3r} = 2 - j$. The fifth one is uniaxial anisotropic, the parameters of which are

$$\bar{\bar{\varepsilon}}_{5r} = \bar{\bar{u}}_{5r} = \begin{bmatrix} 1.5 & 0 & 0 \\ 0 & 1.5 & 0 \\ 0 & 0 & 2 \end{bmatrix}$$

This object is illuminated by an EM plane wave from +z-axis. After discretization, the numbers of triangles and tetrahedrons are 1,024 and 3,626 with 17,576 total unknowns. In this example, the CC of electric flux is attempted to be explicitly enforced on the material-PEC interfaces to reduce the number of volumetric electric unknowns, while 872 volumetric electric unknowns are eliminated due to the use of CC. The numerical results of the bistatic RCS calculated using the VIE-EFIE, VIE-CFIE, and those enforced the CC (denoted by CC-EFIE-VIE and CC-VIE-CFIE) are shown in Fig. 7 for the normal incidence. It is observed that the numerical results from the VIE-EFIE or VIE-CFIE with and without CC are almost in excellent agreement everywhere, indicating that the CC is valid when the bi-anisotropic materials are involved. Table II shows the computational details, containing the number of iterations, the peak memory usage and the total CPU time. It is observed that when the CC is used in either the VIE-EFIE or VIE-CFIE, all of the peak memory usages, the number of iterations and total CPU time are reduced. However, the reduction is very limited, because the eliminated number of unknowns (872) occupies quite a small proportion in the total number of unknowns (17,576). This phenomenon illuminates that the CC is more suitable for the calculation of the thin-coated objects. On the other hand, the CC does not deteriorate the matrix condition for the objects containing complex material blocks. In other words, the CC can always be adopted safely and reliably.

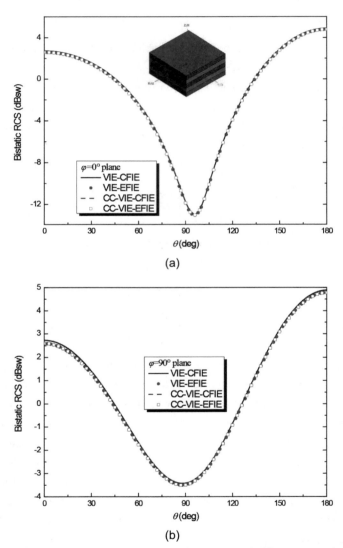

Fig. 7 Bistatic RCS of a multi-tablet containing five layers of different materials, illuminated by a θ-polarized *EM* plane wave from +z-axis

(a) φ=0° plane, (b) φ=90° plane

Table II

Computational details of scattering from five-layer tablet for different implementations with the normal incidence

Equation type	VIE-EFIE	VIE-CFIE	CC-VIE-EFIE	CC-VIE-CFIE
Number of unknowns	17,576		16,704	
Number of iterations	203	74	198	70
Peak memory usage (MB)	2483.7		2365.2	
Total CPU time (min)	10.3	5.1	10.0	4.8

5. Conclusion

In this paper, a second-kind integral equation, called VIE-CFIE, has been proposed and applied in modelling of the EM scattering from the composite objects involving inhomogeneous bi-anisotropic materials and closed PECs. In the process of MoM solution, some new kinds of singularities occur and have been properly handled using the singularity extraction method. The accuracy and efficiency of the proposed VIE-CFIE are demonstrated by the calculation of various objects with different constitutive parameters and geometry structures. Numerical experiment shows that even if the calculated object contains fine structures, complex inhomogeneous materials and multilayers, the VIE-CFIE can always give reliable results and be several times faster than the VIE-EFIE during the iterative solution. The validity of the continuity condition of electric flux explicitly enforced on the bi-anisotropy-PEC interfaces has also been investigated, which can be reliably used to reduce the number of unknowns for the volumetric electric current.

Acknowledgments

This work was supported in part by the National Natural Science Foundation of China under Grant 61701447, Grant 61971384, and Grant 62071436, and in part by the Fundamental Research Funds for the Central Universities under Grant CUC19ZD001.

Data Availability Statement

The data that support the findings of this study are available from the corresponding author upon reasonable request.

Conflict of Interest

All authors declare no conflict of interest.

References

[1] VIITANEN A J, LINDELL I V. Chiral slab polarization transformer for aperture antennas[J]. IEEE Trans. Antennas Propag., 1998, 46 (9): 1395-1397.

[2] YIN W Y, WANG W B, LI P. Guided electromagnetic waves in gyrotropic chirowaveguides[J]. IEEE Trans. Microwave Theory and Tech., 1994, 42(11): 2156-2163.

[3] QIU C W, YAO H Y, LI L W, et al. Backward waves in magnetoelectrically chiral media: propagation, impedance and negative refraction[J]. Phys. Rev. B, 2007, 75: 15.

[4] VARADAN V K, VARADAN V V, LAKHTAKIA A. On the possibility of designing antireflection coatings using chiral media[J]. J. Wave-Mater. Interact., 1987(2): 71-81.

[5] SERDYUKOV A, SEMCHENKO I, TRETYAKOV S, et al. Electromagnetics of bi-anisotropic materials: theory and application[M]. Amsterdam: Gordon and Breach, 2001.

[6] LINDELL I V, SIHVOLA A H, TRETYAKOV S A, et al. Electromagnetic waves in chiral and bi-isotropic media[M]. Norwood, MA: Artech House, 1994.

[7] HARRINGTON R F. Field Computation by moment methods[M]. New York: MacMillan, 1968.

[8] CHEW W C, JIN J M, MICHIELSSEN E, et al. Fast and efficient algorithms in computational electromagnetics[M]. Boston, MA: Artech House, 2001.

[9] KOBIDZE G, SHANKER B. Integral equation based analysis of scattering from 3-D inhomogeneous anisotropic bodies[J]. IEEE trans. antennas propag., 2004, 52(10): 2650-2658.

[10] KOBIDZE G, AYGUN K, SHANKER B. Efficient integral equation based analysis of scattering from PEC-anisotropic bodies[J]. IEEE antennas and propagation society symposium, 2004, 4: 3887-3890.

[11] ZHANG Y, WEI X, LI E. Electromagnetic scattering from three-dimensional bianisotropic objects using hybrid finite element-boundary integral method[J]. J. of Electromagn. Waves and Appl., 2004, 18(11): 1549-1563.

[12] WANG D X, LAU P Y, YUN E K N, et al. Scattering by conducting bodies coated with bi-isotropic materials[J]. IEEE trans. antennas propag., 2007,55,(8): 2313-2319.

[13] WANG D X, YUN E K N, CHEN R S, et al. An efficient volume integral equation solution to EM scattering by complex bodies with inhomogeneous bi-isotropy[J]. IEEE trans. antennas propag., 2007,55(7): 1970-1981.

[14] HU L, LI L W, YEO T S. Analysis of scattering by large inhomogeneous bi-anisotropic objects using AIM[J]. Progress in electromagnetics research, 2009, 99 :21-36.

[15] LU T, ZHAO Y, YANG Y. Application of VSIE method to scattering problem involving conducting and anisotropic bodies[C]. Chengdu, China: International Conference on Microwave and Millimeter Wave Technology, 2010.

[16] MARKKANEN J, YLÄ-OIJALA P, SIHVOLA A. Discretization of volume integral equation formulations for extremely anisotropic materials[J]. IEEE trans. antennas propag., 2012,60(11): 5195-5202.

[17] YLÄ-OIJALA P, MARKKANEN J, JÄRVENPÄÄ S. Current-based volume integral equation formulation for bianisotropic materials[J]. IEEE trans. antennas propag., 2016, 64(8): 3470-3477.

[18] YLÄ-OIJALA P, TASKINEN M. Calculation of CFIE impedance matrix elements with RWG and $n \times$RWG functions[J]. IEEE Trans. Antennas Propag., 2003,51(8): 1837–1846.

[19] LIU J, LI Z, SU J, et al. On the volume-surface integral equation for scattering from arbitrary shaped composite PEC and inhomogeneous bi-isotropic objects[J]. IEEE Access, 2019,7: 85594-85603.

[20] RAO S M, WILTON D R, GLISSON A W. Electromagnetic scattering by surfaces of arbitrary shape[J]. IEEE Trans. Antennas Propag., 1982,30(3): 409-418.

[21] SCHAUBERT D H, WILTON D R, GLISSON A W. A tetrahedral modeling method for electromagnetic scattering by arbitrarily shaped inhomogeneous dielectric bodies[J]. IEEE Trans. Antennas Propag., 1984,32(1) : 77-85.

[22] SAAD Y. Iterative methods for sparse linear systems[M]. 2nd ed. Philadelphia: SIAM, 2003.

[23] WOO A C, WANG H T G, SCHUH M J, et al. EM programmer's notebook-benchmark radar targets for the validation of computational electromagnetics programs[J]. IEEE antennas propag. magazine, 1993,35(1): 84-89.

[24] DUNAVANT D A. High degree efficient symmetrical Gaussian quadrature rules for the triangle[J]. Int. J. Numer. Methods Eng., 1985, 21: 1129-1148.

[25] KEAST P. Moderate-degree tetrahedral quadrature formulas[J]. Computer methods in applied mechanics and engineering, 1986, 55(3): 339-348.

[26] LEVENT GÜREL, ÖZGÜR ERGÜL. Singularity of the magnetic-field integral equation and its extraction[J]. IEEE antennas and wireless propagation letters, 2015, 4: 229-232.

[27] GRAGLIA R D. On the numerical integration of the linear shape functions times the 3-D Green's function or its gradient on a plane triangle[J]. IEEE trans. antennas propag., 1993, 41(10): 1448-1455.

[28] DUFFY M G. Quadrature over a pyramid or cube of integrands with a singularity at a vertex[J]. SIAM J. Numer. Anal., 1982, 19(6): 1260-1262.

Tri-Band Radar Cross-Section Reduction Based on Optimized Multi-Element Phase Cancellation*

1. Introduction

Stealth technology can significantly reduce the detection probability and increase concealment ability in the military. Radar cross section (RCS) is the measurement of a target's ability to reflect radar signals in the direction of the radar receiver. Therefore, RCS reduction is a significant parameter for designing electromagnetic stealth targets.

Metasurfaces, which are two-dimensional counterparts of metamaterials, provide a new type of regulation for electromagnetic wave propagation, such as focusing lens, polarization conversion, RCS reduction, etc. Artificial magnetic conductor (AMC) structure is an example of metasurface with zero reflection phase. Therefore, one important application of the AMC structure is to achieve RCS reduction by controlling the phase of the scattered field. One main concept of the stealth technique is redirecting scattered energy to different directions. In [6], the combination of the perfect electric conductor (PEC) and AMC leads to destructive interference in boresight direction, and the scattered power will be reflected in other directions depending on two structure contributions. However, the bandwidth is narrow. By replacing the AMC and PEC with two different AMC structures, wideband RCS reduction is achieved. A 10-dB RCS reduction can be achieved when the two AMC structures maintain a phase difference (180°±37°) in wideband.

Then, the dual-wideband checkerboard surface is presented in [8], and the surface with 10-dB RCS reduction bandwidths of 61% and 24% by a combination of the two different AMC structures is demonstrated. Another method of reducing RCS is using polarization conversion metasurface. The direction of the electric field of the reflected wave is rotated by 90° compared to the incident wave. Mirror symmetry of the polarization conversion element (PCE) can produce another unit cell with a 180° phase difference. The bandwidth of RCS reduction depends mainly on the polarization conversion rate (PCR) of the PCE.

In the latest literature, the RCS reduction study based on the phase cancellation is mainly divided into three directions: expanding the bandwidth, optimizing the effect of diffuse scattering, and multi-band function. Several layout techniques have been

---

* The paper was originally published in *IET Microwaves, Antennas & Propagation*, 2020, 14 (15), and has since been revised with new information. It was co-authored by Jianxun Su, Hang Yu, Jiayong Yu, Qingxin Guo, and Zengrui Li.

introduced to optimize the effect of diffuse scatterings, such as concave/convex chessboard arrangement [14], sunflower seed inspired distribution, and random coding arrangement. Only a few designs of multi-band RCS reduction metasurfaces have been reported. However, most broadband and multiband RCS reduction are limited to low ratio bandwidth or frequency ratio. A RCS reduction device with large frequency ratio will be needed when a platform is required to work in both centimeter-wave and millimeter-wave bands.

In this paper, due to the consideration of multiple reduction bands that avoid short-circuit frequencies, a tri-band RCS reduction, low-profile metasurface with large frequency ratio is designed and fabricated. The primary AMC unit cell composes of a metal structure, a dielectric layer, the metal bottom ground. Then the simulation results show that the reflection phase between different unit cells is no significant difference at some frequencies. The equivalent circuit model theory is used to explain this phase limited phenomenon. Based on the immutability of the reflection phase blind zones, we adapt the optimized multielement phase cancellation (OMEPC) to achieve a tri-band RCS reduction in our designed frequency band (3-45 GHz). The designed 16-lattice AMC surface can achieve 10-dB RCS reduction in tri-band frequency (3.81–7.33 GHz, 19.16–22.83 GHz, and 35.63–38.37 GHz), which over both centimeter-wave and millimeter-wave bands, under normal incidence for dual-linear polarizations.

2. The Design Of Tri-Band Metasurface

In this section, we propose the AMC structure, which should satisfy the wide reflection phase range with the change of some geometric parameters in multi-frequency bands. By selecting the appropriate thickness and dielectric constant of the substrate, designers can obtain RCS reduction performance in the desired frequency band or multi-band, which avoids short-circuit frequencies. After getting these simulation data of reflection coefficients, the multielement phase cancellation and particle swarm optimization (PSO) algorithm are implemented to determine the geometric parameters of the surface, which can obtain the maximum multiple bandwidths for the backward RCS reduction.

2.1. AMC Unit Structure Design

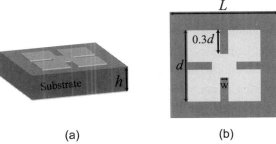

Fig. 1 Configuration of the proposed AMC cell (a) Perspective view, (b) Top view

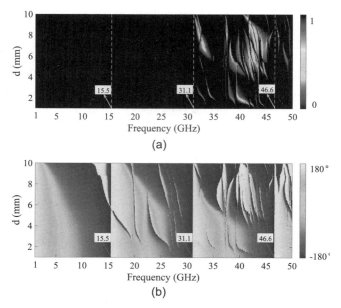

Fig. 2 The simulated fundamental harmonic reflection coefficient of AMC cell with different side length d (a) Amplitude, (b) Phase.

The unit cell is composed of two metallic layers separated by the dielectric substrate of F4B-2 ($\varepsilon_r = 2.65$, $\tan\delta = 0.001$) with a thickness of $h = 5.93$ mm. The top layer consists of a square patch with four slits at the edge of the patch, and the bottom layer is the copper ground. The thickness of the copper layer is 0.035 mm. The configuration of the proposed unit cell is depicted in Fig. 1. The unit cells with a periodicity of $L = 10$ mm, and the parameter of the unit cells is $w = 0.3$ mm. The value of d varies from 1 to 10 mm with a linear step width of 0.1 mm. Numerical simulation is carried out to investigate the reflection phase of the proposed AMC cells by the frequency-domain solver of the CST Microwave Studio®. Periodic boundary conditions are applied to the unit cell boundaries to model an infinite array environment.

As illustrated in Fig. 2(a), the simulated reflection amplitude is approximately uniform ($|\Gamma_{m,n}| \approx 1$) below 30 GHz due to the low-loss substrate and the infinite ground. The AMC is characterized by a periodicity exceeding one wavelength for inducing the excitation of a high number of Floquet harmonics. Since the power scatters toward high order harmonics over 30 GHz, the reflection coefficient of the fundamental harmonic appears less than 1. The simulated reflection phase between any AMC cells with different side lengths are nearly zero at some frequencies, as shown in Fig. 2(b). This phase limited phenomenon can be explained by the equivalent circuit model theory. According to the transmission line theory, the total input impedance (Z_{total}) equal to the parallel connection of an LC circuit

Tri-Band Radar Cross-Section Reduction Based on Optimized Multi-Element Phase Cancellation

and the short-circuited terminal. The input impedance of the short-circuited terminal can be expressed as

$$Z_s(h) = jZ_1\tan(\beta_1 h) = jZ_1\tan(\frac{2\pi\sqrt{\varepsilon_r}h}{c} \times f) \quad (1)$$

Here, Z_1 is the characteristic admittance in the dielectric substrate, β_1 is the phase constant in the dielectric substrate, ϵ_r and h are dielectric constant and thickness of the substrate, c is the speed of light, and f is the frequency. Equation (1) shows that when h and ϵ_r are fixed values, the impedance for the short-circuited line is periodic in frequency, repeating for multiples of $c/2h\sqrt{\varepsilon_r}$.

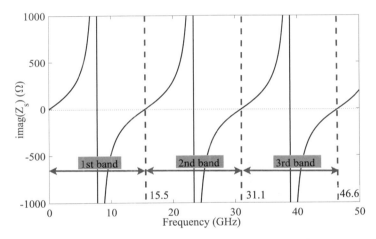

Fig. 3 The input impedance of the short-circuited terminal with substrate constant $\varepsilon_r = 2.65$ and a thickness of h = 5.93 mm.

When $\beta_1 h = n\pi$ or $f_n^{s.c} = \frac{nc}{2h\sqrt{\varepsilon_r}}$ ($n = 1, 2, 3...$), the input impedance of the short-circuited terminal is zero, and the total input impedance is zero ($Z_{total}(h) = 0$). This means any configuration of the upper structure is short-circuited by the metal ground. The reflection coefficients of AMC at these frequency points are the same as the PEC ($\Gamma = -1$). The short-circuited frequencies are decided by the thickness and dielectric constant of the substrate. When $n = 1, 2, and 3$, the short-circuit frequencies of this AMC structure are $f_1^{s.c} = $ 15.5 GHz, $f_2^{s.c} = $ 31.1 GHz, and $f_3^{s.c} = $ 46.6 GHz, respectively. Figure 3 shows the curve of the input impedance of the short-circuited terminal via the frequency. The results of the simulated reflection coefficient in Fig.2 verify this conclusion. Due to the existence of these frequency points, we can divide the reduceable frequency band into the first band (0 - $f_1^{s.c}$), the second band ($f_1^{s.c}$-$f_2^{s.c}$), the third band ($f_2^{s.c}$-$f_3^{s.c}$), etc.

In this way, the phase difference is obtained by changing the upper metallic structure, while the dielectric constant and height of the substrate are fixed; the short-circuit frequencies always occur. Considering the inevitability of short-circuit frequencies, we will optimize the frequency bands where phase cancellation can be performed to obtain multi-band RCS reduction performance.

2.2 Optimized Planar Metasurface Layout

The optimized planar surface is composed of *P* lattices. And each lattice was a sub-array of AMC cells. It is illuminated by the plane wave under normal incidence. According to the principle of the array theory, the combined amplitude of the scattered field is the vector sum of *P* individual lattices. Since zero reflection can only occur at a few discrete frequency points, in the case of a 10 dB reduction, the reflection coefficient should satisfy the following relationship

$$\left|\frac{1}{P}\sum_{p=1}^{P}\Gamma_p\right| \leq \sqrt{0.1} \qquad (2)$$

where Γ_p is the reflection coefficient of the *p*-th AMC lattice. When $P = 2$ and $|\Gamma_p| \approx 1$, this is the traditional opposite phase cancellation (OPC), the phase difference (180°±37°) between the two AMC cells is needed to achieve a 10 dB RCS reduction. When the number of lattices is larger than 2, this situation is called multielement phase cancellation (MEPC), which is proposed in Ref [18], there will be countless cases for achieving destructive interference. Meanwhile, the MEPC is far more powerful to realize destructive interference in the grating lobe region compared to the OPC. The possibility of achieving multi-wideband RCS reduction can obviously be improved. Then, the monostatic RCS reduction can be approximated by [18].

$$\sigma_R(dB) = 20\log_{10}\left|\frac{1}{P}\sum_{p=1}^{P}\Gamma_p\right| \qquad (3)$$

Equation (3) provides a theoretical guideline for predicting the RCS reduction of the multielement AMC surface. According to our numerical optimization experience, as the number of AMC elements gradually increases, the bandwidth increases significantly. However, when the number of lattices is greater than 16, it has little effect on the bandwidth, the number of AMC lattices is $P = 16$. To approximate the periodic boundary condition, each lattice includes a sub-array of M¡ÁM identical AMC cells. The ultimate optimization of the multi-band RCS reduction aim is to obtain the arrangement of 16 lattices. To find the optimal solution in the above reflection phase data, the PSO module is introduced to reduce the calculation of combination cases. The optimized frequency band is 3-45 GHz, where Q (=840) is the number of optimization frequency points $(f_1, f_2, ..., f_Q)$. For RCS reduction surfaces with P AMC lattices, $P \times Q$ values can be obtained from the pre-stored reflection phase table, which has been performed in the simulation of the AMC unit cell. In order to effectively evaluate the results of RCS multielement phase cancellation, the cost function is defined as

$$\text{cost} = \sum_{i=1}^{Q} s(i) \quad (4)$$

and

$$s(i) = \begin{cases} 0, & \text{if } \sigma_R^i \leq -12 \text{ dB} \\ 1, & \text{if } \sigma_R^i > -12 \text{ dB} \end{cases} \quad (5)$$

where σ_R^i is the monostatic RCS reduction value at the i-th frequency point. The 2 dB redundancy is due to the fact that Eq. (3) does not consider the effects of mutual coupling and edge diffraction contributions. Finally, the predicted RCS value of each optimized frequency band forms the basis of the cost function. In this optimization, the smaller value of the cost function, the better effect of RCS reduction.

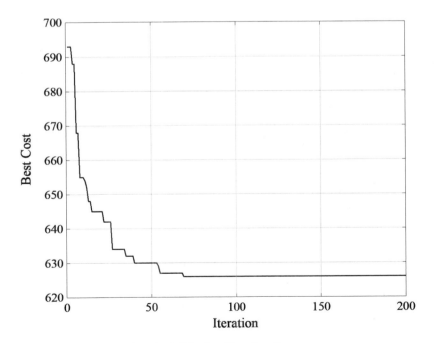

Fig. 4 PSO algorithm iteration curve.

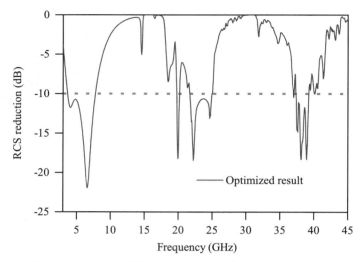

Fig. 5 Optimized RCS reduction by the PSO algorithm.

Table I

The optimized size data of 16 AMC lattices

AMC	1	2	3	4	5	6	7	8
d (mm)	9.6	3.4	8.8	6	3.6	1.3	9.6	1.3
AMC	9	10	11	12	13	14	15	16
d (mm)	9.2	2.1	5.7	9.8	1.3	8.8	9.4	2.0

After 200 iterations, as shown in Fig. 4, we get the optimal geometric parameters of 16 AMC lattices for the surface with the lowest backward RCS over multiple frequency bands. The optimized parameters of 16 basic AMC lattices are shown in Table I, and the optimized predicted RCS reduction result is shown in Fig. 5. The predicted result shows that tri-band RCS reduction is obtained, and the frequency bands of reduction are distributed in the first band, the second band, and the third band.

3. Simulation and Measurement

In this section, the proposed tri-band AMC surface is simulated, fabricated, and measured. The predicted, simulated, and measured results RCS reductions at normal incidence are compared. The bistatic scattering patterns between the proposed metasurface and the equal-sized PEC surface are investigated.

The RCS reduction surface has been designed using 16 optimized AMC lattices which include a sub-array of M × M identical AMC cells to achieve interference cancelation. To determine the number of sub-array, four 16-element RCS reduction surfaces with different lattice size (3×3, 4×4, 5×5, and 6×6 array of AMC cells) are full-wave simulated by the Transient Solver of CST Microwave Studio®. The schematic of the RCS reduction surface with 16 AMC lattices, which consist of a 36 array of AMC cells, is illustrated in Fig. 6.

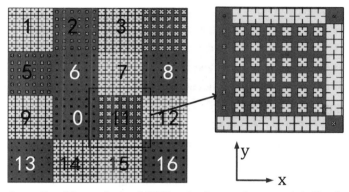

Fig. 6 The schematic of the optimized RCS reduction surface, each lattice includes a sub-array of 6×6 identical AMC cells.

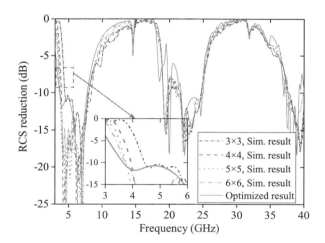

Fig. 7 The simulated and predicted monostatic RCS reduction with different lattice size under normal incidence.

The simulated and predicted monostatic RCS reduction for x-polarization under normal incidence with different lattice size are illustrated in Fig. 7. It can be observed that as the number of cells in the each lattice increases, the low-frequency of RCS reduction moves toward lower frequencies, and the performance is consistent in the high-frequency. Therefore, the lattice size of a 36 array of AMC cells is adopted in this optimized RCS reduction surface. As plotted in Fig. 9, the simulation results show the proposed 16-lattice AMC surface can achieve monostatic 10 dB RCS reduction in tri-band frequency (3.75-7.79 GHz, 21.19-24.85 GHz, and 37.31-39.94 GHz) with bandwidths of 70%, 15.9%, and 6.8% for both x- and y-polarization. The slight deviation between simulated and predicted results can be attributed to the coupling between different AMC lattices and edge effects, which are neglected in the predicted result.

To validate the simulation results, as shown in Fig. 8, a prototype of the AMC surface with dimensions of 240 mm×240 mm is fabricated and measured. The RCS experiment is performed in the compact antenna test range (CATR) system. Since the RCS reduction performance of the surface is independent of polarization, it is possible to measure RCS reduction at x-polarized incidence. The simulated and measured monostatic RCS reduction results are plotted in Fig. 9. The measured 10 dB RCS reduction is in the frequency bands of 3.81–7.33 GHz, 19.16–22.83 GHz, and 35.63–38.37 GHz, with bandwidths of 63%, 17%, and 7.4%, respectively. It can be observed that the measured result of monostatic RCS reduction is in good agreement with the simulation. The reason for the frequency offset at the high-frequency band is the instability of the dielectric constant and the measurement error.

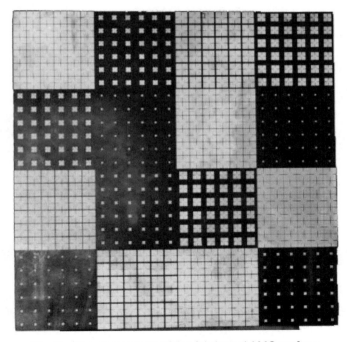

Fig. 8   The photograph of the fabricated AMC surface.

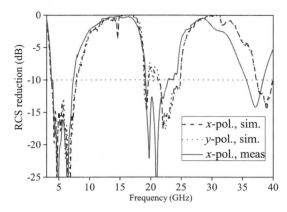

Fig. 9 The simulated and measured monostatic RCS reduction under normal incidence.

1. The simulated results of the normalized 3-D bistatic RCS patterns of the PEC ground and the 16-lattice metasurface for the normally impinging plane wave at 6 GHz, 22 GHz, and 39 GHz are depicted in Fig. 10. It shows that the PEC plate has a strong backscattering, and when the PEC ground is covered with metasurface, diffuse scattering emerges in the free space. Due to the non-uniform distributions of the reflection phase of the surface lattices, the scattered fields are redirected more directions, and the backscattering power is accordingly decreased according to the energy conservation law. Finally, the specular RCS reduction performance of the designed surface is investigated under oblique incidence, as shown in Fig. 11. The simulated results evident that the surface works well with respect to the TE and TM polarizations as well as incident angles up to 40°.

2. A comparison between the previous wideband and multi-band RCS reduction designs and this work is provided in Table II. Obviously, in our work, taking into account the multiple reduction frequency bands which avoid short-circuit frequencies, a large frequency ratio of RCS reduction bands is achieved. To meet the demand for the higher frequency ratio, we can expand the optimized frequency to the fourth band, the fifth band, etc.

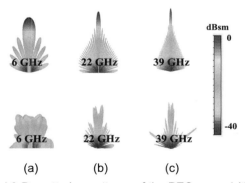

Fig. 10 The normalized 3-D scattering patterns of the PEC ground (the upper part) and the 16-lattice metasurface (the bottom part) at (a) 6 GHz, (b) 22 GHz, and (c) 39 GHz.

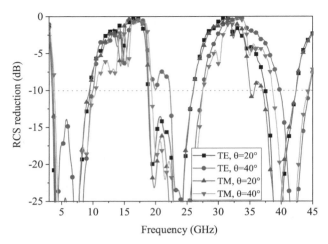

Fig. 11 The simulated specular RCS reduction of the designed surface for different polarizations and incident angles.

Table II

Comparison of our work with previous works

3. Ref.	4. RCSR (dB)	5. Thickness (mm)	6. 1st SCF 7. (GHz)	8. Operating frequency bands (GHz)	9. Frequency ratio
10.[20]	11.10	12.4	13.25.28	14.2.92-3 / 5.57-6.3 / 15.9.5-25.5	16.1 : 2 : 5.9
17.[21]	18.8	19.3	20.30.71	21.3.9-4 / 7.7-19 /-	22.1 : 3.38
23.[8]	24.10	25.6.35	26.15.92	27.3.94-7.4 / 8.41-10.72 /-	28.1 : 1.69
29.[14]	30.10	31.3	32.30.71	33.7.8-23.2 /- /-	34./
35.[13]	36.10	37.11.5	38.12.55	39.3.77-10.14 /- /-	40./
41.Our work	42.10	43.5.93	44.15.5	45.3.81-7.33 / 19.16-22.83 /35.63-38.37	46.1 : 3.77 : 6.64

RCSR: RCS reduction.

1st SCF: The first short-circuited frequency.

Frequency ratio: It is the ratio of the center frequency of each band.

4. Conclusion and Discussion

A tri-band metasurface with large frequency ratio for reducing monostatic and bistatic RCS that combines 16-lattice AMC structures is designed, simulated, fabricated, and measured. While the thickness and dielectric constant of the AMC are fixed, the blind zones of RCS reduction occur periodically regardless of any upper structure. Then, the transmission line theory is applied to analyze the root cause of the bandwidth blind zones, that is, the AMC patch is short-circuited by the ground plane at some frequencies. Theoretically, by selecting the appropriate thickness and dielectric constant of the substrate, designers can obtain RCS reduction performance in the desired frequency band or multi-band, which avoids short-circuit frequencies. Based on the above analysis, we utilize OMEPC to obtain a tri-band RCS reduction surface with 10-dB RCS reduction bandwidths of 70%, 15.9%, and 6.8% under normal incidence. Moreover, the bandwidth of RCS reduction bands is stable for oblique incident angles up to 40° for both TE and TM polarizations. Theoretical prediction, simulation, and measured monostatic results are in good agreement. This paper expands the approach of multi-band RCS reduction.

The AMC with a fixed substrate thickness will limit the bandwidth expansion for RCS reduction because of the short-circuited problem at some frequencies, resulting in the bandwidth blind zone. To overcome the in-phase reflection problem at short-circuited frequencies, multiple substrate thicknesses can be introduced to impose additional phase differences. The common short-circuited frequencies is moved to the high-frequency range.

Acknowledgments

This work was supported by the National Natural Science Foundation of China (61701448 and 61671415) and General Research Programs of Beijing Municipal Education Commission (KM201510015012).

Author Information

J. Su, H. Yu, J. Yu, Q. Guo and Z. Li are with the School of Information and Communication Engineering, Communication University of China, Beijing 100024, China.

References

[1] JUNG T J, HYEON I J, BAEK C W, et al. Circular/linear polarization reconfigurable antenna on simplified RF-MEMS packaging platform in K-band[J]. IEEE transactions on antennas and propagation, 2012, 60(11): 5039-5045.

[2] LI X, XIAO S, CAI B, et al. Flat metasurfaces to focus electromagnetic waves in reflection geometry[J]. Optics letters, 2012, 37(23): 4940-4942.

[3] HUANG L, CHEN X, MÜHLENBERND H, et al. Three-dimensional optical holography using a plasmonic metasurface[J]. Nature communications, 2013, 4(1): 2808.

[4] LI Y, ZHANG J, QU S, et al. Wideband radar cross section reduction using two-dimensional

phase gradient metasurfaces[J]. Applied physics letters, 2014, 104(22): 221110.

[5] PAQUAY M, IRIARTE J C, EDERRA I, et al. Thin AMC structure for radar cross-section reduction[J]. IEEE transactions on antennas and propagation, 2007, 55(12): 3630-3638.

[6] DE COS M E, ALVAREZ-LOPEZ Y, ANDRES F L H. A novel approach for RCS reduction using a combination of artificial magnetic conductors[J]. Progress in electromagnetics research, 2010, 107: 147-159.

[7] CHEN W, BALANIS C A, BIRTCHER C R. Checkerboard EBG surfaces for wideband radar cross section reduction[J]. IEEE transactions on antennas and propagation, 2015, 63(6): 2636-2645.

[8] CHEN W, BALANIS C A, BIRTCHER C R. Dual wide-band checkerboard surfaces for radar cross section reduction[J]. IEEE transactions on antennas and propagation, 2016, 64(9): 4133-4138.

[9] CHEN H, WANG J, MA H, et al. Ultra-wideband polarization conversion metasurfaces based on multiple plasmon resonances[J]. Journal of applied physics, 2014, 115(15): 154504.

[10] AMERI E, ESMAELI S H, SEDIGHY S H. Ultra wideband radar cross section reduction by using polarization conversion metasurfaces[J]. Scientific reports, 2019, 9(1): 478.

[11] SUN S, JIANG W, LI X, et al. Ultrawideband high-efficiency 2.5-dimensional polarization conversion metasurface and its application in RCS reduction of antenna[J]. IEEE antennas and wireless propagation letters, 2019, 18(5): 881-885.

[12] SAIFULLAH Y, WAQAS A B, YANG G M, et al. 4-bit optimized coding metasurface for wideband RCS reduction[J]. IEEE access, 2019, 7: 122378-122386.

[13] SANG D, CHEN Q, DING L, et al. Design of checkerboard AMC structure for wideband RCS reduction[J]. IEEE transactions on antennas and propagation, 2019, 67(4): 2604-2612.

[14] YUAN F, XU H X, JIA X Q, et al. RCS reduction based on concave/convex-chessboard random parabolic-phased metasurface[J]. IEEE transactions on antennas and propagation, 2019, 68(3): 2463-2468.

[15] AL-NUAIMI M K T, HONG W, WHITTOW W G. Aperiodic sunflower-like metasurface for diffusive scattering and RCS reduction[J]. IEEE antennas and wireless propagation letters, 2020, 19(7): 1048-1052.

[16] DAI H, ZHAO Y, YU C. A multi-elements chessboard random coded metasurface structure for ultra-wideband radar cross section reduction[J]. IEEE access, 2020, 8: 56462-56468.

[17] COSTA F, MONORCHIO A, MANARA G. Wideband scattering diffusion by using diffraction of periodic surfaces and optimized unit cell geometries[J]. Scientific reports, 2016, 6(1): 25458.

[18] SU J, LU Y, LIU J, et al. A novel checkerboard metasurface based on optimized multielement phase cancellation for superwideband RCS reduction[J]. IEEE transactions on antennas and propagation, 2018, 66(12): 7091-7099.

[19] EBERHART R, KENNEDY J. A new optimizer using particle swarm theory[C]. MHS'95. Proceedings of the sixth international symposium on micro machine and human science. IEEE, 1995:

39-43.

[20] ZHUANG Y, WANG G, ZHANG Q, et al. Low-scattering tri-band metasurface using combination of diffusion, absorption and cancellation[J]. IEEE access, 2018, 6: 17306-17312.

[21] ZHUANG Y, WANG G, LIANG J, et al. Dual-band low-scattering metasurface based on combination of diffusion and absorption[J]. IEEE antennas and wireless propagation letters, 2017, 16: 2606-2609.

Efficient Triangular Interpolation Methods: Error Analysis and Applications*

1. Introduction

Interpolation plays a very important role in computational electromagnetics (CEM), including finite element method (FEM) and method of moments (MoM). One of other fields using interpolation in CEM is efficient evaluations of the Green's functions (GF) in layered media, periodic Green's function (PGF), and PGF in layered media. In general, because the PGF involves the double summations of infinite series in the spectral or space domain, and Sommerfeld integral if it is the layered media, it is very time-consuming to evaluate the PGF for all observation and source points. Interpolation plays a key component to reduce the CPU time in calculation of the PGF.

Many researchers have made efforts on the efficient and accurate evaluations of the interpolation. However, the commonly used 2D interpolation technique is a product of Lagrange interpolations with polynomials of degree p along each direction to interpolate the function at a given point using $(p+1)2$ pre-calculated data points. Nevertheless, a 2D Lagrange polynomial of degree p contains only $(p+1)(p+2)/2$ terms. For $p>0$, the extra data points required for the interpolation increase the data access and interpolation time. In FEM, $(p+1)(p+2)/2$ points are used for interpolations of all points inside a large triangle formed by these points. In our proposed method, the points used for interpolation are chosen to minimize the interpolation error. In large scale problem, there is more saving in the interpolation time by using triangular interpolation instead of bivariate Lagrange interpolation. The root-mean-square (RMS) errors for lower order, $p=1$ and 2 were reported in [14]. In this letter, the details are presented on choosing the data points to interpolate a function at a given point with lower interpolation error, considering maximum error and RMS error, over triangles and rectangles. The interpolation error analysis and interpolation region selections for the rectangular, right triangular, and equilateral triangular interpolations at first order and second order are presented. The large triangle formed by the $(p+1)(p+2)/2$ points is chosen to keep the interpolation region always in the small triangle

---

* The paper was originally published in *IEEE Antennas and Wireless Propagation Letters*, 2020, 19 (6), and has since been revised with new information. It was co-authored by Wen Luo, Jinbo Liu, Zengrui Li, and Jiming Song.

at the center of the large triangle. Furthermore, the interpolation methods using the regions selected are applied to calculate the 2D singly PGF.

2. Interpolation Region Selection

For arbitrarily complex cases, the interpolation error for p'th order is proportional to the $(p+1)$'th order derivative of the function. But the coefficients of the error depend on the specific function to be interpolated. As is well known, all electromagnetic interactions can be expanded as the combinations of plane waves, which have been applied in this letter to study the interpolation error. The maximum and RMS interpolation errors are used as the error measure.

2.1 1D Interpolation

In 1D interpolation, the plane wave is simply expressed as $f(x) = e^{-jkx}$. The Lagrange interpolation polynomial is given in [15] and [16]. The relative error is expressed as

$$R_p(x) = \left| e^{-jkx} - L_p(x) \right|. \tag{1}$$

where $L_p(x)$ is the Lagrange interpolation polynomial. Here, the maximum (Max) error is the same as the maximum relative error because the amplitude of the plane wave is one. The expression of RMS error is as follows

$$R_p^{RMS} = \sqrt{\int |R_p(x)|^2 \, dx / h}. \tag{2}$$

The p'th order interpolation polynomial is formed by the nearest $p+1$ data points. The interpolation region is between the two adjacent interpolation points for $p=1$ and two half sides of the center interpolation point for $p=2$. The asymptotic maximum error is obtained as

$$R_p^{Max} = (kh/2)^{p+1} (2l-1)!!/(2l)!!, \; l = \text{int}\left[(p+1)/2\right], \tag{3}$$

where int is a function to pick up the integer part. The asymptotic RMS error is derived as

$$R_p^{RMS} = \frac{\left(\dfrac{kh}{2}\right)^{p+1}}{(p+1)!} \left\{ \begin{array}{l} \sqrt{\int_0^1 dx\, x^2 (x-2l)^2 \prod_{i=1}^{l-1} X(i)}, \; p = 2l-1 \\[2ex] \sqrt{\int_0^1 dx\, x^2 \prod_{i=1}^{l} X(i)}, \; p = 2l \end{array} \right. \tag{4}$$

where $X(i) = (x^2 - 4i^2)^2$. The asymptotic maximum and RMS errors are directly proportional to the $(p+1)$'th power of kh.

2.2 2D Interpolation: Linera Interpolation

The 2D Lagrange polynomial interpolation method is used on the rectangular grids. The interpolation function in FEM is applied to the right triangular and equilateral triangular grids. The plane wave is simply expressed as $e^{-j\mathbf{k}\cdot\mathbf{r}}$. Relative interpolation error is defined as

$$R_p(kh,\hat{k},\mathbf{r}) = \left| e^{-j\mathbf{k}\cdot\mathbf{r}} - L_p(kh,\hat{k},\mathbf{r}) \right| \tag{5}$$

where $\mathbf{k} = k\hat{k}$, and $L_p(kh,\hat{k},\mathbf{r})$ is interpolation polynomials with order p.

The RMS error for different directions is

$$R_p^{\text{RMS}}(kh,\hat{k}) = \sqrt{\int d\mathbf{r} \left| e^{-j\mathbf{k}\cdot\mathbf{r}} - L_p(kh,\hat{k},\mathbf{r}) \right|^2 \Big/ \int d\mathbf{r}} \tag{6}$$

and RMS error for different locations is

$$R_p^{\text{RMS}}(kh,\mathbf{r}) = \sqrt{\int d\hat{k} \left| e^{-j\mathbf{k}\cdot\mathbf{r}} - L_p(kh,\hat{k},\mathbf{r}) \right|^2 \Big/ \int d\hat{k}} \tag{7}$$

It is an integral from 0 to 2π for 2D interpolation and over a unit spherical surface for 3D, or in total.

$$R_p^{\text{RMS}}(kh) = \sqrt{\int d\hat{k} \int d\mathbf{r} \left| e^{-j\mathbf{k}\cdot\mathbf{r}} - L_p(kh,\hat{k},\mathbf{r}) \right|^2 \Big/ \left[\int d\hat{k} \int d\mathbf{r} \right]} \tag{8}$$

For the linear interpolation ($p=1$), three data points are used for the triangular interpolation, while four data points for the rectangular interpolation. The distribution of the maximum and RMS interpolation errors for the rectangular, right triangular and equilateral triangular interpolations are shown in Fig. 1. It is found that the right triangular interpolation has the largest maximum and RMS errors. The maximum errors of the equilateral triangular interpolation are the same as the rectangular one, while the former one has the lowest RMS errors.

To further analyze the maximum and RMS errors of the right triangular interpolation, the right triangle is divided into three regions, including one rectangle and two triangles as shown in Figs. 1 (b) and (e). When $\lambda/h = 20$, Areas 1 and 2 have the same maximum error with value of 0.02465, and the RMS errors are 0.00926 and 0.00979, respectively. The numerical results show that the maximum errors in all three regions are the same, but the RMS error of the small rectangle is lower than the two small right triangles as listed. The rectangle has been selected as the interpolation region for the right triangular interpolation. Therefore, for a given point, three nearest data points are used in the right triangular interpolation.

Efficient Triangular Interpolation Methods: Error Analysis and Applications

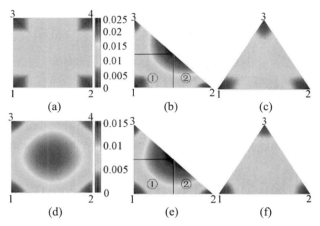

Fig. 1 distribution of maximum and RMS errors,(a) maximum errors, rectangular interpolation; (b) maximum errors, right triangular interpolation, Area 1 (a small rectangle) is the interpolation region; (c) maximum errors, equilateral triangular interpolation; (d) RMS errors, rectangular interpolation; (e) RMS errors, right triangular interpolation; (f) RMS errors, equilateral triangular interpolation.

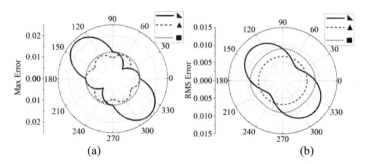

Fig. 2 the interpolation errors of the right triangular (▲), equilateral triangular (▲) and rectangular interpolation (■) for different wave propagation directions, (a) Max error, (b) RMS error.

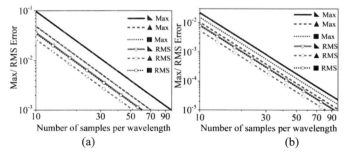

Fig. 3 the interpolation errors of the right triangular (▲), equilateral triangular (▲) and rectangular interpolation (■) for different $\lambda/h : (a) p = 1; (b) p = 2$.

| 305

Using the selected interpolation regions, the asymptotic maximum and RMS errors over different directions are derived. For the linear rectangular interpolation,

$$R_{\blacksquare_1}^{Max}(\alpha) = (kh)^2/8 \tag{9}$$

$$R_{\blacksquare_1}^{RMS}(\alpha) = (kh)^2 \sqrt{1-\sin^2(2\alpha)/12}/(2\sqrt{30}) \tag{10}$$

The asymptotic form of the maximum error is not a function of the propagation directions, while the RMS error is almost a constant. For the first order right triangular interpolation,

$$R_{\blacktriangle_1}^{Max}(\alpha) = \text{Max}\{\cos^2\alpha, \sin^2\alpha, 1-\sin(2\alpha)\}(kh)^2/8 \tag{11}$$

$$R_{\blacktriangle_1}^{RMS}(\alpha) = \sqrt{34-2\cos(4\alpha)-25\sin(2\alpha)}(kh)^2/(16\sqrt{15}) \tag{12}$$

For the first order equilateral triangular interpolation,

$$R_{\blacktriangle_1}^{Max}(\alpha) = \cos^2\left[\text{mod}(\alpha+\pi/6, \pi/3)-\pi/6\right](kh)^2/8 \tag{13}$$

$$R_{\blacktriangle_1}^{RMS}(\alpha) = (kh)^2 \sqrt{3}/(8\sqrt{10}). \tag{14}$$

where mod is the modulo operation. Figures 2 and 3 (a) plot the maximum and RMS errors of the right triangular, equilateral triangular and rectangular interpolations over different propagation directions and number of samples per wavelength using the selection regions, respectively. It is observed that the maximum error of the right triangular interpolation is double of the rectangular interpolation, because the hypotenuse length of the right triangle is $\sqrt{2}$ times of the mesh size. The RMS error of the right triangular interpolation is 5.2% more than the rectangular interpolation, but it uses one point less (3 to 4). Although the sampling points of the equilateral triangular interpolation are 15.5% more than the rectangular grids, the maximum error of the equilateral triangular interpolation is the same as the rectangular interpolation, but the RMS error is 76.7% of the rectangular interpolation.

2.3 2D Interpolation: Quadratic Interpolation

For the quadratic interpolation ($p=2$), the rectangular and triangular interpolations need 9 and 6 data points, respectively. Figure 4 plots the maximum and RMS error distributions of the rectangular, right triangular and equilateral triangular interpolations. It is obvious that the right triangular interpolations have the largest maximum and RMS errors, and the equilateral triangular interpolations have the lowest maximum and RMS errors.

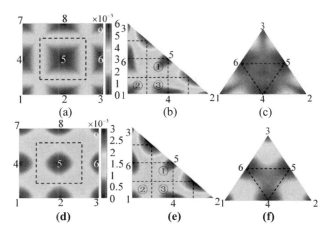

Fig. 4 $p = 2$, distribution of maximum and RMS errors, $\lambda/h = 20$. maximum errors: (a) rectangular interpolation, the rectangle at the center is the interpolation region; (b) right triangular interpolation, Area 1 is the interpolation region; (c) equilateral triangular interpolation, the equilateral triangle at center is the interpolation region; and RMS errors: (d) rectangular interpolation; (e) right triangular interpolation; (f) equilateral triangular interpolation.

The interpolation regions for the rectangular interpolation, equilateral interpolation are shown in Figs. 4 (a) and (c). The selection of the right triangular interpolation region intends to make the error as small as possible. The right triangle formed by nearest six data points is divided into ten parts, including six rectangles and four right triangles. When $\lambda/h = 20$, the numerical maximum errors of Areas 1, 2 and 3 over different propagation directions and locations are 0.00278, 0.00215 and 0.00278, respectively. The numerical RMS errors of Areas 1, 2 and 3 over different propagation directions and locations are 0.00105, 0.00115 and 0.00110, respectively. It is found that Area 2 has the smallest maximum errors, followed by Area 1 and 3, but the RMS error of Area 1 is smaller than Areas 2 and 3. The error distribution in Area 1 is more uniform, so Area 1 is picked as the interpolation region.

The asymptotic RMS errors of the rectangular, right triangular and equilateral interpolations based on the interpolation regions we selected are driven over different propagation directions as follows,

$$R_{\blacksquare 2}^{RMS}(\alpha) = (kh)^3 \sqrt{407} \sqrt{\cos^6 \alpha + \sin^6 \alpha} \Big/ \left(48\sqrt{105}\right) \tag{15}$$

for the rectangular interpolation,

$$R_{\blacktriangle 2}^{RMS}(\alpha) = (kh)^3 \sqrt{Q(\alpha)} \Big/ \left(384\sqrt{210}\right) \tag{16}$$

for the right triangular interpolation, where $Q(\alpha) = 35472 + 16624\cos(4\alpha) - 27405\sin(2\alpha) - 19761\sin(6\alpha)$, and

$$R_{\blacktriangle 2}^{\text{RMS}}(\alpha) \approx (kh)^3 \sqrt{457 - 326\cos(6\alpha)} / \left(96\sqrt{105}\right). \tag{17}$$

for the equilateral triangular interpolation.

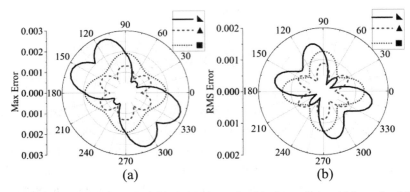

Fig. 5 the interpolation errors of the right triangular (◤), equilateral triangular (▲) and rectangular interpolation (■) for different wave propagation directions, (a) Max error, (b) RMS error.

For the quadratic interpolation, the rectangular interpolation uses the nearest point at first, then eight points are added around the data point to perform the interpolation. For the right/equilateral triangular interpolations, firstly, the three nearest points form a small right/equilateral triangle, then we extend to next three nearest points from the small right/equilateral triangle to form a big one.

Using the selected interpolation region, the maximum and RMS errors of the right triangular, equilateral and rectangular interpolations over different propagation directions and number of samples per wavelength are shown in Figs. 5 and 3(b). The maximum error of the right triangular interpolation is 44% more than the rectangular interpolation, and the RMS error is 104.6% of the rectangular interpolation, but it uses three data points less (6 to 9). The maximum and RMS errors of the equilateral triangular interpolation are the 67.8% and 67.1% of the rectangular interpolation, respectively. For the right triangular interpolation, the RMS error is closer to the rectangular interpolation than $p = 1$. The equilateral triangular interpolation has the lowest RMS error. Both asymptotic and numerical maximum and RMS errors are directly proportional to (kh)p+1 for the linear and quadratic interpolation.

3. Applications of the PGF

The singly PGF of a periodic PEC array in both space domain and spectral domain is given in [17, 18]. The Veysoglu's transformation is used to speed up the convergence of the singly PGF. The selection of interpolation regions is applied to accelerate the evaluation of the singly PGF. Before interpolating, the singularity of PGF is removed. The interpolation performance is evaluated by the interpolation error and CPU time. The relative interpolation error is defined as follows:

$$R_g = \left|\tilde{G} - \tilde{G}_{\text{interpolate}}\right| / \left|\tilde{G}\right| \tag{18}$$

where $\tilde{G}$ is the PGF without the singularity.

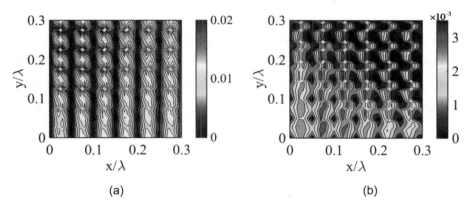

Fig. 6 The relative interpolation errors of the right triangular interpolation: (a) $p = 1$; (b) $p = 2$.

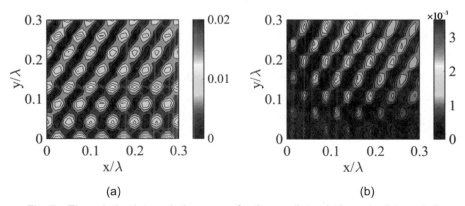

Fig. 7 The relative interpolation errors for the equilateral triangular interpolation: (a) $p = 1$; (b) $p = 2$.

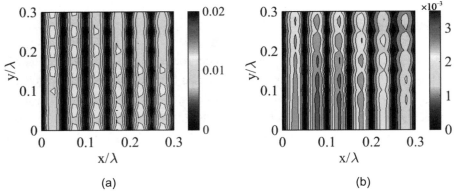

Fig. 8 The relative interpolation errors for the rectangular interpolation: (a) $p=1$; (b) $p=2$.

Figures 6, 7 and 8 show the interpolation relative error distribution for the first order and second order using right triangular, equilateral triangular and rectangular interpolations, respectively, in a region of 0.3λ by 0.3λ. Total of 31 by 31 points of PGF are interpolated from $2p+5$ by $2p+5$ points of sampling pre-calculated for rectangular grids, and $p+7$ by $2p+6$ points for equilateral grids, for $p=1$ and 2 cases. The incident angle is $\varphi=30°$. It is observed that the equilateral triangular interpolation has the smallest interpolation error for both $p=1$ and $p=2$, but the equilateral triangular grids need more pre-calculated sampling points than the rectangular grids. The right triangular interpolation has the largest relative interpolation error. Comparing the first order interpolation, the interpolation errors for the second right triangular interpolation are closer to the rectangular interpolation.

Table I lists the CPU time running on a computer with Intel Core i7-4578U 64-bit 3.00 GHz and 8 GB RAM. When $p=1$, the interpolation CPU times for the right triangular and equilateral triangular interpolations are approximately 54.2% and 30.2% less than the rectangular interpolation, respectively. When $p=2$, the interpolation CPU times for the right triangular and equilateral triangular interpolation are approximately 48.8% and 83.2% of the rectangular interpolation, respectively. It is more time-consuming to apply triangular interpolations in pre-calculation of large scale problem. The total CPU time of the right triangular is 95.1% of the rectangular interpolation for the first order and 95.6% for the second order interpolation. The total CPU time of the equilateral triangular interpolation is 98.7% of the rectangular interpolation for the first order and 97.1% for the second order interpolation. Meanwhile, it has smaller interpolation error.

Table I

CPU Time for the Right triangular (▶), Equilateral Triangular (▲) and Rectangular Interpolation (■)

Order	Direct	Total CPU time		
		■	▶	▲
1	1.2391s	0.3064s	0.2915s	0.3023s
2		0.3669s	0.3509s	0.3563s
—	—	Interpolation CPU time		
1	—	0.0308s	0.0141s	0.0215s
2		0.0422s	0.0206s	0.0351s

4. Conclusion

In this paper, the numerical and asymptotic results of maximum and RMS errors of 1D and 2D interpolations for plane waves are presented. The interpolation errors are directly proportional to the $(p+1)$'th power of kh. The interpolation regions with lower interpolation errors in 2D interpolation are selected to minimize overall errors. The maximum and RMS errors of the right triangular interpolation are closer to the rectangular interpolation for the second order. The interpolation error can be estimated for given mesh size and order of interpolation. The number of samples for the equilateral triangular interpolation is 15.5% more than the rectangular interpolation with the same mesh size. The sampling points of the right triangular grids are the same as the rectangular grids, but the data points used for the right triangular interpolation are only $(p+2)/(2p+2)$ of the rectangular interpolation. By applying interpolation on evaluating the singly PGF, the right triangular interpolation is verified as the most efficient, and the equilateral triangular interpolation is the most accurate.

References

[1] JIN J. The finite element method in electromagnetics[C]. Hoboken: Wiley-IEEE Press, 2014.

[2] SONG J, CHEW W C. Multilevel fast multipole algorithm for solving combined field integral equations of electromagnetic scattering[J]. Micro. Opt. Tech. Let.,1995, 10 (1): 14-19.

[3] CHEW W C, JIN J M, MICHIELSSEN E, et al. Fast and efficient algorithms in computational electromagnetics[M]. Norwood: Artech House, 2001.

[4] CHEW W, XIONG J, SAVILLE M. A matrix-friendly formulation of layered medium Green's function[J]. IEEE antennas & wireless propagation letters, 2006, 5(1): 490-494.

[5] XIONG J L, CHEW W C. A newly developed formulation suitable for matrix manipulation of layered medium Green's functions[J]. IEEE transactions on antennas and propagation, 2010, 58(3): 868-875.

[6] HU F G, SONG J. Integral-equation analysis of scattering from doubly periodic array of 3-D conducting objects[J]. IEEE transactions on antennas and propagation, 2011, 59(12): 4569-4578.

[7] CHEN K, SONG J, KAMGAING T. Accurate and efficient computation of layered medium doubly periodic Green's function in matrix-friendly formulation[J]. IEEE transactions on antennas and propagation, 2015, 63(2): 809-813.

[8] YANG H, YILMAZ A E. A Log-Scale Interpolation Method for Layered Medium Green's Functions[C]. Baston, MA, USA: IEEE International Symposium on Antennas and Propagation; USNC/URSI National Radio Science Meeting, 2018.

[9] ERGUL O, GUREL L. Enhancing the accuracy of the interpolations and anterpolations in MLFMA[J]. IEEE antennas & wireless propagation letters, 2006, 5(1): 467-470.

[10] VOLSKI V, VANDENBOSCH G A E, BACCARELLI P, et al. Interpolation of Green's Functions with 2D Periodicity in Layered Media[C]. Edinburgh: The Second European Conference on Antennas and Propagation (EuCAP 2007), IEEE, 2007: 1-7.

[11] LI L, WANG H G, CHAN C H. An improved multilevel Green's function interpolation method with adaptive phase compensation[J]. IEEE transactions on antennas & propagation, 2008, 56(5): 1381-1393.

[12] SONG J M, LU C C, CHEW W C. MLFMA for electromagnetic scattering from large complex objects[J]. IEEE transactions on antennas and propagation, 1997, 45(10): 1488-1493.

[13] VALERIO G, BACCARELLI P, PAULOTTO S, et al. Regularization of mixed-potential layered-media Green's functions for efficient interpolation procedures in planar periodic structures[J]. IEEE transactions on antennas & propagation, 2009, 57(1): 122-134.

[14] LUO W, LIU J, LI Z, et al. Error Analysis of 2D Triangular Interpolation for Plane Waves[C]. Nanjing, China: 2019 International Applied Computational Electromagnetics Society Symposium, IEEE, 2019: 1-2.

[15] ABRAMOWITZ M, STEGUN I A, ROMAIN J E. Handbook of mathematical functions[J]. Physics today, 1966, 19(1): 120-121.

[16] PRESS W H, TEUKOLSKY S A, VETTERLING W, et al. Numerical recipes[M]. 2nd ed. New York, NY, USA: Cambridge University Press, 1992.

[17] PETERSON A F, SCOTT R L, MITTRA R. Computational methods for electromagnetics[M]. Oxford: Oxford University Press, 1998.

[18] KONG J A. Scattering of electromagnetic waves: numerical simulations[M]. New York, NY, USA: Wiley, 2001.

Ultrawideband Monostatic and Bistatic RCS Reduction for Both Copolarization and Cross Polarization Based on Polarization Conversion and Destructive Interference*

1. Introduction

Radar cross section (RCS) reduction of metallic target plays an important role in the military defense. Metasurface with its advantage of unique characteristic and low profile has been widely utilized to reach this electromagnetic stealth goal.

As to the non-absorptive metasurface, one of the common methods to reduce RCS is based on the destructive interference between two unit cells with equal magnitude and opposite phases. In [3], a combination of perfect electric conductor (PEC) and artificial magnetic conductor (AMC) is used to create destructive interference due to the 180° phase shift of PEC unit and 0° phase shift of AMC. Since the narrow operating frequency of AMC, the 10 dB RCS reduction bandwidth is limited to 6.5%. Therefore, two unit cells, such as artificial magnetic conductors, electromagnetic band gaps etc. are used to keep 180°±37° phase difference in a larger frequency band for 10dB RCS reduction.

In addition, coding metasurfaces are proposed and investigated in recent years. In [9], 2-bit coding metasurfac is presented, which includes four elements with $\pi/2$ phase shift gradient. Unfortunately, this kind of metasurface strictly requires equivalent magnitude and fixed phase difference between elements. Their operating frequency is also limited. Thus, a physical mechanism avoiding the requirement of fixed phase difference is proposed in [11], and low scattering with backscattering coefficient less than 0.2 is achieved in a bandwidth of 112%. In addition to the wideband feature of this mechanism, its flexibility makes it suitable for various applications.

Another method to reduce RCS is based on polarization conversion metasurface (PCM). The PCM can reduce the reflected energy of co-polarization by reflecting most of energy with 90° polarization rotation (cross-polarization). In [15], double arrow unit cell is proposed to rotate the polarization in a bandwidth of 113%. However, its polarization

---

* The paper was originally published in *IEEE Transactions on Antennas and Propagation*, 2019, 67 (7), and has since been revised with new information. It was co–authored by Yao Lu, Jianxun Su, Jinbo Liu, Qingxin Guo, Hongcheng Yin, Zengrui Li, and Jiming Song.

conversion ratio (PCR) is very low. The remaining co-polarized scattering energy leads to the high co-polarized RCS. Thus, PCMs with much higher PCR are investigated in [16-18]. In these literatures, the co-polarized RCS reduction only relies on the PCR of unit cells. However, it is usually hard to design unit cells with high PCR in a very wide frequency band.

Furthermore, the reflected cross-polarized energy becomes much great due to the polarization conversion. The traditional PCMs in [12-15] arrange unit cells uniformly, which has high cross-polarized RCS. To reduce the monostatic RCS of cross polarization, PCMs in [16-17] are composed of original unit cells and their mirror unit cells which are arranged in two orthogonal orientations in four sections, respectively. However, four main beams will appear in the scattering pattern, degrading the bistatic RCS performance.

It can be seen that, when the RCS reduction only relies on either destructive interference or polarization conversion, the magnitude and bandwidth of RCS reduction are limited. Polarization characteristics of scattering field of complex objects are usually very complicated, so the suppression of the total RCS including the cross-polarized component is essential. Since the operating frequency, polarizations and directions of radar receiving antennas are unpredictable, both the co- and cross-polarized RCSs in ultra-wide frequency band are need to be suppressed for both monostatic and bistatic.

In this paper, a novel metasurface based on the combination of polarization conversion and destructive interference is proposed for ultra-wideband total mono- and bistatic RCS reduction. The total RCS is the summation of co- and cross-polarized RCSs and the RCS reduction performance under normal incidence is studied. The proposed metasurface includes 32 original unit cells (O-cells) sharing the same shape of asymmetric double arrow with different arm lengths. The mechanism employing particle swarm optimization (PSO) algorithm and array theory is modified to optimize proper parameters for these O-cells. As a result, a part of reflected energy is rotated to cross-polarization, and the rest of co-polarized energy is reduced by destructive interference produced by these O-cells. Thus, the co-polarized RCS is greatly reduced. In addition, mirror unit cells (M-cells) are introduced. The complete destructive interference occurs between O- and M-cells, leading to extremely low cross-polarized monostatic RCS. The metasurface can achieve the monostatic RCS reductions of larger than 10 dB for both polarization components in an ultra-wide frequency band from 7.4 to 22.6 GHz (101% bandwidth). Moreover, diffusion scattering patterns are produced by carefully designing the random distribution of O- and M-cells to reduce the co- and cross-polarized bistatic RCS. The 10 dB bistatic RCS reduction is obtained from 7.4 to 22.5GHz (101% bandwidth). The design flow of the novel RCS reducer metasurface is depicted in Fig.1.

This paper is organized as follows. Section II describes the design procedure for co-polarized monostatic RCS reduction. In Section III, the cross-polarized monostatic RCS is considered. The bistatic RCS reduction is studied in Section IV. Then, the simulation and measurement results are illustrated in Section V, and this work finalizes with the conclusion in Section VI.

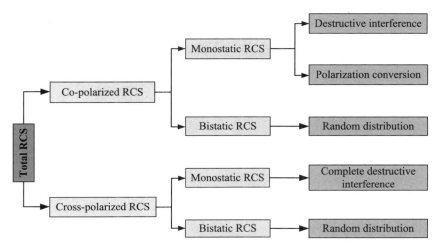

Fig. 1 The proposed design flow for total RCS reduction.

2. Design for Co-Polarized Monostatic RCS Reduction

In the previous researches, the RCS reduction usually refers to the reduction of co-polarized RCS. As shown in Fig. 1, polarization conversion and destructive interference are employed together to reduce the co-polarized monostatic RCS in this paper. The metasurface designed in this paper is composed of 8×8 unit cells including 32 O-cells and their mirror (M-cells). The basic structure of O-cells, the parameter optimization, and the final optimal parameters of O-cells are described as follow.

2.1 The Basic Structure of O-Cells

The O-cell is a single-layer structure. A metallic patch is printed on a grounded Polytetrafluoroethylene Woven Glass substrate (Model: F4B-2, Wangling Insulating Materials, Taizhou, China) with a thickness of 2.93 mm, a dielectric constant of 2.65, and a loss tangent of 0.001. The metal patches and ground are 0.035 mm-thick copper layers. These metallic patches of unit cells share the same shape of asymmetric double arrow but with different geometrical parameters. The basic structure is shown in Fig. 2. The dimensions a=6mm, l_1=4.72mm, l_3=1.16mm, w_1=0.3mm, and w_2=0.24mm are fixed. Dimension l_2 varies from 0.26 to 8.12 mm with an interval of 0.01 mm. When it reaches up to 4.72 mm, it continues increasing with a right angle, as shown in Fig. 2(b). Different unit cells are generated when parameter l_2 changes. Parameter sweep simulation of the basic structure is carried by Frequency Domain Solver of CST microwave studio. Periodic boundary condition is imposed to model an infinite array of unit cells and obtain their reflection coefficients. A part of the co-polarized reflection coefficients are shown in Fig. 3.

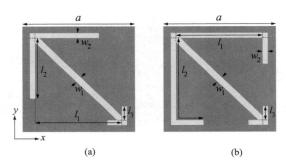

Fig. 2　The basic structure of O-cells. (a) $l_2 < 4.72$ mm. (b) $l_2 > 4.72$ mm.

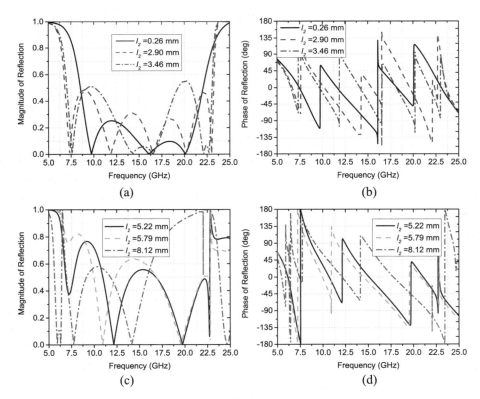

Fig. 3　A part of co-polarized reflection coefficients of O-cells. (a) and (b) $l_2 < 4.72$ mm. (c) and (d) $l_2 > 4.72$ mm.

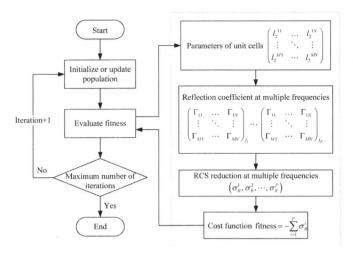

Fig. 4 The optimization flow of O-cells.

2.2 Parameter Optimization of O-Cells

To reduce co-polarized RCS in the wide frequency band, we have to optimize the dimension l_2 of 32 O-cells. To evaluate the RCS reduction of possible combination of l_2, array pattern synthesis (APS) is used to replace the time-consuming software simulation. Array pattern synthesis module is based on the array theory. The monostatic RCS reduction σ_R of the metasurface composed of $M \times N$ lattices under normal incidence can be approximated by

$$\sigma_R = 20\log_{10}\left|\frac{\sum_{m=1}^{M}\sum_{n=1}^{N}|\Gamma_{mn}|e^{j\angle\Gamma_{mn}}}{MN}\right| \quad (1)$$

where, $|\Gamma_{mn}|$, $\angle\Gamma_{mn}$ are reflection magnitude and phase of the (m, n)-th lattice, respectively. The value σ_R can be used to evaluate the RCS reduction performance of each combination. The larger the σ_R is, the larger the monostatic RCS under normal incidence is.

To avoid calculating all possible combinations, PSO module is introduced to reduce the number of combinations that need be evaluated. PSO is an intelligent optimization algorithm through iterations. Each particle in a population is one of possible solutions. The algorithm can automatically adjust particles in population through the performance of last iteration and expect a better performance at next iteration.

Figure 4 shows the optimization flow in our design. Each particle represents a combination of parameters l_2 of 32 O-cells. Firstly, a dimension matrix is generated for each particle by PSO module in the "initialize population" step. Then, the backward RCS reduction of unit cells with these sizes is theoretically calculated by Eq. 1 in the "evaluate fitness" step by APS module. To broaden the frequency band of RCS reduction, the co-polarized RCS reductions of P frequencies covering an ultra-wide frequency band are considered in cost function. Therefore, the RCS reduction performance of each particle in population can be quantified as

$$\text{fitness} = -\sum_{i=1}^{P} \sigma_R^i \qquad (2)$$

Depending on the obtained fitness of all particles of last iteration, a better dimension matrix will be generated for each particle in the "update population" step in next iteration. Therefore, when loop optimization reaches the maximum number of iterations, the optimal dimension matrix will be output.

2.3 Optimization Results

Based on the optimization procedure described before, the optimal results of 32 O-cells are obtained after 1000 iterations, as shown in Table I. A part of the optimal l_2 and their structures are shown in Fig.5. The polarization conversion capability of these O-cells can be represented by PCR,

$$\text{PCR} = |\Gamma_{cross}|^2 / (|\Gamma_{cross}|^2 + |\Gamma_{co}|^2) \qquad (3)$$

where Γ_{co}, Γ_{cross} are the reflection of co- and cross-polarization, respectively. The calculated PCRs of a part of O-cells are depicted in Fig. 6. Without considering about the contribution of destructive interference, 10 dB co-polarized RCS reduction requires the PCR of unit cell to be higher than 0.9. However, it is much difficult to design a simple unit cell with such high PCR in a very large frequency band. In our design, the great co-polarized RCS reduction is unable to be realized only relying on the relatively low PCRs of these O-cells. Therefore, to cancel the co-polarized field component, destructive interference is introduced.

The reflected characteristics and PCRs of O-cells at 8.5GHz and 20GHz are used to illustrate the proposed mechanism of co-polarized RCS reduction. As shown in Fig. 6, most of O-cells resonate at 20 GHz, achieving high PCRs. Co-polarized reflected energy at this frequency is greatly reduced by polarization conversion. However, PCRs of almost all O-cells at 8.5 GHz are less than 0.8, even down to 0.33. Polarization conversion at this frequency is unable to achieve great RCS reduction alone. In this situation, destructive interference plays an important role to help reduce RCS at this frequency, which is owing to the proper phase differences and anomalous magnitudes of different O-cells. Generally,

O-cells are unable to hold the high PCR or good co-polarized destructive interference over an ultra-wide frequency band. Thus, polarization conversion and destructive interference are always combined together to achieve the 10 dB co-polarized RCS reduction in an ultra-wide frequency band. The theoretically calculated co-polarized RCS reduction is shown in Fig. 7. It can be seen that 10 dB monostatic RCS reduction for co-polarization can be obtained from 7.1 to 22.7 GHz (105% bandwidth) based on the optimal dimensions of 32 O-cells.

Table I

The optimization results of l_2 (mm) of O-cells

3.46	8.10	5.22	0.26	8.12	0.26	6.02	8.12
0.26	8.12	0.26	4.94	0.26	5.89	3.45	0.26
0.33	0.26	5.79	5.78	0.26	2.9	0.26	5.79
0.26	0.26	5.21	0.26	3.51	6.10	8.12	0.26

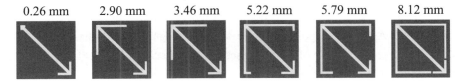

Fig. 5 The dimensions of parameter l_2 and their structures of these O-cells.

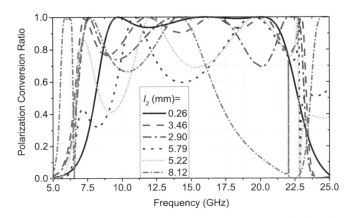

Fig. 6 The polarization conversion ratios of a part of O-cells.

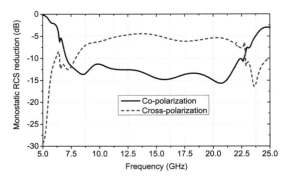

Fig.7 Theoretically calculated monostatic RCS reduction of both co-polarization and cross-polarization.

3. Design for Cross-Polarized Monostatic RCS Reduction

Figure 7 also shows the calculated RCS reduction performance of cross-polarization based on the 32 O-cells. When the co-polarized RCS decreases, the cross-polarized RCS will increases correspondingly. To keep the excellent performance of co-polarized RCS reduction, and simultaneously reduce the cross-polarized RCS, 32 mirror unit cells (M-cells) are introduced. The M-cells are the mirrors of 32 O-cells about the x axis, with the identical geometrical parameters of O-cells.

The co- and cross-polarized reflection characteristics of O- and M-cells with $l_2 = 5.79$ mm are studied. As to the co-polarized reflection, the magnitude and phase of M-cells are the same with those of O-cells. It implies that the introduction of M-cells will not change the monostatic RCS reduction performance for co-polarization. As to the cross-polarized reflection, the magnitude of M-cells is the same with that of O-cells, while the phase of M-cells is exactly 180° different from that of O-cells, as shown in Fig.8 (c) and (d). This feature of cross-polarization highly satisfies the requirement of complete destructive interference and leads to extremely low cross-polarized RCS.

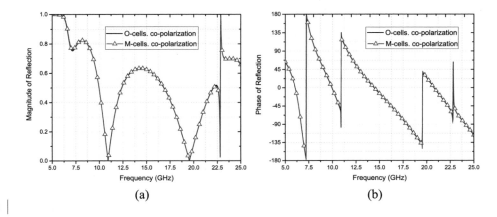

(a)   (b)

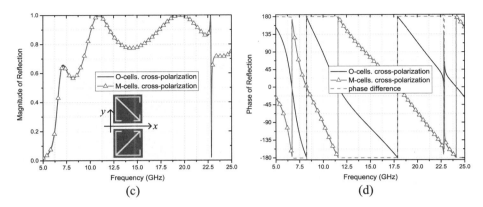

Fig. 8 The reflection coefficient of O- and M-cells with $l_2 = 5.79$mm. (a), (b) co-polarized magnitude and phase. (c), (d) cross-polarized magnitude and phase.

4. Design for Bistatic RCS Reduction

Besides the monostatic RCS reduction, bistatic RCS reduction of the metasurface under normal incidence is also considered. Usually, the bistatic RCS performance depends much on the distribution of unit cells. And anomalous distribution of unit cells is good for generating a diffusion scattering pattern of metasurface. Thus, pseudorandom matrixes with 8×8 elements are generated through the random function in the MATLAB software. Each element in the matrix represents either an O-cell or M-cell. To evaluate the bistatic RCS performance of these matrixes, the scattering patterns are calculated by

$$\sigma(\theta,\varphi) = \left\{ \sum_{m=1}^{M} \sum_{n=1}^{N} |\Gamma_{mn}| e^{j[kd\sin\theta(m\cos\varphi+n\sin\varphi)+\angle\Gamma_{mn}]} \right\}^2 \quad (4)$$

where k is the wave number, and d is the distance between lattices. It is worth noting that the total RCS is the summation of co- and cross-polarized RCSs. Therefore, both the scattering fields of co- and cross-polarization are calculated and summed together to choose the best distribution. After 500 random distributions is performed, the optimal distribution is chosen based on minimum of the sum of maximum values of $\sigma(\theta,\phi)$ at 16 frequency points from 7.4 to 22.5 GHz with an interval of 1.0 GHz, as shown in Fig. 9(a). The finally designed metasurface with a total dimension of $288 \times 288 \text{mm}^2$ is composed of 8×8 lattices and each lattice is formed of 6×6 unit cells, as shown in Fig. 9(b).

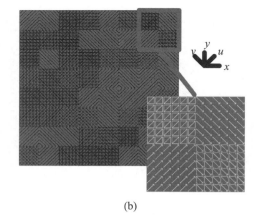

-0.26	0.26	8.10	-5.78	-0.33	5.22	-8.12	0.26
-8.12	-6.10	-8.12	0.26	0.26	-0.26	-0.26	6.02
8.12	0.26	-5.22	8.12	3.46	4.94	-3.51	0.26
-4.94	5.89	3.45	-5.89	-0.26	0.26	-0.26	0.33
8.12	5.79	5.78	-0.26	0.26	-0.26	-0.26	-2.9
-6.02	-5.21	-3.46	2.90	0.26	-3.45	5.79	-5.79
-0.26	0.26	0.26	-5.79	5.21	-8.12	-0.26	0.26
3.51	-0.26	6.10	8.12	-8.10	-0.26	0.26	-0.26

(a)   (b)

Fig. 9   The optimal distribution of the metasurface. (a) Dimensions of unit cells. Numbers with "-" are dimensions of M-cells, otherwise dimensions of O-cells. (b) The schematic view of metsuraface.

5. Simulated and Measured Results

To verify the design through theoretical analysis, the proposed metasurface is simulated, fabricated, and measured. Simulation and measurement results for both monostatic and bistatic RCS reduction are described as follow.

5.1 Monostatic RCS Reduction Performance

The simulated total monostatic RCS reductions under normal incidence of x- and y-polarized plane waves are plotted in Fig. 10. 10 dB RCS reduction can be achieved from 7.4 to 22.6 GHz (101.3% bandwidth) for both x- and y-polarization. This frequency band agrees well with the theoretical calculated monostatic RCS reduction, shown in Fig. 7. It is also can be seen that the RCS performance of the metasurface is polarization-independent.

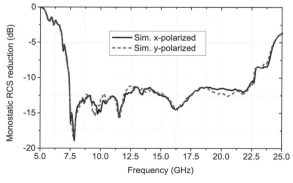

Fig. 10   The simulated monostatic RCS reduction under normal incident x- and y-polarized plane waves.

The total monostatic RCS consists the RCSs of co- and cross-polarization. When the metasurface is illustrated by the x- and y-polarized plane waves, the RCSs of co- and cross-polarization are shown in Fig. 11. Compared with the RCS of equal-sized PEC plate, co-polarized RCS of meatsurface is reduced, due to the polarization conversion and destructive interference between 32 O-cells. In addition, extremely low cross-polarized RCS is obtained, owing to the complete destructive interference between O- and M-cells. This property greatly validates the operating mechanism mentioned in Section II.

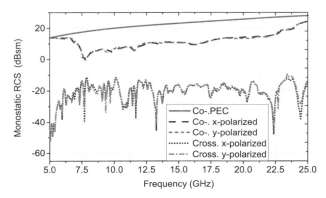

Fig. 11 The monostatic RCS under normal incidence for co- and cross polarization.

The prototype of the metasurface is displayed in Fig. 12. Both the fabricated metsaurface and equal-sized PEC plate are measured using the compact antenna test range (CATR) system of the Science and Technology on Electromagnetic Scattering Laboratory in Beijing, China. Since the RCS performance of the metasurface is polarization-independent, only the RCS reduction under x-polarized incidence is measured. The measured monostatic RCS reduction, normalized to the RCS of equal-sized PEC plate is plotted in Fig. 13. The 10 dB monostatic RCS reduction can be achieved from 7.5 to 22.8 GHz (101% bandwidth). The measurement results agree well with the simulation results. The slight deviation of the measured result is due to the fabrication and experiment error. In summary, the theoretically calculated, simulated and measured RCSs all verify the excellent monostatic RCS reduction property of the proposed metasurface.

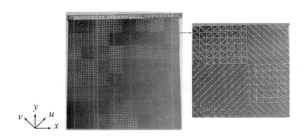

Fig. 12 The fabricated prototype of proposed metasurface.

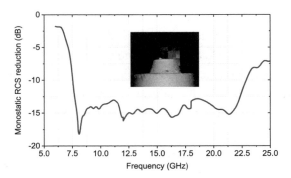

Fig. 13　The measured monostatic RCS reduction under normal incidence.

5.2 Bistatic RCS Reduction Performance

The bistatic RCS under normal incidence is considered in this part. The simulated 3-dimensional (3-D) scattering patterns of the metasurface at 8.5 and 20 GHz are shown in Fig. 14. Comparing the scattering patterns in Fig. 14(a) and (e) with those of Fig. 14(b) and (f), the scattering patterns of PEC with one peak beam are replaced by diffusion patterns of metasurface. The maximum bistatic RCS value of the metasurface is greatly suppressed.

The total bistatic RCS pattern is the summation of the scattering patterns of both co- and cross-polarization. The co- and cross-polarized scattering patterns at 8.5 GHz are illustrated in Fig. 14(c) and (d). Almost half of the reflected energy is converted to cross-polarization. Since the destructive interference between unit cells and random layout of unit cells, diffusion patterns are generated for both co- and cross-polarization. Thus, the total RCS at this frequency is diffused. In addition, the co- and cross-polarized scattering patterns at 20 GHz are illustrated in Fig. 14(g) and (h). There is one beam in the scattering pattern of co-polarization. However, since almost all reflected energy is converted to cross-polarization, the maximum value of this beam is very low. Further, considering the reflected energy of cross-polarization, diffusion pattern is also achieved due to the complete destructive interference between O- and M-cells and the random distribution of these two kinds of unit cells. Therefore, the maximum value of the total RCS at this frequency is much smaller.

In general, polarization conversion splits the reflected energy into two parts with different polarizations. Destructive interference and random distribution of unit cells are employed together to redirect energy of each polarization to more directions, leading to diffusion patterns over the whole operating frequency band. The bistatic RCS reduction (σ_R^{Bi}) defined in [23] is described as

$$\sigma_R^{Bi} = \frac{\max(\text{Bistatic RCS with metasurface})}{\max(\text{Bistatic RCS without metasurface})} \quad (5)$$

The simulated total bistatic RCS reduction under normal incidence is depicted in Fig.

15. 10 dB bistatic RCS reduction can be achieved from 7.4 to 22.5 GHz (101% bandwidth), demonstrating the excellent performance of the proposed metasurface for bistatic RCS reduction.

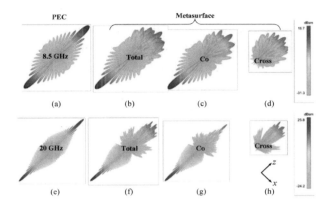

Fig. 14 The 3-D bistatic scattering patterns at 8.5 and 20 GHz. (a) and (e) illustrate the patterns of the same-sized PEC plate at 8.5 and 20 GHz, respectively. (a), (b), and (c) illustrate the total, co- and cross-polarized bistatic RCS of metasurface at 8.5 GHz, respectively. (e), (f), and (g) the total, co- and cross-polarized bistatic RCS of metasurface at 20 GHz, respectively.

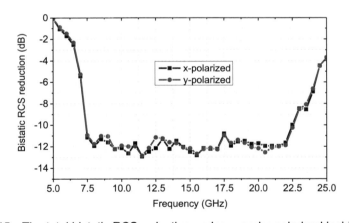

Fig. 15 The total bistatic RCS reduction under x- and y-polarized incidence.

6. Conclusion

A novel metasurface combining polarization conversion and destructive interference is designed, fabricated, and tested for ultra-wideband mono- and bistatic RCS reduction for both co- and cross-polarization. The metasurface is composed of 32 O-cells and their mirror cells (M-cells) with asymmetric double arrow patches. These patches share the same simple shape but with different geometrical parameters. Due to the special shape of unit cells, polarization conversion occurs, splitting the reflected energy to co- and cross-polarization. The parameters l_2 of 32 O-cells are optimized using the PSO and APS modules together to reduce the remaining co-polarized monostatic RCS based on the destructive interference. 32 M-cells are introduced to reduce the cross-polarized monostatic RCS based on the complete destructive interference between O- and M-cells. In addition, the random arrangement of these unit cells is designed to diffuse the reflected energy, leading to an ultrawide bistatic RCS reduction. The theoretical calculated, simulated and measured results are in good agreement and validate that the proposed metasurface is able to achieve 10 dB RCS reduction from 7.4 to 22.5 GHz (101% bandwidth) for both monostatic and bistatic RCS.

Acknowledgments

This work was supported by the National Natural Science Foundation of China under Grant 61701448, Grant 61671415, and Grant 61331002.

Author Information

Y. Lu, J. Liu, Q. Guo, and Z. Li are with the School of Information Engineering, Communication University of China, Beijing 100024, China.

J. Su and H. Yin are with the School of Information Engineering, Communication University of China, Beijing 100024, China, and also with the Science and Technology on Electromagnetic Scattering Laboratory, Beijing 100854, China.

J. Song is with the Department of Electrical and Computer Engineering, Iowa State University, Ames, IA 50011 USA, and also with the School of Information Engineering, Communication University of China, Beijing 100024, China.

References

[1] KNOTT E F, SCHAEFFER J F, TULLEY M T. Radar cross section[M]. Raleigh, NC, USA: SciTech Publishing, 2004.

[2] SINGH H, JHA R M. Active radar cross section reduction[M]. Cambridge: Cambridge University Press, 2015.

[3] PAQUAY M, IRIARTE J C, EDERRA I, et al. Thin AMC structure for radar cross-section reduction[J]. IEEE transactions on antennas and propagation, 2007, 55(12): 3630-3638.

[4] DE COS M E, ALVAREZ-LOPEZ Y, ANDRES F L H. A novel approach for RCS reduction using a combination of artificial magnetic conductors[J]. Progress in electromagnetics research, 2010, 107: 147-159.

[5] GALARREGUI J C I, PEREDA A T, DE FALCON J L M, et al. Broadband radar cross-

section reduction using AMC technology[J]. IEEE transactions on antennas and propagation, 2013, 61(12): 6136-6143.

[6] CHEN W, BALANIS C A, BIRTCHER C R. Checkerboard EBG surfaces for wideband radar cross section reduction[J]. IEEE transactions on antennas and propagation, 2015, 63(6): 2636-2645.

[7] ESMAELI S H, SEDIGHY S H. Wideband radar cross-section reduction by AMC[J]. Electronics letters, 2016, 52(1): 70-71.

[8] ZHANG H, LU Y, SU J, et al. Coding diffusion metasurface for ultra-wideband RCS reduction[J]. Electronics letters, 2017, 53(3): 187-189.

[9] GAO L H, CHENG Q, YANG J, et al. Broadband diffusion of terahertz waves by multi-bit coding metasurfaces[J]. Light: science & applications, 2015, 4(9): e324-e324.

[10] WANG K, ZHAO J, CHENG Q, et al. Broadband and broad-angle low-scattering metasurface based on hybrid optimization algorithm[J]. Scientific reports, 2014, 4(1): 5935.

[11] SU J, LU Y, ZHANG H, et al. Ultra-wideband, wide angle and polarization-insensitive specular reflection reduction by metasurface based on parameter-adjustable meta-atoms[J]. Scientific reports, 2017, 7(1): 42283.

[12] YAN D, GAO Q, WANG C, et al. A novel polarization convert surface based on artificial magnetic conductor[C]. Suzhou, China: IEEE Asia-Pacific Microwave Conference Proceedings, 2005.

[13] ZHU X C, HONG W, WU K, et al. A novel reflective surface with polarization rotation characteristic[J]. IEEE antennas and wireless propagation letters, 2013, 12: 968-971.

[14] GRADY N K, HEYES J E, CHOWDHURY D R, et al. Terahertz metamaterials for linear polarization conversion and anomalous refraction[J]. Science, 2013, 340(6138): 1304-1307.

[15] CHEN H, WANG J, MA H, et al. Ultra-wideband polarization conversion metasurfaces based on multiple plasmon resonances[J]. Journal of applied physics, 2014, 115(15): 154504.

[16] JIA Y, LIU Y, GUO Y J. et al. Broadband polarization rotation reflective surfaces and their applications to RCS reduction[J]. IEEE transactions on antennas and propagation, 2015, 64(1): 179-188.

[17] JIA Y, LIU Y, GUO Y J, et al. A dual-patch polarization rotation reflective surface and its application to ultra-wideband RCS reduction[J]. IEEE transactions on antennas and propagation, 2017, 65(6): 3291-3295.

[18] LI S J, CAO X Y, XU L M, et al. Ultra-broadband reflective metamaterial with RCS reduction based on polarization convertor, information entropy theory and genetic optimization algorithm[J]. Scientific reports, 2016, 6(1): 37409.

[19] WILLIS N J. Bistatic radar[M]. Raleigh, NC, USA: SciTech Publishing, 2005.

[20] BALANIS C A. Antenna theory: analysis and design[M]. Hoboken: John wiley & sons, 2016.

[21] EBERHART R, KENNEDY J. A new optimizer using particle swarm theory[C]. Nagoya, Japan: MHS'95. Proceedings of the sixth international symposium on micro machine and human science. IEEE, 1995: 39-43.

[22] BOERINGER D W, WERNER D H. Particle swarm optimization versus genetic algorithms for phased array synthesis[J]. IEEE Transactions on antennas and propagation, 2004, 52(3): 771-779.

[23] EDALATI A, SARABANDI K. Wideband, wide angle, polarization independent RCS reduction using nonabsorptive miniaturized-element frequency selective surfaces[J]. IEEE transactions on antennas and propagation, 2013, 62(2): 747-754.

Dual-Polarization Absorptive/Transmissive Frequency Selective Surface Based on Tripole Elements[*]

1. Introduction

Frequency selective surface (FSS) has been extensively investigated in the past decades, and many outstanding FSSs with spatial filtering characteristics have been presented for stealthy radomes. In recent years, a new type of FSS, which have characteristics of being transparent to incident electromagnetic waves in certain passband and absorptive outside of the passbands, has attracted growing attention. It is often called as rasorber, which is a combination of the words radome and absorber. It is also called as absorptive/ transmissive frequency selective surfaces (ATFSS), because it has both passband and absorption band.

Although the conceptual design of an absorptive/transmissive radome has been proposed in 1995, the complete conceptual rasorber design of multilayer structures was not presented until 2009. Some ATFSSs which use two or three layers of FSS with certain spacing have been reported. Some designs in the literature have the absorption band above the transmission window. Some have the absorption band below the transmission window, and others have two absorption bands on both sides of a passband. Some basic FSS elements or their deformed structures were utilized for rasorber design: dipole-based element loaded with lumped resistors were presented in [6]-[8]; four-legged loaded element were used in [9]; square-loop and hybrid resonator were used in [10] and [11]. More recently, the concept of a 3-D frequency selective rasorber (FSR) based on a 2-D periodic array of multimode cavities was proposed in [12]. After that, several 3-D rasorbers were presented. Compared to a multilayer structure, the structure of the 3-D FSS is more complicated, although high selectivity and angular stable performance can readily be achieved.

This letter presents an ATFSS with a transparent window between two absorption bands based on the design of cascading of an absorber and a filter, which were realized with tripole loops and tripole slots elements, respectively. Periodic structure was designed as hexagonal structure. Equivalent circuit, impedances and current distributions of the FSSs at different resonant frequencies were used to illustrate the operating principle. A prototype

---

[*] The paper was originally published in *IEEE Antennas and Wireless Propagation Letters*, 2019, 18 (5), and has since been revised with new information. It was co-authored by Qingxin Guo, Zengrui Li, Jianxun Su, Lamar Y. Yang, and Jiming Song.

of the designed ATFSS was fabricated and measured after it was elaborately designed. The results showed that a passband with a center frequency of 5 GHz was obtained between a lower absorption band from 2.66 to 4.5 GHz and a higher absorption band from 5.66 to 8.56 GHz. Good agreement was observed between simulated and measured results.

2. Analysis and Design of the ATFSS

2.1 Structure, Response and Equivalent Circuit

An ideal absorber absorbs all the incident waves at the absorption band, namely, there is no incident power to be reflected or transmitted. As a result, $|S_{11}| = 0$ and $|S_{21}| = 0$. An ideal filter enables the in-band incident waves to pass through with low insertion loss but reflect most of the out-of-band signal, so that, $|S_{11}| = 0$ and $|S_{21}| = 1$ in the passband, but $|S_{11}| = 1$ and $|S_{21}| = 0$ in the stopband. But for an ATFSS, its response can be shown in Fig. 1(a), in which the S_{11} (the solid red line) is different from that of a traditional bandpass FSS (the dotted magenta line). There are two absorption bands around f_1 and f_3 which are located on both sides of a transmission band with low insertion loss in the middle (around f_2). The coefficients of both reflection and transmission are small outside of the passband due to the absorption introduced. Ideally, the design objects include: $|S_{11}| = 0$ and $|S_{21}| = 1$ at f_2, while $|S_{11}| = 0$ and $|S_{21}| = 0$ at f_1 and f_3, However, it is more practical for the $|S_{11}|$ to be continuously less than a certain value, such as -10 dB.

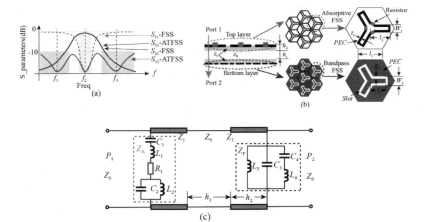

Fig. 1 Absorptive/transmissive frequency selective surface. (a) Response. (b) Structure. (c) Equivalent circuit. (l_1 = 14 mm, l_2 = 13.2 mm, l_3 =10 mm, W_1 = 2 mm, W_2 = 4 mm, t_1 = g_1 = 0.5 mm, h_1 = 10 mm, h_2 = 0.5 mm, L_1 = 9.32 nH, L_2 = 2.06 nH, L_3 = 4.38 nH, L_4 = 1.36 nH, C_1 = 0.10 pF, C_2 = 0.48 pF, C_3 = 0.55 pF, C_4 = 0.103 pF, R_1 = 300 Ω, Z_0 =377 Ω, $Z_1 = Z_0 / \sqrt{\varepsilon_r}$).

In order to meet the above condition, an ATFSS can be realized by cascading one layer of lossy FSS and another layer of lossless FSS. The lossy FSS acts as an absorber and the lossless FSS functions as a bandpass filter. Two layers of FSS are separated by a spacer, whose thickness is about $\lambda_0/4$ at f_2. Fig. 1(b) shows the configuration of the presented ATFSS. The cells of both FSSs were arranged as a hexagonal structure. The absorber consists of one tripole loop element and three lumped resistors. Each corner near the center of the three legs was broken and connected with one lumped resistor for an absorptive feature. The loop needs to parallel resonate at f_2. The total length of each leg ($2l_1+W_1$) was initially set as a half wavelength ($\lambda_0/2$) at f_2. Tripole slots were chosen for the bandpass FSS. The total length of $2l_3+W_2$ was initially set as $\lambda_0/2$ at f_2, but it is necessary to optimize the length and the width in consideration of the passband and stopband. Each FSS was supported with a substrate of 0.5 mm thickness (h_2). The relative permittivity and loss tangent of the substrate were 2.2 and 0.002, respectively. The air spacer with a thickness (h_1) of 10 mm was used to separate two FSSs. The electrical length between two FSSs is approximately equal to $h_1 + 2h_2\sqrt{\epsilon_r}$, which is slightly less than $\lambda_0/4$ at f_2 after considering the performances of the ATFSS.

The equivalent circuit model (ECM) is shown in Fig. 1(c). The loop FSS was modelled by a series-parallel resonance circuit which consists of a series circuit (R_1, L_1 and C_1) and a parallel circuit (L_2 and C_2), and the slot FSS was modelled by another parallel circuit consisting of L_3, C_3 and series of L_4 and C_4. Z_A and Z_F can be expressed by (1) and (2), respectively.

$$Z_A = R_1 + j\frac{\left(\omega^2 L_1 C_1 - 1\right)\left(\omega^2 L_2 C_2 - 1\right) - \omega^2 L_2 C_1}{\omega C_1 \left(\omega^2 L_2 C_2 - 1\right)} \quad (1)$$

$$Z_F = j\frac{-\omega L_3 \left(\omega^2 L_4 C_4 - 1\right)}{\left(\omega^2 L_3 C_3 - 1\right)\left(\omega^2 L_4 C_4 - 1\right) - \omega^2 L_3 C_4} \quad (2)$$

At the resonance frequency ω_2, where

$$\omega_2 = 1/\sqrt{L_2 C_2} \quad (3)$$

Im(Z_A) $\rightarrow \infty$, which means the signal coming from P_1 cannot pass through Z_A but transmits forward to Z_F. At frequencies ω_1 or ω_3 given by

$$\omega^2_{1\ or\ 3} = \frac{B \pm \sqrt{B^2 - 4L_1 C_1 L_2 C_2}}{2L_1 C_1 L_2 C_2} \quad (4)$$

where $B = L_1 C_1 + L_2 C_2 + L_2 C_1$, Im Im($Z_A$) = 0 and RE($Z_A$)=$R_1$. Part of the incident power coming from port P_1 is consumed by R_1 because the current inevitably traverses the resistor. The RE(Z_A) always equals to R_1, which means Z_A always consumes power in the wideband except ω_2.

No real part in (2) indicates that Z_F is lossless. If L_4 and C_4 series resonate, that is to say,

$$\omega_4 = 1/\sqrt{L_4 C_4} \tag{5}$$

Z_F is shorted, then a transmission zero can be obtained. The incident power coming from Z_A will be reflected back and is absorbed by R_1 again. At frequencies ω_2 or ω_5 given by

$$\omega_{2\text{ or }5}^2 = \frac{D \pm \sqrt{D^2 - 4L_3 C_3 L_4 C_4}}{2L_3 C_3 L_4 C_4} \tag{6}$$

where $D = L_3 C_3 + L_4 C_4 + L_3 C_4$, $\text{Im}(Z_F) \longrightarrow \infty$, the signal coming from Z_A transmits directly to port P_2. As a result, two passbands at ω_2 and ω_5 are obtained.

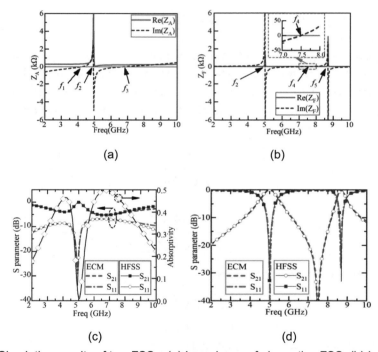

Fig. 2 Simulation results of two FSSs. (a) Impedance of absorptive FSS. (b) Impedance of bandpass FSS. (c) S parameter and absorptivity of absorptive FSS. (d) S parameter of bandpass FSS.

2.2 Simulation Results and Discussion

Fig. 2(a) and (b) show the simulated impedance of the absorptive FSS and the bandpass FSS, respectively. The simulation results were obtained by using the HFSS. Fig. 2(a) shows that the Im(Z_A) is equal to 0, but the Re(Z_A) is not zero at f_1 and f_3 (4.1 and 6.8 GHz); and both the Re(Z_A) and Im(Z_A) approach the maximum at f_2 (5 GHz). The results indicate that the loop with resistors is parallel resonant at f_2, but is absorptive at other frequencies. Fig. 2(b) shows that two parallel resonances locate at f_2 and f_5 (5 and 8.7 GHz) since both parts of Z_F change rapidly and reach the maximum. Within the range of 2 to 10 GHz except 5 and 8.7 GHz, not only the Re(Z_A) but also the Im(Z_A) are close to zero indicates that the bandpass FSS is equivalent to a ground plane. It is worth mentioning that one transmission zero will be obtained around f_4 because $Z_F = 0$. Both Z_A and $Z_F \to \infty$ at f_2 means that one passband can be obtained. The frequencies f_1 to f_5 are formulated with (3) to (6).

Fig. 2(c) and (d) show the comparison results of S parameters which were simulated by using ECM and HFSS. The results of absorptive FSS shown in Fig. 2 (c) exhibit that a passband with a low insertion loss is obtained at 5 GHz, and at out-of-band, $|S_{21}|$ ranges from -5 to -2 dB. The absorptivity, which is calculated by using $A = 1 - |S_{11}|^2 - |S_{21}|^2$, is also plotted in the figure. It seems that only part of incident power is absorbed by the FSS. Nearly 50% is absorbed at 4.1 and 6.8 GHz. Two passbands, one of which locates at 5 GHz and the other at 8.7 GHz, and one transmission zero at 7.5 GHz are shown in Fig. 2(d). The results from ECM agree very well with that obtained from HFSS.

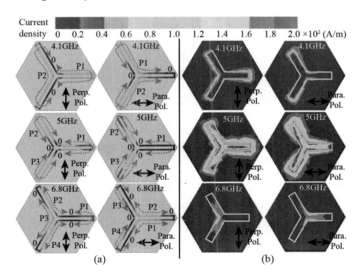

Fig. 3 Surface current distribution at resonance frequencies. (a) Absorptive FSS (0: Zero point of the current). (b) Bandpass FSS.

We also investigated the surface current distribution of both FSSs to verify the presented concept. The surface current on the tripole loop excited by the incident field is shown in Fig. 3. At 5GHz, for both perpendicular and parallel polarizations, three current paths are induced, and each pole has one path. As shown in the center row of Fig. 3(a), zero points of these current paths almost fixed and exist in the positions where the lumped resistors were placed. The absorber exhibits lossless characteristic because no current pass through the lumped resistors. While at first series resonance frequency, 4.1 GHz, only two current paths are excited. The currents inevitably traverse the lumped resistors and the incident power is consumed. Similar phenomenon appears at 6.8 GHz, the second series resonance frequency, there are four paths, among which at least two paths pass through the lumped resistors. As a result, the tripole loops become lossy at 4.1 and 6.8 GHz. Fig. 3(b) shows the surface current distribution of tripole slot. At the resonance frequency of 5 GHz, strong current is excited along the slot, so the incident wave can pass through the slots under low insertion loss. However, at other two frequencies, very low surface current is induced, which means that the tangential electric field on the surface of the bandpass FSS is very weak. The filtering FSS become a ground plane and reflect almost all incident power backward to the absorptive FSS.

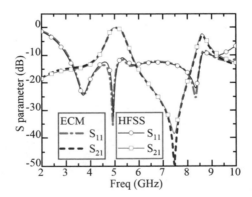

Fig. 4　Comparison of S parameter obtained from ECM and HFSS.

Fig. 4 shows the simulated S parameter of the ATFSS obtained from the ECM and HFSS. They agree with each other very well below 9.5 GHz. It shows that a transmission band of around 5 GHz with a minimum insertion loss of 0.15 dB was obtained. One deep valley is shown at 7.5 GHz, just as the prediction that one transmission zero will be obtained at f_4. The band of $|S_{11}| < -10$ dB covered from 3 to 8.6 GHz, which is much wider than the passband. The result indicates that the out-of-band electromagnetic wave incident from port P_1 to port P_2 is absorbed because both the reflection and transmission coefficients are small.

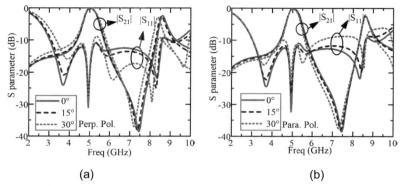

Fig. 5 Simulated performance at oblique incidence. (a) Perpendicular polarization. (b) Parallel polarization.

The S parameters of oblique incidence for this structure were further studied. The oblique incidence angle (θ_{in}) scans from 0° to 30° with a step of 15° for both the perpendicular and parallel polarization wave. As shown in Fig. 5, the passband was very stable as the θ_{in} is increased, although the bandwidth of the passband expanded slightly under perpendicular polarization but decreased a little bit under parallel polarization. For perpendicular polarization, $|S_{11}|$ was getting worse in low absorption band and getting better in high absorption band with increasing θ_{in}, while for parallel polarization, it was getting worse in both bands.

2.3. Experimental Results

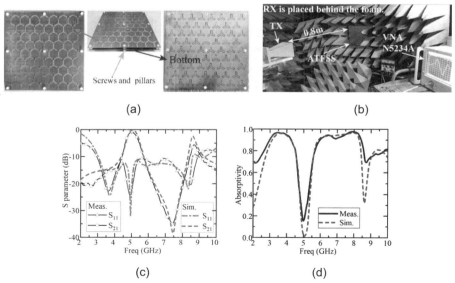

Fig. 6 Experimental results. (a) Photo of ATFSS. (b) Photo of measurement system. (c) S parameter. (d) Absorptivity.

To verify the designs and simulation results discussed above, a prototype of the proposed ATFSS structure was fabricated using standard printed circuit board (PCB) technology. Each FSS was etched on the F4B substrate whose thickness and relative permittivity are 0.508 mm and 2.2, respectively. The overall size of the fabricated structure was $4\lambda_0 \times 4\lambda_0 \times 0.183\lambda_0$, where λ_0 is the free-space wavelength at 5 GHz. Plastic pillars and screws were used to fix the essential 10 mm air gap between the two layers. The photo of the prototype is shown in Fig. 6(a).

The measurement system is shown in Fig. 6(b). The ATFSS was measured in an anechoic chamber by using Keysight N5234A with the time-domain gating. The aperture of two horn antennas for testing is 75 mm × 75 mm. The space between transmitting antenna (TX) or receiving antenna (RX) and the ATFSS was 0.8 m. Due to the measurement set-up limitation, the far-field condition was hard to meet for the entire frequency range. Our prototype was positioned on the far field of the TX antennas so that the transmission can be considered similar to the simulation. As a result, some discrepancy could arise concerning the reflection effects at the higher frequencies. The ATFSS was surrounded with absorbent foam to reduce the effect of the edge diffraction. Time-domain gating was used during measuring the reflection coefficient for reducing the multipath interference.

Fig. 6(c) shows the simulation and measurement S parameters under normal incidence. A passband with a minimum insertion loss of 0.35 dB at 5 GHz was obtained. The measurement bandwidth of the passband with $|S_{21}| \geq -3$ dB was from 4.68 to 5.36 GHz. The $|S_{11}|$ was below -10 dB in the range of 3 to 8.8 GHz. It is seen that not only the measured band of $|S_{11}| < -10$ dB was larger than the simulated one but also the measured insertion loss was higher than the simulated one, too. The main reason for this difference is because of the parasitic effect of lumped resistors, which was not taken into consideration in the full-wave simulations, for the sake of a simplified model.

A comparison of simulation and measurement of the absorptivity under the normal incidence is shown in Fig. 6(d). It is noted that two absorption bands located at two sides of the passband are seen. The measured absorption bandwidth with an absorption rate higher than 80% were 51.4% (2.66 to 4.5 GHz) and 40.8% (5.66 to 8.56 GHz) in lower and higher bands, respectively.

Table I lists the performance comparisons between the proposed ATFSS and others in the literatures.

Table I

Performance comparison of the state-of-the-art

Ref.	Absorptive Band1		f_c2 (GHz)	IL3 (dB)	Pol.4	Thickness5	
	Lower	Upper				λ_c	λ_{Am}
[4]	N.A.	7-18	1	1.9	Single	0.025	0.197

Table I Continued

Ref.	Absorptive Band1		f_c2 (GHz)	IL3 (dB)	Pol.4	Thickness5	
	Lower	Upper				λ_c	λ_{Am}
[5]	N.A.	10-18	4.6	0.3	Dual	0.07	0.166
[6]	3-9	N.A.	10.2	0.15	Single	0.41	0.12
[7]	3-9	N.A.	10	0.2	Dual	0.4	0.12
[8]	3.76-8.7	12-16.1	10.3	0.3	Dual	0.24	0.088
[11]	1.9-5.1	7.1-9.8	6.3	0.6	Dual	0.345	0.104
This work	2.76-4.44	5.64-8.42	5	0.35	Dual	0.183	0.098

1 Unit (GHz), the bandwidths are of 80% absorption rate.
2 f_c: Center frequency of passband.
3 IL: The minimum insertion loss of passband.
4 Pol.: Type of polarization, single or dual.
5 λ_c and λ_{Am}: Wavelength of f_c and the lowest frequency of absorption band.

3. Conclusion

By cascading a tripole loop FSS which acts as an absorber and a tripole slot FSS which functions as a bandpass filter, we designed an ATFSS with a transparent band in between two absorption bands. The tripole loops were loaded with lumped resistors to obtain one lossless passband and two lossy absorptive bands. The equivalent circuit model, the equivalent impedances of both FSSs, and the surface current distribution at different resonance frequencies were utilized to investigate the working principle. Simulation results showed that stable performances were observed under oblique incidence with the incident angles increased to 30°. After the ATFSS was elaborately designed, one prototype was fabricated using printed circuit board technology to verify the design. Measured results agreed well with the simulated ones. The measured results showed that a passband of 3 dB bandwidth covered from 4.68 to 5.36 GHz, in which the minimum insertion loss was 0.35 dB. Below and above the passband, two absorptive bands with an absorption rate higher than 80% were also achieved.

References

[1] MUNK B A. Frequency selective surfaces: theory and design [M]. New York: Wiley, 2000.
[2] MUNK B A. Metamaterials: critique and alternatives [M]. Hoboken, NJ, USA: Wiley, 2009.
[3] ARCENEAUX W S, AKINS R D, MAY W B. Absorptive transmissive radome [P]. 1995-5-21.
[4] MOTEVASSELIAN A, JONSSON B L G. Design of a wideband rasorber with a polarisation

sensitive transparent window[J]. IET Microw. Antenna Propag., 2012,6(7): 747-755.

[5] COSTA F, MONORCHIO A. A frequency selective radome with wideband absorbing properties[J]. IEEE Trans. Antennas Propag., 2012,60(6): 2740-2747.

[6] CHEN Q, LIU L, CHEN L, et al. Absorptive frequency selective surface using parallel LC resonance[J]. Electron. Lett., 2016,52(6):418-419.

[7] CHEN Q, YANG S, BAI J, et al. Design of absorptive/transmissive frequency-selective surface based on parallel resonance[J]. IEEE Trans. Antennas Propag., 2017,65(9): 4897-4902.

[8] CHEN Q, SANG D, GUO M, et al. Frequency-selective rasorber with interabsorption band transparent window and interdigital resonator[J]. IEEE Trans. Antennas Propag., 2018,66(8): 4105-4114.

[9] XIU X, CHE W, HAN Y, et al. Low-profile dual-polarization frequency-selective rasorbers based on simple-structure lossy cross-frame elements[J]. IEEE antennas wireless propag. lett., 2018,17(6): 1002-1005.

[10] SHANG Y, SHEN Z, XIAO S. Frequency-selective rasorber based on square-loop and cross-dipole arrays[J]. IEEE trans. antennas propag., 2014,62(11): 5581-5589.

[11] HUANG H, SHEN Z. Absorptive frequency-selective transmission structure with square-loop hybrid resonator[J]. IEEE antennas wireless propag. lett., 2017,16(11): 3212-3215.

[12] SHEN Z, WANG J, LI B. 3-D frequency selective rasorber: concept, analysis, and design[J]. IEEE trans. microw. theory techn., 2016,64(10): 3087–3096.

[13] YU Y, SHEN Z, DENG T, et al. 3-D frequency-selective rasorber with wide upper absorption band[J]. IEEE trans. antennas propag., 2017,65(8): 4363-4367.

[14] ZHANG Y, LI B, ZHU L, et al. Frequency selective rasorber with low insertion loss and dual-band absorptions using planar slotline structures[J]. IEEE antennas wireless propag. lett., 2018,17: 633-36.

Ultrawideband Radar Cross-Section Reduction by a Metasurface Based on Defect Lattices and Multiwave Destructive Interference*

1. Introduction

In physics, the interference is a phenomenon in which two or more waves of same type superpose to form a resultant wave of greater, lower, or the same amplitude. The interference usually refers to the interaction of waves that are correlated or coherent with each other. The principle of superposition of waves states that when two or more propagating waves traverse into the same space, the resultant amplitude is the vector sum of the amplitudes of the individual waves. Interference effects can be observed with all types of waves, for example, light, radio, acoustic, surface water waves or elastic waves.

Metasurfaces, which are essentially two-dimensional (2D) metamaterials with low fabrication cost, are capable of generating abrupt interfacial phase changes and providing a unique way of fully controlling the local wave front at the subwavelength scale. One of the potential applications of metasurface is to reduce the scattering field of a metal object. Generally, the RCS reduction can be achieved by two ways. One is electromagnetic (EM) wave absorption, and the other is phase cancellation or destructive interference. For the absorptive method, the radar absorbing metamaterial can also be used for RCS reduction by transforming electromagnetic energy into heat. However, metamaterial absorbers usually operate in the vicinity of resonance frequency. For the phase cancellation method, the scattered energy could be redirected away from the source direction.

As a traditional phase cancellation method, opposite phase cancellation has been widely implemented to achieve the RCS reduction. A 180°phase reversal is generated for two reflection coefficients, resulting in destructive interference.Since the frequency and direction of incoming waves are unpredictable in reality, bandwidth and oblique incidence performance are two important factors of stealth technology. In 2007, Paquay *et al.* combine the perfect electric conductors (PEC) and artificial magnetic conductors (AMC) to

---

* The paper was originally published in *Physical Review Applied*, 2019, 11 (4), and has since been revised with new information. It was co–authored by Jianxun Su, Huan He, Yao Lu, Hongcheng Yin, Guanghong Liu, and Zengrui Li.

design a checkerboard surface. The backscattered field is successfully suppressed near the broadside direction of the surface. However, the RCS reduction occurs over a very narrow bandwidth owing to the limited in-phase reflection characteristics of the AMC. Then, a planar broadband checkerboard structure formed by combining two AMC cells is presented for wideband RCS reduction in [16]. The 41% fractional bandwidth (FBW) is obtained for 10 dB RCS reduction. In [17], a 2D phase gradient metasurface designed using a square combination of 49 split-ring sub-unit cells is proposed to reduce RCS. In the frequency range of 7.8-17 GHz (74.19% FBW), RCS reduction is larger than 10 dB. In [18], a new AMC structure is designed to achieve ultra-wideband RCS reduction. This three layer structure composed of two trefoil type AMC unit cells with different heights and relative permittivities in a chessboard-like configuration. The 10 dB bandwidth spans from 13.1-44.5 GHz (109% FBW). In [19], a planar low cost and thin metasurface is composed of two different AMC unit cells which have a Jerusalem cross pattern with different thicknesses. The metasurface reduces RCS more than 10 dB from 13.6-45.5 GHz (108% FBW) for both TE and TM polarizations. Different from previous metasurface designs that mainly focus on one dimension, a shared aperture metasurface design is divided into two dimensions. Square patch is selected as metasurface elements to construct the metasurface, printed on both sides of the dielectric substrate. And an air substrate is added between dielectric substrate and ground plane. This metasurface can achieve 10 dB RCS reduction almost from 4.8 GHz to 17.5 GHz (113.9% FBW) under normal incidence. Furthermore, many other metasurfaces have been presented in previous researches. These checkerboard surfaces of periodic phase arrangement can create four strong scattering beams, which are bad for the bistatic RCS.

Recently, novel designs of coding diffuse metasurface, which can provide a more flexible way for the manipulation of reflected waves, are highly desirable. In [24], coding, digital and programmable metamaterials with excellent abilities for manipulating EM waves have firstly been presented, and 10 dB RCS reduction bandwidth of 66.67% is achieved. Then two metasurfaces using the windmill-shaped unit have been developed. The metasurface in [25] realizes 10 dB RCS reduction from 7-13 GHz, and the other shows 10 dB RCS reduction within the whole X band. A single-layer terahertz metasurface is proposed in [27], which produces untralow reflections from 1 to 1.8 THz. Based on a three-layer polarization convertor, an ultra-broadband reflective metamaterial is presented in [28]. The bandwidth of 10 dB RCS reduction covers from 5.21 to 15.09 GHz (94.38% FBW) under normal incidence. In [29], a broadband and wide-angle 2-bit coding metasurface is designed and characterized at terahertz (THz) frequencies. The ultrathin metasurface is composed of four digital elements based on a metallic double cross line structure, resulting in 60% FBW for 10 dB RCS reduction. In [30], a metasurface composed by three kinds of simply patterned elements with different resonant properties is designed. The metasurface shows excellent backward scattering from 0.77 THz to 1.97 THz. A broadband and broad-angle polarization-independent random coding metasurface is investigated in [31]. The metasurface gets 10 dB RCS reduction bandwidth of 84.75%. In [32], two kinds of AMC unit cells constitute a coding diffuse metasurface, which realizes 10 dB RCS reduction from 5.4 to 7.4 GHz. Then, a coding phase gradient metasurface constructed by phase gradient metasurface as the coding elements is proposed in [33]. The 10 dB RCS reduction bandwidth of 64.18% is realized. In [34], a random combinatorial gradient metasurface is proposed for broadband, wide-angle, and polarization-independent diffusion scattering. The metasurface suppresses the RCS by more than 10 dB within the frequency band from 7.1 to 15.6 GHz (74.89% FBW). Compared to the checkerboard metasurfaces with periodic phase arrangement, the reflection energy is distributed into more directions away from the source direction by the coding diffuse metasurfaces, which makes the lower bistatic RCS.

The objective of this work is to develop a defect lattice structure and a new physical mechanism of MWDI for breaking the bandwidth constraints of traditional phase cancellation methods such as opposite phase cancellation. The defect lattices, which destroy the periodicity of the finite array, are capable of generating amplitude-phase manipulated waves. And the defect lattices themselves can achieve primary destructive interference, leading to RCS reduction. The interference between the local backscattered waves generated by the defect lattices is manipulated and optimized simultaneously by the superposition principle of waves and PSO to obtain second destructive interference. The proposed metasurface can achieve 10 dB RCS reduction from 6.16 to 41.63 GHz with a ratio bandwidth (RBW) of 6.76:1 under normal incidence. A comparison between other recent researches and this work is provided in Table I. Obviously, our results are far beyond the state of the art of the performance in terms of bandwidth of this type of structures.

Table I

Comparison of our work and previous researches

Article	σ_R (dB)	OFB (GHz)	FBW (%)	RBW (f_H / f_L)
[16]	10	14.5-21.8	41	1.50:1
[17]	10	7.8-17	74.19	2.18:1
[24]	10	7.5-15	66.67	2.00:1
[28]	10	5.21-15.09	94.38	2.90:1

Table I Continued

Article	σ_R (dB)	OFB (GHz)	FBW (%)	RBW (f_H / f_L)
[29]	10	700-1300	60	1.86:1
[21]	10	3.8-10.7	95	2.82:1
[22]	10	6.1-17.8	98	2.92:1
[31]	10	17-42	84.75	2.47:1
[33]	10	9.83-19.12	64.18	1.95:1
[23]	10	5.4-14.2	89.8	2.63:1
[18]	10	13.1-44.5	109	3.40:1
[19]	10	13.6-45.5	108	3.35:1
[20]	10	4.8-17.5	113.9	3.65:1
This Work	10	6.16-41.63	148.4	6.76:1

OFB: The operating frequency band

This paper is organized as follows. Section II shows the new physical mechanism of multi-wave destructive interference. The design process of ultra-wideband RCS reducer metasurface is shown in section III. The simulated results for normal and oblique incidences are illustrated in Section IV, and the measured monostatic RCS reduction under normal incidence is shown in Section V. The conclusions are drawn in Section VI.

2. Multi-Wave Destructive Interference

A metasurface consists of a two-dimensional array of M×N defect lattices. Fig. 1 shows the design flowchart for RCS reducer metasurface. First, the lattice structure is designed to produce amplitude-phase manipulated waves and overcome the short circuit problem over ultra-wide frequency band. Second, the combination of the new mechanism of multi-wave destructive interference and PSO algorithm is adopted to optimize and determine the geometric parameters of the defect lattices for achieving ultra-wideband monostatic RCS reduction. Third, the defect lattices are arranged randomly to consist the metasurface. Fourth, EM simulation is carried out by the transient solver of CST Microwave Studio® to verify the performance of monostatic and bistatic RCS reductions. At last, the high-precision RCS measurement is conducted to validate the theoretical design and simulation.

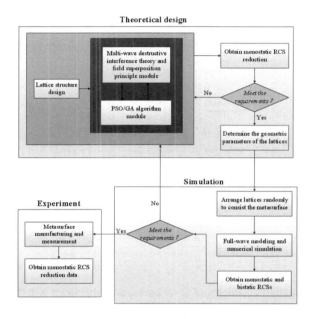

Fig. 1　The design procedure of the metasurface.

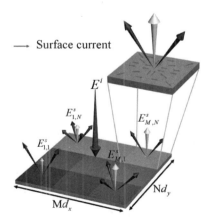

Fig. 2　Plane wave is illuminating on a metasurface consisted of $M \times N$ defect lattices. $M \times N$ amplitude-phase manipulated waves are produced locally by the defect lattices. The yellow arrows indicate backscattered waves while the blue arrows represent side-scattered waves/ side lobes.

The main theory for the metasurface design is the defect lattice and the combination of multi-wave destructive interference and PSO algorithm. The $M \times N$ defect lattices are uniformly spaced with dx in the x direction and dy in the y direction, as shown in Fig. 2. The defect structures, which destroy the periodicity of the finite array, are capable of generating reverse current, leading to primary destructive interference. $M \times N$ backscattered waves (yellow arrows) with full control of amplitude and phase produced locally by the defect lattices are superimposed and optimized to achieve second destructive interference, as shown in Fig.3.

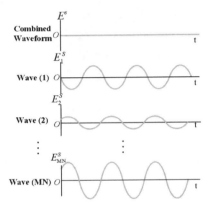

Fig. 3 The combined waveform of $M \times N$ backscattered waves with full control of the reflection amplitude and phase.

The RCS reduction (σ_R) of the metasurface for second destructive interference, compared to an equal-sized PEC surface, can be represented by [35].

$$\sigma_R = 10\log_{10}\left[\frac{|E^s_{MS}|^2}{|E^s_{PEC}|^2}\right] = 20\log_{10}\left|\frac{\sum_{m=1}^{M}\sum_{n=1}^{N}|E^s_{m,n}|e^{j\angle E^s_{m,n}}}{MN|E^i_{m,n}|e^{j\angle E^i_{m,n}}}\right| \quad (1)$$

where E^s_{MS} and E^s_{PEC} are the scattering fields of the metasurface and equal-sized PEC surface, respectively. $E^s_{m,n}$ and $E^i_{m,n}$ are the scattering field and incident field of the (m, n)-th lattice, respectively. Based on the measurement method of frequency selective surface, the reflection coefficient of the (m, n)-th defect lattice is defined as

$$\Gamma_{m,n} = |\Gamma_{m,n}|e^{j\angle\Gamma_{m,n}} = \frac{|E^s_{m,n}|}{|E^i_{m,n}|}e^{j\angle E^s_{m,n}} \quad (2)$$

Thus, the total RCS reduction can be approximated by

$$\sigma_R = 20\log_{10}\frac{\left|\sum_{m=1}^{M}\sum_{n=1}^{N}|\Gamma_{m,n}|e^{j\angle\Gamma_{m,n}}\right|}{MN} \quad (3)$$

Eliminating the phase term in Eq. (3), the RCS reduction for primary destructive interference just produced by the defect lattices themselves is described as

$$\sigma_R^{1st} = 20\log_{10}\frac{\sum_{m=1}^{M}\sum_{n=1}^{N}|\Gamma_{m,n}|}{MN} \quad (4)$$

Complete destructive interference requires

$$\sum_{m=1}^{M}\sum_{n=1}^{N}|E_{m,n}^s|e^{j\angle E_{m,n}^s} = 0 \quad (5)$$

which is an indefinite equation. The amplitude and phase of $M \times N$ backscattered waves can be manipulated independently and optimized to achieve ultra-wideband destructive interference. It is noted that Eq. (5) has countless solutions. Obviously, opposite phase cancellation and coding metamaterials are just two special solutions. If there are two electric fields, complete destructive interference requires two electric fields with equal amplitude and opposite phase, as shown in Fig. 4. If the number of electric fields is larger than 2, there will be countless case for achieving complete destructive interference. These cases shown in Fig. 5, which are the solutions of the above indefinite equation, can also achieve complete phase cancellation. The phase difference between the electric fields is variable. However, opposite phase cancellation and coding metamaterials would exclude these cases. For opposite phase cancellation and coding metamaterials, due to the fixed number of unit cells and the fixed phase difference between them, the ability to realize phase cancellation is greatly reduced, leading to a narrow bandwidth of RCS reduction.

In this paper, a metasurface based on defect lattices and a new physical mechanism of MWDI is proposed for ultra-wideband RCS reduction. Due to the presence of reverse current, the defect lattices can generate amplitude and phase manipulated waves and has the capacity of primary destructive interference and RCS reduction. As depicted in Fig. 3, multiple backscattered waves from the defect lattices are superimposed and optimized to achieve second destructive interference. More defect lattices and, in particular, full control of both the amplitude and phase of the reflection between them greatly increase the ability to control EM waves and achieve maximum bandwidth of RCS reduction.

$|E_1|e^{j\varphi_1} \longleftarrow \bullet \longrightarrow |E_2|e^{j\varphi_2}$

Fig. 4 Opposite phase cancellation of two electric fields.

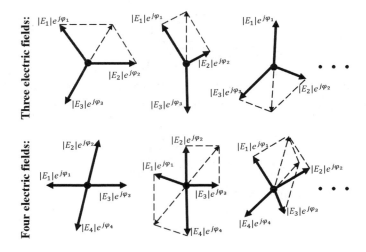

Fig. 5 Complete destructive interference for 3 and 4 electric fields.

3. Metasurface Design

The design process of ultra-wideband RCS reducer metasurface is shown in this section.

3.1 Defect Lattice and Its Reflection Characteristics

The basic lattice used in this work is a defect structure, which is different from the traditional lattice of a finite array. The traditional lattice with uniform current distribution just has the capacity of phase manipulation, whose reflection amplitude is approximately equal to unity. Based on the traditional lattice consisted of a finite array with 7×7 square ring patches, 9 square ring patches in the middle part are replaced by a cross patch to constitute the defect lattice. The geometry structure of the defect lattice is depicted in Fig. 6. The metallic patches are printed on the surface of Polytetrafluoroethylene Woven Glass (Model: F4B-2, Wangling Insulating Materials, Taizhou, China) substrate with a dielectric constant $\varepsilon_r = 2.65$ and loss tangent tan δ=0.001. The back of the substrate is the metallic ground plane. The side length of the square ring is L. The length and width of the cross strip are $5L$ and L, respectively.

The defect structures, which destroy the periodicity of the finite array, are capable of generating amplitude and phase manipulated waves. Wave path-difference is produced between tiles with different thickness. This additional phase difference can extend the coverage range of reflected phase over an ultra-wide frequency band. The range of the reflection phase changes is large enough with the change of geometrical parameter L, and the reflection amplitude can be less than unity. It means that the defecct lattice has the capacity of amplitude-phase manipulation.

The lattices are simulated by the transient solver of CST Microwave Studio®. In this simulation, the periodicity a of the unit cell and the width w of square ring are fixed. Therefore, two geometrical parameters are controlled to tailor EM wave reflection characteristics. The geometrical parameter L varies from 0.8 to 7.8 mm with a step size of 0.1 mm, while there are three choices of layer thickness for the dielectric substrate: 2 mm, 4 mm and 6 mm. A part of the reflection amplitude and phase curves are plotted in Fig. 7. The lattice of single layer structure will limit the bandwidth for RCS reduction because of the short circuit problem at some frequencies. For the thickness h=6 mm, the reflection phase around 15.35 GHz and 30.71 GHz hardly depend on the value of L due to short circuit. Transmission line theory can better explain this condition. The equivalent circuit model is shown in Fig. 8. The infinite metal ground is equivalent to the short circuit terminal. When the layer thickness $h = n\lambda_g / 2$ ($n = 1, 2, 3 \ldots$), the patch is short circuited by the metal ground. $\lambda_g (= \lambda_0 / \sqrt{\epsilon_r})$ is waveguide wavelength. The substrate thickness (h=6 mm) is equal to $0.5\lambda_g$ and $1.0\lambda_g$ at 15.35 GHz and 30.71 GHz, respectively. Therefore, the reflection phases at these two frequencies do not change with the size L of the patch, as shown in Fig. 7(f). However, defect lattices with various layer thicknesses can effectively overcome the short circuit problems.

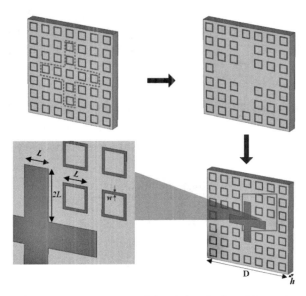

Fig. 6 Geometrical structure of the defect lattice. Dimensions are:
a=8, w=0.3, h= 2, 4, 6 in mm.

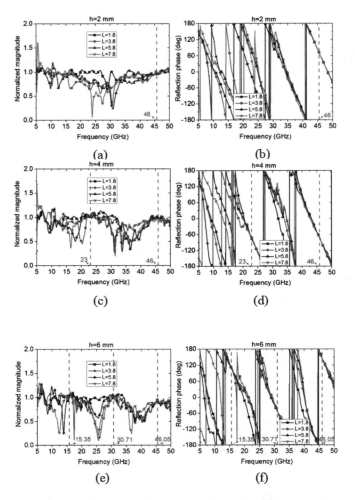

Fig. 7 The reflection coefficients of the basic defect lattices. (a, c, e) The normalized amplitudes and (b, d, f) the reflection phases with the change of geometer parameter L for the layer thickness of 2, 4, and 6 mm, respectively.

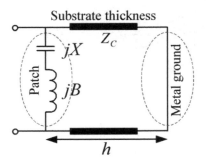

Fig. 8 Equivalent circuit model.

Full and free control of both the amplitude and phase of the local waves would greatly increase the ability of ultra-wideband manipulation of EM waves. A comparison of the current distribution and the scattering pattern between defect lattice and traditional lattice is presented in Fig. 9. Due to the presence of reverse current, the defect lattice produces primary destructive interference with several scattered beams, leading to RCS reduction and reflection amplitude-phase manipulation.

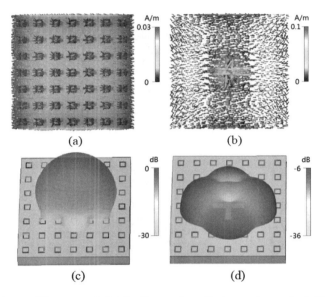

Fig. 9 Comparison of the current distribution and scattering pattern of a lattice with h=6 mm and L=3.1 mm at 5.4 GHz. (a), (c) traditional lattice. (b), (d) defect lattice.

3.2 Optimization Design of the Basic Defect Lattices

This is the most important step for achieving ultra-wideband RCS reduction. The metasurface presented in this research is composed of 4×4 defect lattices. The backscattering field is only related to the reflection coefficient of the basic lattices but independent of their distribution under normal incidence. Thus, a combination of PSO and MWDI is utilized to optimize the geometrical parameters of 16 defect lattices for achieving the maximum bandwidth of RCS reduction.

The PSO is a new class of stochastic optimizers based on the collective behavior of nature, which is very easy to understand, easy to implement and highly robust. It is similar in some ways to genetic algorithms, but requires less computational bookkeeping. It has been applied in conjunction with Eq. (3) to optimize the geometrical parameter L and layer thickness h of 16 defect lattices. The detailed flowchart for the PSO algorithm with

MWDI module to optimize the design of the basic defect lattices is depicted in Fig. 10. The geometrical parameters ($L_{1,1},\cdots L_{1,4},\cdots L_{4,4}$) and ($h_{1,1},\cdots h_{1,4},\cdots h_{4,4}$) of 16 defect lattices are put into MWDI module. The parameters L are initialized to be in the range of [0.8, 7.8] mm, while the thickness h is a discrete value chosen from the set {2, 4, 6} mm. Iteration number (Nmax) is set to 500 enough in the optimization.

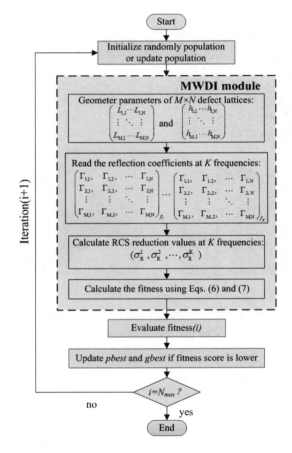

Fig. 10 Optimization flowchart of the basic defect lattices for ultra-wideband RCS reduction.

In MWDI module, the reflection coefficients at K frequencies can be read from the pre-calculated reflection amplitude and phase table. Based on Eq. (3), the RCS reduction (σ_R) can be calculated. Taking into consideration that a cancellation error occurs when some frequencies have a very large σ_R value and others with a very small σ_R value, the fitness function used to assess the wideband RCS reduction performance is defined as

$$\text{fitness} = \sum_{i=1}^{K} s(i) \qquad (6)$$

and

$$s(i) = \begin{cases} 0, & \text{if } |\sigma_R^i + 12| \leqslant 1 \text{ dB} \\ 0.5, & \text{if } 1 \text{ dB} \leqslant |\sigma_R^i + 12| \leqslant 2 \text{ dB} \\ 1.0, & \text{if } |\sigma_R^i + 12| > 2 \text{ dB} \end{cases} \qquad (7)$$

where σ_R^i is the backward RCS reduction under normal incidence at i-th frequency. The optimized target value for RCS reduction is set to 12 dB. The redundancy of 2 dB is considered for the approximation error by Eq. (3) owing to the coupling between the lattices and their edge effects. To get better RCS reduction performance, the fitness should be as low as possible.

Then, the RCS reduction values are scored by Eq. (6). The output fitness score will be evaluated at each iteration. The individual optimal location (*p*best) and the global optimal location (*g*best) will be updated if fitness score is lower. After 500 iterations, the optimal parameters ($L_{1,1}, \cdots L_{1,4}, \cdots L_{4,4}$) and ($h_{1,1}, \cdots h_{1,4}, \cdots h_{4,4}$) are obtained, which are listed in Table II.

The proposed metasurface can be constituted by the optimized 16 defect lattices with optimal parameters shown in Table II. Fig. 11 gives the predicted monostatic RCS reduction of the metasurface with a plane wave normally impinging. RCS reduction values more than 10 dB can be realized from 6.16 to 41.63 GHz, which is about 148.4% FBW and 6.76:1 RBW.

Table II

Optimal results of geometrical parameter *L* and layer thickness *h*

Lattice :	1	2	3	4	5	6	7	8
L (mm)	1.4	6.9	6.5	1.5	7.4	6.3	7.0	2.4
h (mm)	6	2	6	6	6	4	6	2
Lattice :	9	10	11	12	13	14	15	16
L (mm)	6.2	2.0	6.7	3.1	3.7	2.2	7.3	7.2
h (mm)	2	4	2	6	2	6	2	2

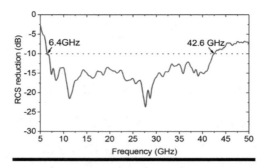

Fig. 11 The predicted monostatic RCS reduction under normal incidence.

The estimated results indicate that the metasurface based on defect lattices and MWDI can greatly expand the bandwidth of RCS reduction. The reflection amplitudes and phases of 16 defect lattices are shown in Fig. 12 (a) and (b), respectively. The reflection amplitudes and phases are functions of frequency, but they can always satisfy

$$|\sum_{m=1}^{M}\sum_{n=1}^{N}\Gamma_{m,n}| \leqslant MN\sqrt{0.1} \tag{8}$$

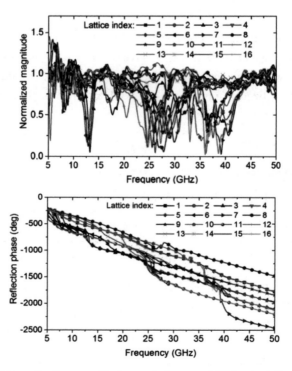

Fig. 12 The reflection amplitudes and phases of 16 basic defect lattices.

The RCS reduction produced by defect lattices themselves is shown in Fig. 13. The presented MWDI together with PSO algorithm is used to control and optimize the amplitudes and phases of 16 defect lattices in an ultra-wide frequency band simultaneously to achieve good destructive interference.

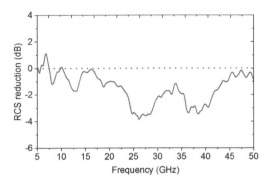

Fig. 13   The RCS reduction produced by the defect lattices themselves.

4. Ultra-Wideband RCS Reduction Performance

The metasurface is composed of these 16 basic defect lattices. The scattering pattern $E^s(\theta,\varphi)$ in space is determined by the arrangement of the basic defect lattices, except for the backward direction $(\theta=0, \varphi=0)$. The diffusion scattering of EM wave is caused by randomized lattice distribution, leading to a low bistatic RCS. To facilitate manufacturing of the metasurface, the lattices with the equal layer thickness are put together. The random distribution of 16 basic defect lattices and the full structure of the metasurface are depicted in Fig. 14. The model of the optimized RCS reducer metasurface is occupied by 4×4 defect lattices with an overall dimension of 224×224 mm².

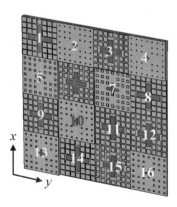

Fig. 14   The full CST model of the ultra-wideband RCS reducer metasurface. The inserted index is the lattice number shown in Table II.

To investigate the RCS reduction of the proposed metasurface, the monostatic RCS of the metasurface as well as that of an equal-sized PEC surface are simulated by the transient solver of CST Microwave Studio® under normal incidence. The RCS reduction of the metasurface normalized to the equal-sized PEC surface with a plane wave normally illuminating is shown in Fig. 15. Almost the same results achieved for both polarizations exhibit the polarization-independent feature of the metasurface. Significant RCS reduction can be observed in the frequency band from 5 to 50 GHz. It can be seen that a 10 dB RCS reduction is achieved in an ultra-wide frequency band from 6.16 to 41.63 GHz with a RBW of 6.76:1. The proposed defect lattice and MWDI have shown to greatly expand the bandwidth of RCS reduction. The simulation results are in good agreement with the estimated results, so the RCS reduction can be approximated by Eq. (3), which does not include edge effects, but provides a good guideline for RCS reduction of the metasurface compared to an equal-sized PEC surface.

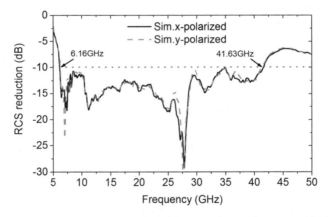

Fig. 15 The simulated monostatic RCS reduction under normal incidence.

Taking a further step, the 3D bistatic RCS patterns of the designed metasurface and the equal-sized PEC surface under normal incidence are depicted in Fig. 16 at 7 GHz, 12 GHz and 20 GHz, respectively. It can be observed that the bistatic RCS is dramatically decreased compared to the equal-sized PEC surface. The designed metasurface is capable of generating abrupt interfacial amplitude and phase changes. Due to the non-uniform distributions of the reflection amplitudes and phases of 16 defect lattices, the scattered fields are redirected to other directions and the specular reflection is effectively suppressed.

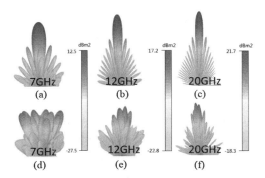

Fig. 16 Comparison of bistatic scattering patterns between the proposed metasurface and equal-sized PEC surface under x-polarized wave normal incidence at 7 GHz, 12 GHz and 20 GHz, respectively. (a)-(c) Specular reflection by the PEC surface. (d)-(f) Diffusion scattering by metasurface.

For the case of oblique incidence, the simulated RCS reduction of the metasurface for TE and TM polarized waves, as a function of frequency at different incident angles, is shown in Fig. 17. The results indicate that a remarkable RCS reduction is achieved across the ultra-wide frequency band (5-50 GHz) for both polarizations. Thus, for wide-angle incidences, the proposed metasurface still performs well in the operating frequency band.

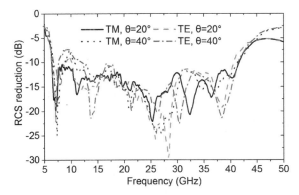

Fig. 17 The RCS reductions under wide-angle oblique incidences for both TE and TM polarizations. TE/TM: The direction of the electric/magnetic field is perpendicular to the plane of incidence.

5. Experimental Verification

To validate the effectiveness of the proposed RCS reduction structure, a sample manufactured using printed circuit board technology is measured to investigate the performance of the proposed metasurface. A 224 mm × 224 mm square sample is shown in Fig. 18. Three pieces of the metasurface board with different thicknesses are processed separately and then stitched together according to our metasurface design. The top and bottom copper layers with thickness of 0.035 mm are printed on a Polytetrafluoroethylene Woven Glass substrate with a dielectric constant $\varepsilon_r = 2.65$ (loss tangent $\tan\delta = 0.001$). A high-precision RCS measurement is conducted using the compact antenna test range system of the Science and Technology on Electromagnetic Scattering Laboratory in Beijing, China, as depicted in Fig. 19. The spherical waves emitted by the horn antenna are reflected by the parabolic metal reflector and become a plane wave. The short test distance between the metasurface sample and the reflector makes it easy to meet the far field conditions. Two horn antennas placed adjacently serve as a transmitter and a receiver, respectively. Due to the limitations of our equipment, the measurement could not be conducted in the frequency range higher than 40 GHz. The entire measured band from 6 GHz to 40 GHz is covered by five pairs of standard linearly polarized horn antennas, which work at 6-8 GHz, 8-12 GHz, 12-18 GHz, 18-26.5 GHz and 26.5-40 GHz, respectively. The antennas are connected to a vector network analyzer, which has the function of a time domain gating. The RCS of the metasurface and equal-sized copper ground of the metasurface are separately measured first. Then, subtraction is made between their values to obtain the RCS reduction.

Fig. 18   The photograph of the fabricated metasurface.

Ultrawideband Radar Cross-Section Reduction by a Metasurface Based on Defect Lattices and Multiwave Destructive Interference

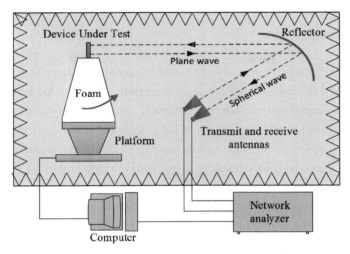

Fig. 19 The schematic of compact antenna test range system for the monostatic RCS measurement.

Fig. 20 shows the measured monostatic RCS reduction of the sample under normal incidence. The RCS reduction maxima is more than 25 dB at 22.7 GHz, while the RCS of the sample is reduced by more than 10 dB within the frequency band of 6.28-40 GHz for both x- and y- polarized incident waves. The measured and the simulated RCS reduction results have excellent agreements over the entire frequency band. The value deviations can be attributed to manufacturing and measurement errors. Second destructive interference is achieved in an ultra-wide frequency band. Overall, the ultra-wideband RCS reduction and diffusion scattering by the proposed metasurface based on defect lattices and MWDI are confirmed.

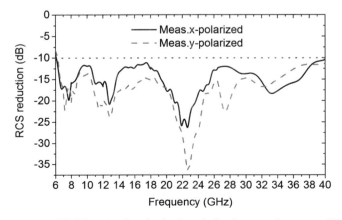

Fig. 20 Measured RCS reduction for both polarizations under normal incidence.

6. Conclusion

In this paper, a novel metasurface, based on defect lattices and a novel physical mechanism of MWDI, has been investigated for greatly expanding the bandwidth of RCS reduction. The proposed metasurface can effectively obtain an ultra-wideband, wide-angle, polarization independent monostatic and bistatic RCS reduction. Sixteen defect lattices are employed to construct the metasurface, each of which is composed of an aperiodic array of square rings with an embedded cross with the capacity of RCS reduction and amplitude-phase manipulation. The primary destructive interference can be realized by defect lattice themselves. More basic lattices and full and free control of both the amplitude and phase of the reflection greatly improve the destructive interference performance. Interference between multiple local backscattered waves produced by the basic defect lattices at multiple frequencies sampled in an ultra-wide frequency band is manipulated and optimized simultaneously by the superposition principle of waves and PSO to obtain second destructive interference. The simulated results demonstrate that the monostatic RCS reduction value exceeds 10 dB almost from 6.16 to 41.63 GHz compared with the RCS of the equal-sized PEC surface, with a ratio bandwidth (RBW) of 6.76:1 for both polarizations under normal incidence. Moreover, the random lattice distribution has realized the diffusion scattering of EM waves, which contributes to a low bistatic RCS simultaneously. The estimated, simulated and measured results match quite well, and validate that the proposed metasurface is significant for bandwidth expansion of RCS reduction.

Acknowledgments

This work was supported by the National Natural Science Foundation of China (61701448 and 61671415), and the High Quality Cultivation Project of CUC (CUC18A007-1).

Author Information

J.X. Su and H.C.Yin are with the School of Information Engineering, Communication University of China, Beijing 100024, China and also with the Science and Technology on Electromagnetic Scattering Laboratory, Beijing 100854, China (e-mail: sujianxun_jlgx@163.com).

H.He, J. Y. Lu and Z. R. Li are with the School of Information Engineering, Communication University of China, Beijing 100024, China (e-mail: hehuan_cuc@163.com, zrli@cuc.edu.cn).

G.H.Liu is with the Information Science Academy, China Electronics Technology Group Corporation (CETC), Beijing 100015, China.

References

[1] BECKER R. Electromagnetic field and interactions[M]. New York: Dover Publications, 1982.

[2] PAIN H J, RANKIN P. Introduction to vibrations and waves[M]. 1st ed. New York: Wiley, 2015.

[3] NURFAIZEY A H, STANGER J, TUCKER N, et al. Manipulation of electrospun fibres in flight: the principle of superposition of electric fields as a control method[J]. Sci.,2012, 47 (3): 1156-1163.

[4] LI J Y, QI Y X, ZHOU S G. Shaped beam synthesis based on superposition principle and Taylor method[J]. IEEE trans. antennas propagat., 2017, 65 (11): 6157-6160.

[5] KOOPMANN G H, SONG L M, FAHNLINE J B. A method for computing acoustic fields based on the principle of wave superposition[J]. J. Acoust. Soc. Am., 1989, 86 (3): 2433-2438.

[6] PENDRY J B, SCHURIG D, SMITH D R. Controlling electromagnetic fields[J]. Science, 2006, 312 (5781): 1780-1782.

[7] ZHANG L, MEI S, HUANG K, et al. Advances in full control of electromagnetic waves with metasurfaces[J]. Adv. Optical Mater., 2016, 4 (6): 818-833.

[8] LIU R, JI C, MOCK J J, et al. Broadband ground-plane cloak[J]. Science, 2009, 323 (5912): 366-369.

[9] WANG X H, CHEN F, SEMOUCHKINA E. Implementation of low scattering microwave cloaking by all-dielectric metamaterials[J]. IEEE microw. wireless compon. lett., 2013, 23 (2): 63-65.

[10] MICHELI D, PASTORE R, APOLLO C, et al. Broadband electromagnetic absorbers using carbon nanostructure-based composites[J]. IEEE trans. microw. theory tech., 2011, 59 (10): 2633-2646.

[11] WATTS C M, LIU X, PADILLA W J. Metamaterial electromagnetic wave absorbers[J]. Adv. Mater., 2012, 24 (23): OP98-120, OP181.

[12] YOO M, KIM H K, LIM S. Angular- and polarization-insensitive metamaterial absorber using subwavelength unit cell in multilayer technology[J]. IEEE antennas wireless propag. lett., 2016, 15: 414-417.

[13] BASKEY H B, AKHTAR M J. Design of flexible hybrid nano-composite structure based on frequency selective surface for wideband radar cross section reduction[J]. IEEE trans. microw. theory tech., 2017, 65 (6): 2019-2029.

[14] SUKHAREVSKY O I. Electromagnetic wave scattering by aerial and ground radar objects[M]. 1st ed. Boca Raton, FL, USA: CRC Press, 2014.

[15] PAQUAY M, IRIARTE J C. EDERRA I, et al. Thin AMC structure for radar cross-section reduction[J]. IEEE trans. antennas propagat., 2007, 55 (12): 3630-3638.

[16] IRIARTE GALARREGUI J C, PEREDA A T, MATÍNEZ DE FALCÓN J L, et al. Broadband radar cross-section reduction using AMC technology[J]. IEEE trans. antennas propagat., 2013, 61 (12): 6136-6143.

[17] LI Y F, ZHANG J Q, QU S B, et al. Wideband radar cross section reduction using two-dimensional phase gradient metasurfaces[J]. Appl. Phys. Lett.,2014, 104 (22): 221110.

[18] AMERI E, ESMAELI S H, SEDIGHY S H. Ultra wide band radar cross section reduction using multilayer artificial magnetic conductor metasurface[J]. J. Phys. D: Appl. Phys.,2018, 51 (28): 285304-285304.

[19] AMERI E, ESMAELI S H, SEDIGHY S H. Low cost and thin metasurface for ultra wide band and wide angle polarization insensitive radar cross section reduction[J]. Appl. Phys. Lett., 2018, 112 (20): 201601-201601.

[20] ZHENG Y J, CAO X Y, GAO J, et al. Shared aperture metasurface with ultra-wideband and wide-angle low-scattering performance[J]. Optical materials express, 2017, 7 (8): 2706-2706.

[21] HAJI-AHMADI M J, NAYYERI V, SOLEIMANI M, et al. Pixelated checkerboard metasurface for ultra-wideband radar cross section reduction[J]. Sci. Rep., 2017, 7 (1-4): 11437.

[22] JIA Y T, LIU Y, GUO Y J, et al. A dual-patch polarization rotation reflective surface and its application to ultra-wideband RCS reduction[J]. IEEE trans. antennas propagat., 2017, 65 (6): 3291-3295.

[23] XUE J J, JIANG W, GONG S X. Chessboard AMC surface based on quasi-fractal structure for wideband RCS reduction[J]. IEEE antennas wireless propag. lett., 2018, 17 (2): 201-204.

[24] CUI T J, QI M Q, WAN X, et al. Coding metamaterials, digital metamaterials and programmable metamaterials[J]. L. Sci. Appl., 2014, 3 (10): e218-e218.

[25] WANG K, ZHAO J, CHENG Q, et al. Broadband and broad-angle low-scattering metasurface based on hybrid optimization algorithm[J]. Sci. Rep., 2014, 4 (1): 5935.

[26] CHEN J, CHENG Q, ZHAO J, et al. Reduction of radar cross section based on a metasurface [J]. Progress in electromagnetics research, 2014, 146: 71-76.

[27] DONG D S, et al. Terahertz broadband low-reflection metasurface by controlling phase distributions[J]. Advanced optical materials, 2015, 3 (10): 1405-1410.

[28] LI S J, CAO X Y, XU L M, et al. Ultra-broadband reflective metamaterial with RCS reduction based on polarization convertor, information entropy theory and genetic optimization algorithm[J]. Sci. Rep., 2016, 5 (1): 37409.

[29] LIANG L J, WEI M G, YAN X, et al. Broadband and wide-angle RCS reduction using a 2-bit coding ultrathin metasurface at terahertz frequencies[J]. Sci. Rep., 2016, 6 (1): 39252.

[30] ZHAO J, et al. Fast design of broadband terahertz diffusion metasurfaces[J]. Optics Express, 2017, 25 (2): 1050-1061.

[31] SUN H Y, GU C Q, CHEN X L, et al. Broadband and broad-angle polarization-independent

metasurface for radar cross section reduction[J]. Sci. Rep., 2017, 7 (1-4): 40782.

[32] LIU X, GAO J, XU L M, et al. Coding diffuse metasurface for RCS reduction[J]. IEEE antennas wireless propag. lett., 2017, 16: 724-727.

[33] ZHENG Q Q, LI Y F, ZHANG J Q, et al. Wideband, wide-angle coding phase gradient metasurfaces based on Pancharatnam-Berry phase[J]. Sci. Rep., 2017, 7 (1-4): 4184-4187.

[34] ZHUANG Y Q, WANG G M, LIANG J G, et al. Random combinatorial gradient metasurface for broadband, wide-angle and polarization-independent diffusion scattering[J]. Sci. Rep., 2017, 7 (1-4): 16560.

[35] CHEN W G, BALANIS C A, BIRTCHER C R. Checkerboard EBG surfaces for wideband radar cross section reduction[J]. IEEE trans. antennas propag., 2015, 63 (6): 2636-2645.

[36] BOERINGER D W, WERNER D H. Particle swarm optimization versus genetic algorithms for phased array synthesis[J]. IEEE trans. antennas propag., 2004, 52 (3): 771-779.

[37] GAO L H, CHENG Q, YANG J, et al. Broadband diffusion of terahertz waves by multi-bit coding metasurfaces[J]. L. Sci. Appl., 2015, 4 (9): e324-e324.

A Novel Checkerboard Metasurface Based on Optimized Multielement Phase Cancellation for Superwideband RCS Reduction*

1. Introduction

Metamaterial and metasurfaces are artificial structures designed for controlling electromagnetic and acoustic waves or fields. They exhibit exceptional, unexpected physical properties from their chemical constituents. These properties lead to some fascinating phenomena, such as negative refraction, subwavelength imaging, field enhancement, and anomalous tunneling effects, etc. Novel devices based on these ideas, such as ultrathin metalenses or perfect lenses, invisibility cloak, plasmonic waveguide, and polarization converter, have been fabricated and tested over the past few years. One of the potential applications of metasurfaces is to reduce the scattering field of a metal object.

To effectively reduce the radar cross-section (RCS) of the target is challenging. Four approaches are usually used to reduce the RCS of a target in electromagnetism. The first approach is to apply radar-absorbing materials (RAM) which transform the electromagnetic energy into heat. However, most RAMs usually operate in the vicinity of resonance frequency; and the narrow working bandwidth limits its applications. The second approach is to alter the appearance of the target (shaping) to reduce the scattered field along source directions, but that may destroy the aerodynamic layout and increases the complexity of the shape design. The third approach is the use of transformation electromagnetics and optics. Electromagnetic waves propagate around the target surface, and the backscattered field is suppressed. The last approach is the opposite phase cancellation (OPC). As an effective method of suppressing the vector fields, it has been widely used in the fields of electromagnetics, optics, and acoustics.

Opposite phase cancellation (OPC) is a traditional method of achieving the RCS reduction. The basic idea is to exploit the cancellation effects arising from the well-known 180° phase difference between the corresponding reflection coefficients. Since the frequency and direction of incoming waves are unpredictable in reality, bandwidth

---

* The paper was originally published in *IEEE Transactions on Antennas and Propagation*, 2018, 66 (12), and has since been revised with new information. It was co-authored by Jianxun Su, Yao Lu, Jiayi Liu, Yaoqing (Lamar) Yang, Zengrui Li, and Jiming Song.

and oblique incidence performance are two important factors of stealth technology. In 2007, based on a combination of artificial magnetic conductors (AMC) and perfect electric conductors (PEC) in a chessboard like configuration, Paquay et al proposed a planar structure for RCS reduction. The backscattered field can be effectively canceled by redirecting it along other angles. However, the narrow in-phase reflection bandwidth of the AMC restricts the RCS reduction frequency range. In [22], a planar monolayer chessboard structure is presented for broadband RCS reduction using AMC technology. Fractional bandwidth (FBW) of more than 40% is obtained with a monostatic RCS reduction larger than 10 dB. These rectangular checkerboard surfaces of periodic phase arrangement create four scattering beams and bistatic RCS reduction of about 8.1 dB. In 2015, Balanis et al proposed a hexagonal checkerboard surface of periodic phase arrangement, with a 10 dB monostatic RCS reduction bandwidth of about 61%, which can create six bistatic RCS lobes, leading to further bistatic RCS reduction. A chessboard AMC surface composed of saltire arrow and four-E-shaped unit cells has a bandwidth of 85% for 10 dB RCS reduction. Then, the dual wideband checkerboard surfaces are presented in [27]; and the bandwidths of 10 dB RCS reduction in the frequency bands of 3.94-7.40 GHz and 8.41-10.72 GHz is about 61% and 24% by utilizing two dual-band electromagnetic bandgap (EBG) structures. In 2017, Haji-Ahmadi et al proposed a pixelated checkerboard metasurface for ultra-wideband RCS reduction . Two unit cells are designed based on the pixelation and optimization of a square patch and achieve 95% measured bandwidth for 10 dB RCS reduction. In [29], the polarization conversion metasurface is utilized to realize wideband RCS reduction. Phase differences of cross-polarized field components between the polarization conversion element and its mirror element are exactly 180° , resulting in opposite phase cancellation. In [30], the metasurface, composed of a square and L-shaped patches, can convert the polarization of the incident wave to its cross-polarized direction. A 10 dB RCS reduction is achieved over an ultra-wideband of 98%.

Recently, coding and digital metamaterials have been proposed for wideband RCS reduction through the design of coding sequences using digital elements "0" and "1", which possess opposite phase responses. RCS reduction is essentially based on opposite phase cancellation. In [31], 1-bit and 2-bit coding metasurfaces composed of digital elements were proposed; and 10 dB RCS reduction bandwidth of 66.67% was achieved. A 3-bit coding metasurface based on multiresonant polarization conversion elements was presented in [32]. The bandwidth of 10 dB RCS reduction was 89.9%. A broadband and broad-angle polarization-independent, random-coding metasurface for RCS reduction has been proposed in [33]. The 10 dB RCS reduction bandwidth of 84.75% was realized.

Previous research focused mainly on the design of unit cells with a fixed phase difference of approximately 180°for opposite phase cancellation or coding metamaterials. However, bandwidth expansion for RCS reduction is extremely difficult. This work was aimed at breaking the bandwidth constraints of opposite phase cancellation and coding metamaterial.

Our research is focused primarily on the development of novel phase cancellation methods. A metasurface based on the new physical mechanism of optimized multielement phase cancellation (OMEPC) is proposed for super-wideband RCS reduction. More basic meta-particles and variable phase differences between them greatly enhance the ability to achieve phase cancellation. Super-wideband manipulation of multiple electromagnetic (EM) waves is realized by adjusting the side length of the square ring patch and the thickness of the dielectric layer of the basic meta-particles. The geometric parameters of the basic meta-particles are optimized and determined by the field superposition principle and particle swarm optimization (PSO) algorithm to achieve good RCS reduction in a super-wide frequency band. The metasurface can achieve more than 10 dB RCS reduction in a super-wide frequency band ranging from 5.5 to 32.3 GHz with a ratio bandwidth (f_H / f_L) of 5.87:1 under normal incidence for both polarizations. Furthermore, the RCS reduction is larger than 8 dB from 5.4 to 40 GHz with a ratio bandwidth of 7.4:1. The metasurface also has an excellent performance under wide-angle oblique incidences. The analysis, simulation and measurement results demonstrate that our proposed physical mechanism greatly expands the RCS reduction bandwidth.

The paper has been structured as follows. The novel physical mechanism, optimized multielement phase cancellation (OMEPC), is introduced in Section II. Section III describes design process of the super-wideband RCS reducer metasurface in detail in three steps. The simulations and measurements of the monostatic and bistatic scattering characteristic for normal and oblique incidences are discussed in Sections IV. Finally, this work is summarized in Section V.

2. Optimized Multielement Phase Cancellation

The phenomenon of interference between waves is based on the field superposition principle. When two or more waves from different local areas of a metasurface travel into the same space, the net amplitude at each point is the sum of the amplitudes of the individual waves. A metasurface consists of a two-dimensional array of $M \times N$ tiles, which are uniformly spaced with d_x in the x direction and d_y in the y direction, as shown in Fig.1 (a). Based on planar array theory, the scattering field $E^s(\theta, \phi)$ of the metasurface under normal incidence can be synthesized accurately by [34].

$$E^S = \mathrm{EP} \times \mathrm{AF} \tag{1}$$

where EP is the pattern function of a tile, which is fixed in our model. AF represents the array factor which can be described as

A Novel Checkerboard Metasurface Based on Optimized Multielement Phase Cancellation for Superwideband RCS Reduction

$$\text{AF}(\theta,\varphi) = \sum_{m=1}^{M}\sum_{n=1}^{N} A_{m,n} \exp\{j[2\pi \sin\theta(\cos\varphi \cdot md_x \\ + \sin\varphi \cdot nd_y)/\lambda + \phi_{m,n}]\} \quad (2)$$

where θ and φ are the elevation angle and azimuth angle of an arbitrary scattering direction, respectively. $\Gamma_{m,n} = A_{m,n}e^{j\varnothing_{m,n}}$ is the reflection coefficient of the (m, n)-th basic tile. Especially in the backward direction, array factor is simplified to

$$\text{AF}(0,\varphi) = \sum_{m=1}^{M}\sum_{n=1}^{N} A_{m,n} e^{j\phi_{m,n}} \quad (3)$$

It is worth noting that the backscattering field is independent of the tile arrangement but only dependent on their reflection coefficients.

The RCS reduction (σ_R) of the metasurface, compared to an equal-sized PEC surface, can be represented by [25].

$$\sigma_R = 10\log \frac{|E^s|^2}{|E^i|^2} \quad (4)$$

For a multi-element checkerboard surface with $P(= M \cdot N)$ tiles, the RCS reduction can be derived based on Eq. (5) in [25], which can be approximated by

$$\sigma_R = 10\log\left|\frac{\text{AF}(0,\varphi)}{P}\right|^2 = 10\log\left|\frac{\sum_{i=1}^{P} A_i e^{j\phi_i}}{P}\right|^2 \quad (5)$$

The reflection amplitudes are unity ($A_1 \approx A_2 \approx \cdots A_P \approx 1$) due to a lossless ground surface. To achieve a 10 dB RCS reduction, the reflection phases of basic meta-particles need to satisfy the follow relationship

$$\left|\sum_{i=1}^{P} e^{j\phi_i}\right| \leq P\sqrt{0.1} \quad (6)$$

which is a multivariate exponential inequality. Completely phase cancellation requires

$$\left|\sum_{i=1}^{P} e^{j\phi_i}\right| = 0 \quad (7)$$

which is an exponential indefinite equation. It is noted that Eq. (6) and (7) have many solutions. Obviously, opposite phase cancellation and coding metamaterials are just two special solutions. However, more solutions will be excluded. For opposite phase cancellation and coding metamaterials, due to the fixed number of unit cells and the fixed phase difference between them, the ability to realize phase cancellation is greatly reduced, leading to a narrow bandwidth of RCS reduction.

Here, a metasurface based on a new physical mechanism of optimized multielement phase cancellation is proposed for super-wideband RCS reduction. Multiple reflected waves are superimposed in space to achieve phase cancellation, as shown in Fig.1 (b). The phase control of the basic meta-particles is arbitrary. More basic meta-particles and, in particular, variable phase difference between them greatly increase the ability to achieve super-wideband phase cancellation. The design process of the metasurface for super-wideband RCS reduction is divided into the following three steps:

2.1 Unit Cell and Its Reflection Characteristics

The choice of unit cell shape should satisfy that reflection phase range with the change of some geometric parameters of unit cell is large enough in a super-wide frequency band. This phase feature guarantees the super-wideband manipulation of EM waves.

2.2 Optimization Design of Basic Meta-Particles

This is the most important step. From Eq. (3), we know that the backscattering field is only related to the reflection phase of the basic meta-particles but independent of their distribution under normal incidence. Optimized multi-element phase cancellation (OMEPC) combining Eq. (5) and PSO algorithm is used to optimize and determine the geometric parameters of basic meta-particles, which can obtain the maximum bandwidth for the monostatic/backward RCS reduction.

2.3 Surface Meta-Particle Layout

Based on planar array theory, random meta-particle distributions were performed to find the optimal surface phase layout for diffuse scattering of EM waves in a super-wide frequency band, leading to a low bistatic RCS simultaneously.

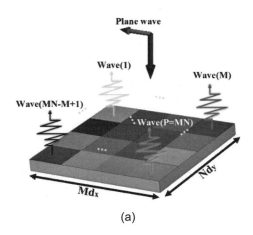

(a)

A Novel Checkerboard Metasurface Based on Optimized Multielement Phase Cancellation for Superwideband RCS Reduction

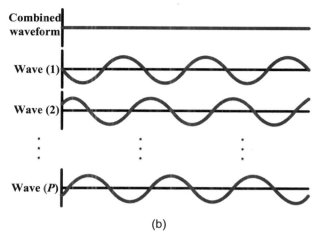

(b)

Fig. 1 Plane wave incident on a metasurface consisted of $P(=M \cdot N)$ tiles. (a) P scattering waves with a different reflection phase were produced locally by the tiles. (b) The combined waveform in the far field is the superposition of these p scattering waves.

3. Metasurface Design

In this section, the design process of the super-wideband RCS reducer metasurface is described in three steps.

3.1 Unit Cell and Its Reflection Characteristics

The square ring patch was chosen as the basic meta-particle of the metasurface for its reflection phase change characteristics. The range of reflection phase change with the change of side length is large enough in a super-wide frequency band. The unit cells were printed on the surface of Polytetrafluoroethylene Woven Glass (Model: F4B-2, Wangling Insulating Materials, Taizhou, China) substrate with a dielectric constant $\epsilon_r = 2.65$ and loss tangent $tan\delta = 0.001$. The back of the substrate is a PEC ground plane. The geometry structure of the basic meta-particle is illustrated in Fig. 2 (a). The basic meta-particles were simulated by Frequency Domain Solver (finite element method) of CST Microwave Studio®. The periodic boundary condition (PBC) is imposed on unit cells to create infinite structure and obtain their reflection coefficients. In this simulation, side length L of the meta-particles varied from 1.2 to 7.6 mm with a step size of 0.02 mm, while there were three choices of layer thickness for the dielectric substrate: 2 mm, 4 mm, and 6 mm. The periodicity a of the unit cell and the width w of square ring were fixed. A part of the reflection phase curves are plotted in Fig. 2 (b)-(d), respectively. The available phase coverage by tuning the side length of the square ring and dielectric layer thickness was larger than 250° at frequencies ranging from 5 to 28 GHz. This phase feature guarantees the possibility of super-wideband manipulation of EM waves.

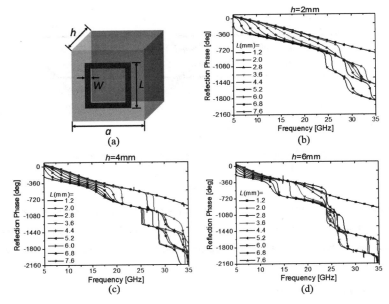

Fig. 2 Geometry of the basic meta-particle and its reflection phase properties. (a) Geometry structure of the square ring element. Dimensions are: a=8, w=0.4, h=2, 4, 6 in mm. (b), (c), and (d) The reflection phase of basic meta-particles with the change of side length L of square rings for the dielectric layer thickness of 2, 4, and 6 mm, respectively.

3.2 Optimization Design of Basic Meta-Particles

In this research, the RCS reducer metasurface consisted of 4×4 tiles. Each tile was a sub-array of basic meta-particle, i.e., the number of basic meta-particles was $p=16$. Effective manipulation of EM waves in an ultra-wide frequency band by adjusting the geometric parameter of the basic meta-particles was first proposed and proved in [35]. The schematic diagram for super-wideband control of RCS reduction by adjusting the side length of square ring and the thickness of dielectric layer is depicted in Fig. 3, where M is the number of optimization frequencies. The reflection phase versus frequency plotted in Fig. 2 were pre-stored in a table. For a combination of geometric parameters for basic meta-particles, M combinations of reflected phases at the optimization frequencies $(f_1, f_1 ... f_M)$ can be read from the reflection phase table. These M phase combinations resulted in M values of RCS reduction. Finally, these M values were used to evaluate the corresponding fitness for this combination of geometric parameters.

The main objective of this work was to control 16 local waves produced by 16 basic meta-particles to achieve phase cancellation for a 10 dB RCS reduction in the super-wide frequency band. However, this process is very challenging. Particle swarm optimization (PSO) together with Eq. (5) was used to optimize the side length L and layer thickness h of 16 basic meta-particles. In optimization, RCS reduction (σ_R) values at M optimization frequencies sampled in a super-wide frequency band needed to be evaluated. A cancellation error occurs when some frequencies have a very large σ_R value and others with a very small σ_R value. Therefore, we could not directly sum up the σ_R values at all of the optimization frequencies to evaluate the fitness. Thus, the fitness function is defined as

$$\text{fitness} = \sum_{i=1}^{M} s(i) \tag{8}$$

and

$$s(i) = \begin{cases} 1, & \text{if } \sigma_R^i > -10 \text{ dB} \\ 0, & \text{if } \sigma_R^i \leqslant -10 \text{ dB} \end{cases} \tag{9}$$

where σ_R^i is the RCS reduction under normal incidence at i-th frequency. The smaller the fitness, the better RCS reduction.

The detailed flow-chart of the optimized multielement phase cancellation for super-wideband RCS reduction is presented in Fig. 4. The initial side length of the basic meta-particle is a random value chosen from a uniform distribution between 1.2 mm and 7.6 mm. Layer thickness h is a discrete value chosen from 2 mm, 4 mm, and 6 mm. When 1000 iterations were finished, we got the optimal geometric parameters of 16 basic meta-particles for the metasurface with the lowest backward RCS in a desired super-wide frequency band.

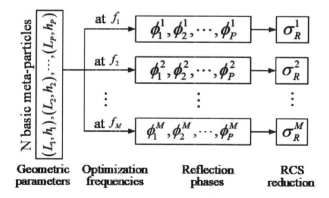

Fig. 3 Schematic diagram for super-wideband manipulation of RCS reduction by geometric parameter adjustment. The side length L and dielectric layer thickness h are two adjustable geometric parameters.

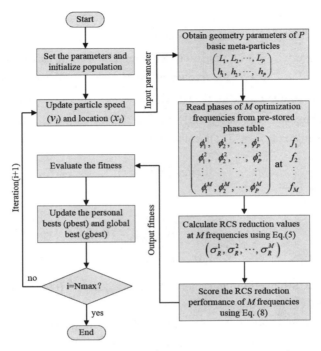

Fig. 4 Flowchart of the basic meta-particle optimization process for super-wideband RCS reduction.

The optimized results of side length L and layer thickness h are listed in Table I. The predicted RCS reduction is shown in Fig. 5(a). In a super-wide frequency band from 5.08 to 27.74 GHz, the RCS reductions are larger than 10 dB. Their corresponding reflection phases are shown in Fig. 5(b). For different frequencies, the basic meta-particles have different reflection phase responses, but they can always satisfy the phase cancellation condition for 10 dB RCS reduction depicted in Eq. (6). Compared with opposite phase cancelation (OPC) and coding metamaterial, the advantage of this approach is that more basic meta-particles and variable phase differences between them greatly increase the ability for super-wideband manipulation of EM waves and realizing super-wideband phase cancellation.

Table I

The optimized results of 16 basic meta-particles

Meta-particle	1	2	3	4	5	6	7	8
L(mm)	7.1	1.2	4.0	6.8	6.0	4.2	7.3	7.6
h(mm)	4	6	4	4	2	2	4	2

Table I Continued

Meta-particle:	9	10	11	12	13	14	15	16
L(mm)	4.3	6.7	4.3	6.8	2.5	7.6	6.7	6.1
h(mm)	2	6	6	4	6	2	2	6

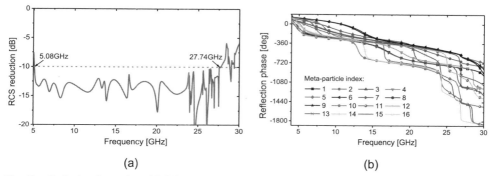

Fig. 5 Optimized results of RCS reduction. (a) The predicted monostatic RCS reduction. (b) The reflection phase of these 16 basic meta-particles.

3.3 Surface Meta-Particle Layout

The metasurface consisted of these 16 basic meta-particles. According to Eqs. (2) and (3), the scattering pattern $E^s(\theta,\phi)$ in space is determined by the arrangement of the basic meta-particles, except for the backward direction ($\theta=0, \phi=0$). We need to find the optimal meta-particle distribution with a lowest bistatic RCS in the super-wide frequency band. The bistatic RCS reduction (σ_R^{Bi}) defined in [23] is described as

$$\sigma_R^{Bi} = \frac{\max\left(Bistatic\ RCS\ with\ metasurface\right)}{\max\left(Bistatic\ RCS\ without\ metasurface\right)} \quad (10)$$

For a checkerboard surface with P basic tiles, the bistatic RCS reduction can be approximated by

$$\sigma_R^{Bi} = 10\log\left[\frac{\max|\mathrm{AF}(\theta,\varphi)|}{P}\right]^2 \quad (11)$$

The arithmetic mean of bistatic RCS reductions at $M=24$ frequency points from 5 to 28 GHz with an interval of 1 GHz are used to evaluate the figure of merit for bistatic scattering characteristics (FMB). Thus, FMB can be described as

$$\text{FMB} = \frac{20}{M}\sum_{i=1}^{M}\log\left[\frac{\max|\text{AF}_i(\theta,\varphi)|}{P}\right] \quad (12)$$

The diffuse scattering of EM wave, which can effectively reduce the bistatic RCS, is caused by random meta-particle distribution through a certain computer-generated pseudo-random matrix. In order to facilitate manufacturing of the metasurface, the tiles with the equal layer thickness are put together. After 500 random distributions were performed, an optimal surface meta-particle layout with minimum FMB is chosen to build the RCS reducer metasurface. The FMB value versus distribution index is plotted in Fig. 6. For the optimal distribution, the average of bistatic RCS reductions from 5 to 28 GHz is 7 dB. The optimal distribution of 16 basic meta-particles and the full structure of the metasurface are depicted in Fig. 7. To approximate the periodic boundary condition (PBC) used in simulation, each tile is occupied by a subarray of 7×7 meta-particles. The model of the optimized RCS reducer metasurface has an overall dimension of $224\times 224 mm^2$.

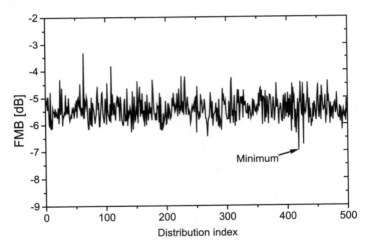

Fig. 6   The random process of FMB.

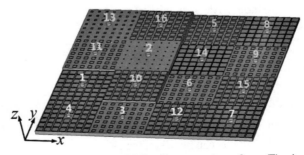

Fig. 7   The full model of the optimized RCS reducer metasurface. The inserted index is the meta-particle number shown in Table I.

4. Simulation and Measurement

4.1 Simulated Results

A full structure of the metasurface was full-wave simulated by the transient solver of CST Microwave Studio®. The RCS of the metasurface normalized to that of an equal-sized PEC surface with a plane wave normally impinging is shown in Fig. 8. The metasurface can achieve 10 dB RCS reduction in a super-wide frequency band from 5.4 to 33.0 GHz with a ratio bandwidth of 6.11:1. The proposed physical mechanism of optimized multielement phase cancellation (OMEPC) has shown to greatly expand the bandwidth of RCS reduction.

The 16 meta-particles of the metasurface present different phase responses, which is critical in disturbing the equiphase reflection. Due to the nonuniform distributions of the phase gradient between neighboring tiles, the specular reflections no longer dominate within the whole scattered waves. As a result, the metasurface generates a diffusion scattering pattern in the far-field with suppressed amplitude. The scattering patterns between the proposed metasurface and the equal-sized PEC surface under normal incidence are compared in Fig. 9 at 6 GHz, 12 GHz and 24 GHz, respectively. The bistatic RCS of the metasurface is much lower than that of the equal-sized PEC surface. It is worth noting that, the periodicity of the array factor (AF) decreases as the frequency increases, leading to close concentrations of the scattering beams. For different frequencies, the phase layout response of the metasurface is different, resulting in different diffusion scattering patterns.

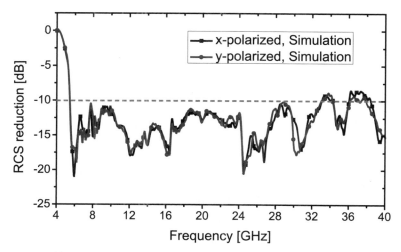

Fig. 8   The simulated RCS reduction versus frequency.

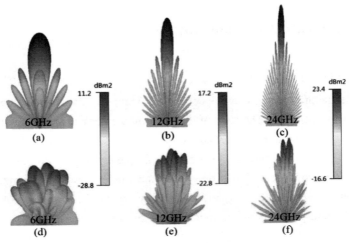

Fig. 9　Comparison of the bistatic scattering patterns between the proposed metasurface and equal-sized PEC surface under normal incidence at 6 GHz, 12 GHz and 24 GHz, respectively. (a)-(c) For equal-sized PEC surface. (d)-(f) For the metasurface.

In addition, the scattering properties of the metasurface under oblique incidences for both transverse-electric (TE) and transverse-magnetic (TM) polarizations are also provided in Fig.10. Significant RCS reduction is obtained from 5.5 to 40 GHz under wide-angle incidences for both polarizations. Simulation results illustrate the excellent performance of RCS reduction under oblique incidences.

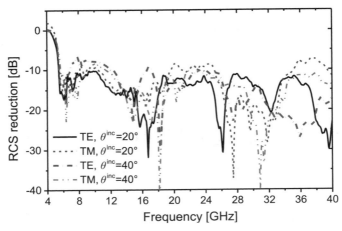

Fig. 10　The RCS reductions in the specular direction under oblique incidence for TE and TM polarizations. TE/TM: The direction of the electric/magnetic field is perpendicular to the plane of incidence.

4.2 Measured Results

To validate the predicted performance of the proposed metasurface, a sample of the metasurface was fabricated and measured. The sample was manufactured using printed circuit board (PCB) technology. Three pieces of the metasurface board with different thicknesses were processed separately, and then stitched together and fixed on the metal ground with Teflon™ bolts, as per our metasurface design. The dielectric substrate is Polytetrafluoroethylene Woven Glass substrate with a dielectric constant $\epsilon_r = 2.65$ (loss tangent $tan\delta = 0.001$). The metal patches and ground are 0.035 mm-thick copper layers. The diameter of the Teflon bolt is 3 mm and the dielectric constant is 2.1. A photo of the fabricated sample is shown in Fig. 11(a). A high-precision RCS measurement was conducted using the compact antenna test range (CATR) system of the Science and Technology on Electromagnetic Scattering Laboratory in Beijing, China. An equal-sized copper ground of the proposed metasurface is also measured as a reference. Fig.11 (b) shows the measured results of the RCS reductions versus frequency under normal incidence. The metasurface can realize an RCS reduction of larger than 8 dB in a super-wide frequency band from 5.4 to 40 GHz with a ratio bandwidth of 7.4:1 for both x- and y-polarizations. The RCS reductions are larger than 10 dB from 5.5 to 32.3 GHz with a ratio bandwidth of 5.87:1. We note that the measurements agree well with the simulations shown in Fig. 8. The value deviations can be attributed to manufacture and measurement error. Good phase cancellation is achieved in a super wide frequency band. A comparison between the previous researches and this work is provided in Table II. Obviously, our work has an overwhelming advantage in bandwidth expansion of RCS reduction. Overall, the bandwidth maximization for RCS reduction and diffusion scattering by the proposed metasurface based on optimized multielement phase cancellation is confirmed.

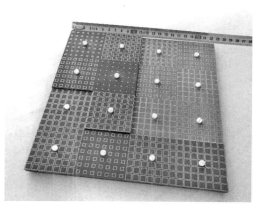

(a)

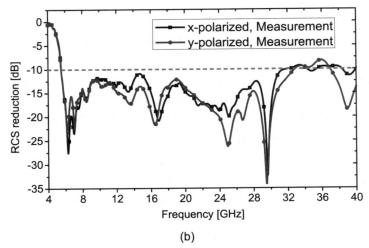

(b)

Fig. 11 The fabrication and measurement of the proposed RCS reducer metasurface. (a) The photograph of the fabricated metasurface. (b) Measured results of RCS reduction for the normal incidences.

Table II

Comparison of our work and previous researches

Article	σ_R (dB)	OFB (GHz)	FBW (%)	RBW (f_H / f_L)
[22]	10	14.5-21.8	41	1.50:1
[23]	10	9.3-15.5	50	1.67:1
[25]	10	4.1-7.59	60	1.85:1
[26]	10	9.40-23.28	85	2.48:1
[27]	10	10.5-18	52.63	1.71:1
[32]	10	7.9-20.8	89.9	2.63:1
[33]	10	17-42	84.75	2.47:1
[30]	10	6.1-17.8	98	2.92:1
[28]	10	3.8-10.7	95	2.82:1
This Work	10	5.5-32.3	141.8	5.87:1
	8	5.4-40	152.4	7.4:1

σ_R: RCS reduction

OFB: Operating frequency band

FBW: The fractional bandwidth ($FBW = (f_H - f_L) / f_c$, $f_c = (f_H + f_L) / 2$)

RBW: The ratio bandwidth ($RBW = f_H / f_L$)

5. Conclusion

A novel checkerboard metasurface based on optimized multielement phase cancellation (OMEPC) was designed, fabricated, and tested for super-wideband RCS reduction. More basic meta-particles and variable phase differences between them greatly increase the ability to achieve phase cancellation. Multiple local waves produced by the basic meta-particles at multiple frequencies sampled in a super-wide frequency band are manipulated and optimized simultaneously to achieve phase cancellation by field superposition principle and particle swarm optimization (PSO) algorithm. The metasurface can achieve more than 10 dB RCS reduction in a super-wide frequency band ranging from 5.5 to 32.3 GHz with a ratio bandwidth of 5.87:1 under normal incidence for both x- and y-polarized waves. Furthermore, the RCS reduction is larger than 8 dB from 5.4 to 40 GHz with a ratio bandwidth of 7.4:1. The diffuse scattering of EM waves is caused by the random meta-particle distribution, leading to a low bistatic RCS simultaneously. The analysis, simulation, and test results show that the new physical mechanism can effectively break the bandwidth constraints of traditional phase cancellation method and greatly expanding the bandwidth of RCS reduction.

Acknowledgments

This work was supported by the National Natural Science Foundation of China (61671415 and 61701448), and the key projects of engineering planning of Communication university of China (3132016XNG1604).

Author Information

J. X. Su is with the School of Information Engineering, Communication University of China, Beijing 100024, China and also with the Science and Technology on Electromagnetic Scattering Laboratory, Beijing 100854, China (e-mail: sujianxun_jlgx@163.com).

Y. Lu, J. Y. Liu and Z. R. Li are with the School of Information Engineering, Communication University of China, Beijing 100024, China (e-mail: zrli@cuc.edu.cn).

Y. Q. Yang is with the Department of Electrical and Computer Engineering, University of Nebraska-Lincoln, NE, 68182, USA.

J. M. Song is with the School of Information Engineering, Communication University of China, Beijing 100024, China and also with the Department of Electrical and Computer Engineering, Iowa State University, Ames, IA 50011, USA (e-mail: jisong@iastate.edu).

References

[1] CUI T J, SMITH D R, LIU R. Metamaterials: Theory, Design, and Applications[C]. New York: Springer Science & Business Media, 2009.

[2] SMITH D R, PENDRY J B, WILTSHIRE M C. Metamaterials and negative refractive index[J]. Science, 2004,305 (5685): 788-792.

[3] BARNES W L, DEREUX A, EBBESEN T W. Surface plasmon subwavelength optics[J]. Nature, 2003, 424 (6950): 824-830.

[4] LITCHINITSER N M, MAIMISTOV A I, GABITOV I R, et al. Metamaterials: electromagnetic enhancement at zero-index transition[J]. Opt. Lett., 2008, 33 (20): 2350-2352.

[5] ALÙ A, ENGHETA N. Pairing an epsilon-negative slab with a mu-negative slab: resonance, tunneling and transparency[J]. IEEE trans. antennas propag, 2003, 51 (10) :2558-2571.

[6] PENDRY J B. Negative refraction makes a perfect lens[J]. Phys. Rev. Lett., 2000, 85 (18): 3966-3969.

[7] SCHURIG D, MOCK J J, JUSTICE B J, et al. Metamaterial electromagnetic cloak at microwave frequencies[J]. Science, 2006, 314 (5801): 977-980.

[8] LIU R, JI C, MOCK J J, et al. Broadband ground-plane cloak[J]. Science, 2009, 323 (5912): 366-369.

[9] MA H F, CUI T J. Three-dimensional broadband ground-plane cloak made of metamaterials[J]. Nat. Comm., 2010, 1 (3).

[10] DAVOYAN A R, LIU W, MIROSHNICHENKO A E, et al. Mode transformation in waveguiding plasmonic structures[J]. Photonics and nanostructures, 2011, 9 (3): 207-212.

[11] YIN J Y, WAN X, ZHANG Q, et al. Ultra wideband polarization-selective conversions of electromagnetic waves by metasurface under large-range incident angles[J]. Sci. Rep, 2015, 5: 12476.

[12] WATTS C M, LIU X, PADILLA W J. Metamaterial electromagnetic wave absorbers[J]. Adv. Mater, 2012, 24 (23): OP98-OP120.

[13] YANG H, CAO X, GAO J, et al. Low RCS metamaterial absorber and extending bandwidth based on electromagnetic resonances[J]. Progress in electromagnetics research M, 2013, 33: 31-44.

[14] YOO M, KIM H K, LIM S. Angular- and polarization-insensitive metamaterial absorber using subwavelength unit cell in multilayer technology[J]. IEEE antennas wireless propag. lett., 2016, 15: 414-417.

[15] LEE D S, GONZALEZ L F, SRINIVAS K, et al. Aerodynamic/RCS shape optimisation of unmanned aerial vehicles using hierarchical asynchronous parallel evolutionary algorithms[R]. San Francisco: AIAA, 2007.

[16] KNOTT E F. Radar Cross Section Measurements[M]. Berlin: Springer, 2012.

[17] ZHANG Y, MUELLER J H, MOHR B, et al. A multi-frequency multi-standard wideband fractional-N PLL with adaptive phase-noise cancellation for low-power short-range standards[J]. IEEE

Trans. Microw. Theory Techn., 2016, 64 (4): 1133-1142.

[18] HU J M, ZHOU W, FU Y W, et al. Uniform rotational motion compensation for ISAR based on phase cancellation[J]. IEEE Geosci. Remote Sens. Lett., 2011, 8 (4): 636-640.

[19] WEAR K A. The effect of phase cancellation on estimates of calcaneal broadband ultrasound attenuation in Vivo[J]. IEEE Trans. Ultrason. Ferroelect. Freq. Contr., 2007, 54 (7): 1352-1359.

[20] SUAREZ J, Prucnal P R. System-level performance and characterization of counter-phase optical interference cancellation[J]. J. Lightwave Technol 2010, 28 (12): 1821-1831.

[21] PAQUAY M, IRIARTE J C, EDERRA I, et al. Thin AMC structure for radar cross-section reduction[J]. IEEE Trans. Antennas Propag., 2007, 55 (12): 3630-3638.

[22] GALARREGUI J C I, PEREDA A T, FALCÓN J L M D, et al. Broadband radar cross-section reduction using AMC technology[J]. IEEE Trans. Antennas Propag., 2013, 61 (12): 6136-6143.

[23] EDALATI A, SARABANDI K. Wideband, wide angle, polarization independent RCS reduction using nonabsorptive miniaturized-element frequency selective surfaces[J]. IEEE Trans. Antennas Propag., 2014, 62 (2): 747-754.

[24] ZHENG Y J, GAO J, CAO X Y, et al. Wideband RCS reduction of a microstrip antenna using artificial magnetic conductor structures[J]. IEEE Antennas Wireless Propag. Lett., 2015, 14: 1582-1585.

[25] CHEN W, BALANIS C A, BIRTCHER C R. Checkerboard EBG surfaces for wideband radar cross section reduction[J]. IEEE Trans. Antennas Propag., 2015, 63 (6): 2636-2645.

[26] ESMAELI S H, SEDIGHY S H. Wideband radar cross-section reduction by AMC[J]. Electron. Lett., 2016, 52 (1): 70-71.

[27] PAN W B, HUANG C, PU M B, et al. Combining the absorptive and radiative loss in metasurfaces for multi-spectral shaping of the electromagnetic scattering[J]. Sci. Rep., 2016, 6: 21462.

[28] HAJI-AHMADI M J, NAYYERI V, SOLEIMANI M, et al. Pixelated checkerboard metasurface for ultra-wideband radar cross section reduction[J]. Sci. Rep., 2017, 7 (1): 11437.

[29] JIA Y T, LIU Y, GUO Y J, et al. Broadband polarization rotation reflective surfaces and their applications to RCS reduction[J]. IEEE Trans. Antennas Propag, 2016, 64 (1): 179-188.

[30] JIA Y T, LIU Y, GUO Y J, et al. A dual-patch polarization rotation reflective surface and its application to ultra-wideband RCS reduction[J]. IEEE Trans. Antennas Propag., 2017, 65 (6): 291-3295.

[31] CUI T J, QI M Q, WAN X, et al. Coding metamaterials, digital metamaterials and programmable metamaterials[J]. L. Sci. Appl., 2014, 3 (10): e218.

[32] SU P, ZHAO Y J, JIA S L, et al. An ultra-wideband and polarization-independent metasurface for RCS reduction[J]. Sci. Rep., 2016, 6: 20387.

[33] SUN H Y, GU C Q, CHEN X L, et al. Broadband and broad-angle polarization-independent metasurface for radar cross section reduction[J]. Sci. Rep., 2017, 7: 40782.

[34] BALANIS C A. Antenna theory: analysis and design[M]. 3rd ed. New York: Wiley, 2005.

[35] SU J X, LU Y, ZHANG H, et al. Ultra-wideband, wide angle and polarization-insensitive specular reflection reduction by metasurface based on parameter-adjustable meta-atoms[J]. Sci. Rep., 2017, 7: 42283.

[36] BOERINGER D W, WERNER D H. Particle swarm optimization versus genetic algorithms for phased array synthesis[J]. IEEE trans. antennas propag., 2004, 52 (3): 771–779.

[37] GAO L H, CHENG Q, YANG J, et al. Broadband diffusion of terahertz waves by multi-bit coding metasurfaces[J]. L. Sci. Appl., 2015, 4 (9): e324.